# SEMANTICAL INVESTIGATIONS IN HEYTING'S INTUITIONISTIC LOGIC

SYNTHESE LIBRARY

STUDIES IN EPISTEMOLOGY,

LOGIC, METHODOLOGY, AND PHILOSOPHY OF SCIENCE



VOLUME 148

DOV M. GABBAY
*Professor of Logic, Bar-Ilan University*

# SEMANTICAL INVESTIGATIONS IN HEYTING'S INTUITIONISTIC LOGIC

D. REIDEL PUBLISHING COMPANY
DORDRECHT : HOLLAND / BOSTON : U.S.A.
LONDON : ENGLAND

**Library of Congress Cataloging in Publication Data**

CIP

Gabbay, Dov M. 1945–
Semantical investigations in Heyting's intuitionistic logic.

(Synthese library; v. 148)
Bibliography: p.
Includes index.
1. Intuitionistic mathematics. 2. Heyting, Arend, 1898– . I. Title.
QA9.47.G32 511'.2 80-27460

ISBN 978-90-481-8362-3

---

Published by D. Reidel Publishing Company
P.O. Box 17, 3300 AA Dordrecht, Holland

Sold and distributed in the U.S.A. and Canada
by Kluwer Boston Inc.,
190 Old Derby Street, Hingham, MA 02043, U.S.A.

In all other countries, sold and distributed
by Kluwer Academic Publishers Group
P.O. Box 322, 3300 AH Dordrecht, Holland

D. Reidel Publishing Company is a member of the Kluwer Group



Printed in The Netherlands

# PREFACE

From the point of view of non-classical logics, Heyting's implication is the smallest implication for which the deduction theorem holds. This book studies properties of logical systems having some of the classical connectives and implication in the neighbourhood of Heyting's implication. I have not included anything on entailment, although it belongs to this neighbourhood, mainly because of the appearance of the Anderson-Belnap book on entailment.

In the later chapters of this book, I have included material that might be of interest to the intuitionist mathematician. Originally, I intended to include more material in that spirit but I decided against it. There is no coherent body of material to include that builds naturally on the present book. There are some serious results on topological models, second order Beth and Kripke models, theories of types, etc., but it would require further research to be able to present a general theory, possibly using sheaves. That would have postponed publication for too long.

I would like to dedicate this book to my colleagues, Professors G. Kreisel, M.O. Rabin and D. Scott. I have benefited greatly from Professor Kreisel's criticism and suggestions. Professor Rabin's fundamental results on decidability and undecidability provided the powerful tools used in obtaining the majority of the results reported in this book. Professor Scott's approach to non-classical logics and especially his analysis of the Scott consequence relation makes it possible to present Heyting's logic as a beautiful, integral part of non-classical logics.

TABLE OF CONTENTS

# INTRODUCTION

In the course of development of intuitionistic mathematics, several logical systems were put forward for the purpose of formalising intuitionistic notions, the main one among which is Heyting's predicate calculus. It soon became apparent that these logical systems have an independent interest from the point of view of non-classical logics, and they were studied for their own sake as a branch of intensional logics.

This book presents semantical methods in the study of these logics, including topics that are of interest from the point of view of non-classical logics as well as topics that have some applications to problems considered in intuitionistic mathematics. The book mainly contains the author's research, but related research of other authors is also included in order to present a more complete picture.

Results that are of related interest but are not in the mainstream of the book are presented as exercises.

One formulation of the classical predicate calculus (CPC) as a Hilbert system is in a language with $f$, $\wedge$, $\vee$, $\rightarrow$, $\forall$, $\exists$ with the following axioms and rule schemas:

*Axiom Schemas*

(1) $A \rightarrow (B \rightarrow A)$

(2) $(A \rightarrow (B \rightarrow C)) \rightarrow ((A \rightarrow B) \rightarrow (A \rightarrow C))$

(3) $((A \rightarrow B) \rightarrow A) \rightarrow A$

(4) $A \wedge B \rightarrow A; \quad A \wedge B \rightarrow B$

(5) $A \rightarrow A \vee B; \quad B \rightarrow A \vee B$

(6) $A \rightarrow (B \rightarrow A \wedge B)$

(7) $(A \rightarrow C) \rightarrow ((B \rightarrow C) \rightarrow (A \vee B \rightarrow C))$

(8) $f \rightarrow A$

(9) $A(x) \rightarrow \exists y A(y)$

(10) $\forall y\, A(y) \rightarrow A(x)$

*Rules*

(11) $$\frac{\vdash A, \vdash A \to B}{\vdash B}$$

(12) $$\frac{\vdash A(x) \to B}{\vdash \exists x A(x) \to B}; \qquad \frac{\vdash B \to A(x)}{\vdash B \to \forall x A(x)}$$

$x$ not free in $B$.

Heyting's predicate calculus, HPC, can be formulated as the subsystem with all the above axioms and rules except axiom (3). That is:

$$\text{HPC} + (3) = \text{CPC}.$$

We examine HPC and compare it with CPC from several points of view, among which the main ones are:

(a) A study of the proof theoretical properties of HPC.
(b) Semantical interpretations for HPC.
(c) A study of various other logics in the neighbourhood of HPC and CPC.
(d) Study of systems arising from taking various axioms $A$ and comparing HPC + $A$ to CPC + $A$.

We begin in Chapter 1 with the general question on the nature of a logical system and the meaning of a certain connective in a given logical system when it is described as truth-functional. This interests us because we ask (since HPC is obtained from CPC by weakening), which of the connectives of CPC are no longer truth-functional in HPC. We also try to introduce HPC independently of CPC, via some general considerations regarding logical systems. This can be done as follows: Suppose we begin with a binary connective $\to$ and demand the minimal axioms allowing the deduction theorem to hold for $\to$;

$$(\text{i.e.,} \quad A_i, B \vdash C \quad \text{iff} \quad A_i \vdash B \to C)$$

then this yields the pure $\to$ part of HPC. Suppose we want to add to $\to$ the connectives $f$, $\wedge$, $\vee$ of CPC, and that we want them to retain their truth-functional meaning, (the concept is made precise in Chapter 1), then we get propositional HPC with $f$, $\wedge$, $\vee$, $\to$ denoted by $h$. (Propositional CPC is denoted by $c$).

Thus we conclude that the propositional part of HPC, $h$, can be regarded as being composed of axioms making $f$, $\wedge$, $\vee$ truth-functional and minimal in yielding the deduction theorem for $\rightarrow$.

Chapter 2 introduces HPC and more of its properties are studied. Among them the disjunction and existence properties for HPC (i.e., $\vdash A \vee B$ iff $\vdash A$ or $\vdash B$ and $\vdash \exists x A(x)$ iff for some $y \vdash A(y)$).

Two intermediate logics are introduced. The system MH, which is the smallest extension of HPC for which the following holds:

$$\text{CPC} \vdash A \quad \text{iff} \quad \text{MH} \vdash (A \rightarrow f) \rightarrow f$$

and the system *CD* which is HPC + the scheme $\forall x\ (A \vee B(x)) \rightarrow A \vee \forall x B(x)$, $x$ not free in $A$.

It appears that in HPC we can regard either $f$, $\wedge$, $\forall$, $\exists$ as classical (retaining the same meaning as they have in CPC) or $f$, $\wedge$, $\vee$, $\exists$ as classical. But we cannot regard all of $f$, $\wedge$, $\vee$, $\exists$, $\forall$ as classical because in this case we get the logic CD. Thus CD is obtained by taking the weakest logical system with $\rightarrow$, $f$, $\wedge$, $\vee$, $\forall$, $\exists$ in which $\rightarrow$ satisfies the deduction theorem and all the other connectives and quantifiers are classical (as in CPC). Of course, the propositional parts of HPC, MH and CD are all equal to $h$.

Chapter 3 introduces three semantical interpretations for HPC. The most useful one is the Kripke interpretation (the others are similar). Kripke structures are of the form $(T, <, A_t)$, $t \in T$, $(T, <)$ being a partially ordered set and $A_t$ is the classical structure associated with $t$. For $t \leqslant s$, $A_t$ is homomorphic into $A_s$. Any wff which holds at $t$ holds also at any $s$, $t \leqslant s$. Thus, $(T, <)$ can be regarded as a branching future flow of time in which the truth of various wffs become known. The meaning of $A \rightarrow B$ is that whenever $A$ becomes true, so is $B$. This interpretation is useful especially for AE sentences. For general wffs a sheaf interpretation is more natural. But we shall not go into sheaves in this book.

Chapters 4–7 resume the study of other neighbouring systems. We mainly consider propositional extensions or weakenings of $h$. We look for intermediate logics (i.e., logics between $h$ and $c$) satisfying various kinds of conditions. It appears that different axioms correspond to different semantical conditions on $(T, <)$. Also the notion of the finite model property (f.m.p.) is discussed, i.e., whether a logic is complete for a class of $(T, <)$ with finite $T$'s.

The interest in such logics stems from trying to characterise $h$ as the strongest logic below $c$ satisfying certain properties (e.g., the disjunction property, as Łukasiewicz thought). Thus, for example, various intermediate logics with the disjunction property are found, and a logic without f.m.p. is described.

Chapters 6 and 7 study further special properties of intermediate propositional logics. The lattice of all propositional formulae of $h$ with one atom is described (it is infinite). Various possibilities of additional connectives to $h$ are studied (for example $\#p$ saying $p$ is true in all the immediate future points).

Chapter 8 proves the interpolation theorem for HPC and some extensions, including CD and MH.

Since the manuscript of the book was completed more results have been obtained in this area. It has been shown by Y. Komari that there are many intermediate logics for which interpolation fails. In fact L. Maksimova shows that there are only eight intermediate logics which allow interpolation (in the propositional part).

Chapter 9 studies the propositional calculus supplemented with quantifiers over atomic propositions. In the classical case the system reduces to the propositional calculus itself but in the intuitionistic case we get a system which is essentially as strong as HPC itself.

Chapters 10–13 study theories formulated in HPC such as groups, fields, etc. We mentioned that a structure for HPC can be obtained by partially ordering classical structures of CPC. Thus, if we take a set $\tau$ of axioms (e.g., groups), we can regard it as a theory $\tau_C$ in CPC (the theory of groups) or as a theory $\tau_H$ in HPC (with the underlying logic being HPC). Can we put together classical structures of $\tau_C$ (i.e. groups) to get the intuitionistic structures for $\tau_H$? For AE sets of axioms $\tau$ with matrix of the form ($\wedge$ atoms $\rightarrow$ $\vee$ atoms) the results are readily available, as partially ordered sets with certain homomorphisms. For more complicated axioms nothing can be said and the more natural interpretation is probably through sheaves.

However, enough can be done for AE theories with matrix of the form ($\wedge$ atoms $\rightarrow$ $\vee$ atoms) to have a base to settle the decision problem for intuitionistic theories treated in the last two chapters. Given $\tau$, we again look at the $\tau_C$ and $\tau_H$. If $\tau_C$ is undecidable, so is $\tau_H$. We can ask, for a decidable $\tau_C$ what is the situation for $\tau_H$? We obtain mainly undecidability results for various $\tau_H$. We give sufficient conditions for undecidability. These conditions depend on the exact logical complexity of the axioms in $\tau$ and also on the classical model theoretic properties of $\tau_C$ (e.g., model completeness, homomorphism, existence of certain kinds of models etc.). The results of decidability are secured using Rabin's theorem on the decidability of Monadic second order theory *SwS*. A class of structures is found which is expressible in *SwS* and for which $\tau_H$ is complete. For example, for $\tau$ = abelian groups (usual additive formulation (with $\circ, +, -, =$) $\tau_H$ is undecidable. Add the axioms, $\forall x(mx = 0)$, $m$ fixed, and $\forall x(x = 0 \vee x \neq 0)$, then the resulting theory is undecidable iff for some prime $p$, $p^2$ divides $m$. Decidability also depends on the underlying intermediate logic. For example, there exists an intermediate logic $X$ such that the theory of one unary predicate $P$ is decidable but that of two is not! We treat algebraic and real closed fields, unary functions, equality and other theories.

Chapters 13 and 14 consider semantic interpretations for HPC in recursive structures and in Post systems. In such structures we get often that the set of valid wff is not arithmetical. Such structures are of interest especially from the intuitionistic point of view. Recently, intuitionistically presented interpretations for HPC have been given. We do not treat these in this book.

The manuscript of this book was submitted to the publisher in 1976. For technical reasons, production did not start until 1980. The most important related development in the meantime is, to my belief, the topological sheaf models for intuitionistic logic developed by D. Scott, M. Fourman, and R. Grayson (see References). The subject is one which deserves a separate volume to itself. A book by Dummett, on intuitionistic mathematics, has also appeared.

## Notation

We use the usual notions of wffs, free and bound occurrences of variables, substitution instances, etc.

$\varphi, \psi$ denote *finite* sets of wffs. $\varphi'$ is said to be a substitution instance of $\varphi$ iff $\varphi'$ is obtained from $\varphi$ by uniform substitution into the elements of $\varphi$.

Wffs are denoted by $A$, $B$, $C$, sometimes by $\alpha$, $\beta$, $\gamma$.

$\omega$ is the set of natural numbers. * denotes concatenations and $\Lambda$ is the empty sequence. $\emptyset$ denotes the empty set.

## CHAPTER 1

# LOGICAL SYSTEMS AND SEMANTICS

This chapter discusses the notions of a logical system, a semantics for a logical system, and the notion of what is a classical connective in a logical system. Examples are given, to prepare the background for the introduction of the Heyting systems in the next chapter.

## 1. SCOTT AND TARSKI SYSTEMS

DEFINITION 1. Let $L$ be a language, and let $\varphi$, $\psi$ denote finite, possibly empty sets of wffs. Let $\emptyset$ denote the empty set. A binary relation $\Vdash$ on sets $\varphi$, $\psi$ is called *Scott consequence relation* iff the following conditions hold:

(a) $\varphi \Vdash \varphi$, for $\varphi \neq \emptyset$

(b) if $\varphi \Vdash \psi$ then $\varphi \cup \varphi' \Vdash \psi \cup \psi'$ for any $\varphi'$, $\psi'$.

(c) (Cut rule): if $\varphi, A \Vdash \psi$ and $\varphi \Vdash \psi, A$ then $\varphi \Vdash \psi$

DEFINITION 2. We write $\varphi, A \Vdash \psi, B$ instead of $\varphi \cup \{A\} \Vdash \psi \cup \{B\}$. Similarly, we use $\varphi, \varphi' \Vdash \psi, \psi'$ and $\varphi, A_1, \ldots, A_n \Vdash \psi, B_1, \ldots, B_k$, instead of $\varphi \cup \varphi' \Vdash \psi \cup \psi'$ and $\varphi \cup \{A_1, \ldots, A_n\} \Vdash \psi \cup \{B_1, \ldots, B_k\}$ respectively.

DEFINITION 3. (a) A (Scott) consequence relation is said to be *consistent* iff $\emptyset \nVdash \emptyset$.

(b) A Scott consequence relation $\Vdash$ is said to be a *Scott system* (S.S) iff $\Vdash$ is closed under substitution.

DEFINITION 4. Let $\Delta$, $\Theta$ be sets of wffs, and let $\Vdash$ be a Scott consequence relation. We write $\Delta \Vdash \Theta$ iff for some $\varphi \subseteq \Delta$, $\psi \subseteq \Theta$, $\varphi \Vdash \psi$.

**Exercise 5.** Let $\Delta$, $\Theta$ be sets of wffs such that $\Delta \nVdash \Theta$. Define a relation $\Vdash^*$ by $\varphi \Vdash^* \psi$ iff $\Delta, \varphi \Vdash \Theta, \psi$. Show $\Vdash^*$ is a Scott consequence relation.

LEMMA 6. *For any Scott consequence relation* $\Vdash$, *if* $\varphi, A_i \Vdash \psi$ *for* $1 \leq i \leq n$ *and* $\varphi \Vdash \psi, A_1, \ldots, A_n$ *then* $\varphi \Vdash \psi$.

*Proof.* By induction on $n$. For $n = 1$ this is the cut rule. For $n = k + 1$, notice that $\varphi, A_i \Vdash \psi, A_{k+1}$ for $1 \leqslant i \leqslant k$ and $\varphi \Vdash \psi, A_{k+1}, A_1, \ldots, A_k$ and so by the induction hypothesis $\varphi \Vdash \psi, A_{k+1}$. Now since $\varphi, A_{k+1} \Vdash \psi$ we conclude $\varphi \Vdash \psi$.

DEFINITION 7. A *Tarski consequence relation* (for $L$) is a binary relation containing pairs of the form $(\varphi, \psi)$ (written $\varphi \vdash \psi$) with $\bar{\bar{\psi}} = 1$, satisfying the following properties. (We use the conventions of Definition 2 for $\vdash$ as well.)

(a) $A \vdash A$

(b) if $\varphi \vdash \psi$ then $\varphi, \varphi' \vdash \psi$

(c) if $\varphi, C \vdash \psi$ and $\varphi \vdash C$ then $\varphi \vdash \psi$ (cut rule)

DEFINITION 8. (a) A Tarski consequence relation $\vdash$ is called a *Tarski system* (T.S) iff $\vdash$ is closed under substitution.

(b) $\vdash$ is said to be consistent iff for some $\varphi, A, \varphi \nvdash A$.

**Exercise 9.** Let $\Vdash$ be a Scott consequence relation. For $\bar{\bar{\psi}} = 1$ let $\varphi \vdash \psi$ iff (def) $\varphi \Vdash \psi$; show that $\vdash$ is a Tarski consequence relation.

LEMMA 10. *If* $\varphi \vdash A_i$, $1 \leqslant i \leqslant n$ *and* $\varphi, A_1, \ldots, A_n \vdash \psi$ *then* $\varphi \vdash \psi$.

*Proof.* Let $\varphi' \subseteq \{A_1, \ldots, A_n\}$ and let $\bar{\bar{\varphi}}' = n - k$. We show by induction on $k$ that $\varphi \cup \varphi' \vdash \psi$. The lemma will follow for the case $n = k$.

*Case* $k = 1$: Let $\{A_1, \ldots, A_n\} = \varphi' \cup \{A\}$. $A \notin \varphi'$. Then $\varphi \cup \varphi' \vdash A$ and $\varphi, \varphi', A \vdash \psi$ and therefore by the cut rule, $\varphi, \varphi' \vdash \psi$.

*Case* $k$: Let $\varphi'' = \varphi' \cup \{A\}$, $\bar{\bar{\varphi}}'' = n - k$, with $A \notin \varphi'$, $\varphi'' \subseteq \{A_1, \ldots, A_n\}$.

By the induction hypothesis $\varphi, \varphi' \vdash \psi$ but also $\varphi, \varphi' \vdash A$ and therefore by the cut rule, $\varphi, \varphi' \vdash \psi$. This proves Lemma 10.

THEOREM 11. *Let* $\vdash$ *be a Tarski consequence relation and let* $\Vdash^-$ *be defined by* $\varphi \Vdash^- \psi$ *iff (def) for some* $B \in \psi$, $\varphi \vdash B$, *then* $\Vdash^-$ *is a Scott consequence relation.*

*Proof.* Clearly conditions (a) and (b) of Definition 1 are satisfied. We verify the cut rule. Assume that $\varphi, C \Vdash^- \psi$ and $\varphi \Vdash^- \psi, C$. By definition, for some $B \in \psi$ and $A \in \psi \cup \{C\}$ we have that $\varphi, C \vdash B$

and $\varphi \vdash A$. If $A \in \psi$ we are finished. If $A \notin \psi$, then $A = C$, and thus $\varphi \vdash C$, and $\varphi, C \vdash B$ and so $\varphi \vdash B$ and again we are finished.

DEFINITION 12. Let $\vdash$ be a Tarski consequence relation and let $\Vdash$ be a Scott consequence relation (for the same language $L$). $\Vdash$ and $\vdash$ are said *to agree* iff for all $\varphi$, $\psi$, $\bar{\bar{\psi}} = 1$, $\varphi \vdash \psi$ iff $\varphi \Vdash \psi$.

THEOREM 13 (*Scott*). *Let $\vdash$ be a Tarski consequence relation, then there exist two Scott consequence relations $\Vdash^+_{\vdash}$ and $\Vdash^-_{\vdash}$ that agree with $\vdash$ and such that for any $\Vdash$ that agrees with $\vdash$ we have $\Vdash^-_{\vdash} \subseteq \Vdash \subseteq \Vdash^+_{\vdash}$.*

*Proof.* (a) For $\Vdash^-_{\vdash}$ take the Scott consequence relation defined in Theorem 11. Assume that $\Vdash$ is any Scott consequence relation that agrees with $\vdash$. Then if $\varphi \Vdash^-_{\vdash} \psi$ then for some $A \in \psi$, $\varphi \vdash A$ and therefore $\varphi \Vdash A$ and hence $\varphi \Vdash \psi$.

(b) We define $\Vdash^+_{\vdash}$.

Let $\varphi \Vdash^+_{\vdash}$ iff for some finite set $\Delta$, property $(*)$ below holds, Where:

$(*)$ For any partition $(\Delta_1, \Delta_2)$ of $\Delta$ (i.e. $\Delta_1 \cup \Delta_2 = \Delta$, $\Delta_1 \cap \Delta_2 = \emptyset$) there exists a Scott consequence relation $\Vdash$ that agrees with $\vdash$ such that $\varphi, \Delta_1 \Vdash \psi, \Delta_2$.

First we show that $\Vdash^+_{\vdash}$ is a consequence relation

(a) Clearly if $\varphi \neq \emptyset$ then $\varphi \Vdash^+_{\vdash} \varphi$ take $\Delta = \emptyset$

(b) If $\varphi \Vdash^+_{\vdash} \psi$ let $\Delta$ be such that $(*)$ holds, then for any $\varphi'$, $\psi'$, $\varphi \cup \varphi' \Vdash^+_{\vdash} \psi \cup \psi'$ as the same $\Delta$ is adequate.

(c) Assume $\varphi, A \Vdash^+_{\vdash} \psi$ and $\varphi \Vdash^+_{\vdash} \psi, A$. Let $\Delta$, $\Delta^*$ resp. be the two sets having the properties $(*)$ in the definition of $\Vdash^+_{\vdash}$. Regard $\Delta' = \Delta \cup \Delta^* \cup \{A\}$, we claim $\varphi \Vdash^+_{\vdash} \psi$ since $\Delta'$ has the property $(*)$ required in the definition.

We now have to show that $\Vdash^+_{\vdash}$ agrees with $\vdash$. Assume $\varphi \Vdash^+_{\vdash} A$ we want to show that $\varphi \vdash A$. Since $\varphi \Vdash^+_{\vdash} A$, there exists a $\Delta$ with property $(*)$. We show by induction on $n$, that for any $\Theta \subseteq \Delta$, $\bar{\bar{\Theta}} = n$ we have that $\varphi \cup (\Delta - \Theta) \vdash A$. For $n = 1$, let $B \in \Theta$, be arbitrary. So $\varphi \cup (\Delta - \{B\}) \Vdash_1 A, B$ for some $\Vdash_1$, that agrees with $\vdash$, because property $(*)$ holds. Also for some $\Vdash_2$ that agrees with $\vdash$, $\varphi, \Delta \Vdash_2 A$. Since $\Vdash_1$, $\Vdash_2$ agree with $\vdash$ we get that $\varphi, \Delta \Vdash_1 A$ and so by cut, $\varphi \cup (\Delta - \{B\}) \Vdash_1 A$ and so $\varphi \cup (\Delta - \{B\}) \vdash A$.

Now assume that for any $\Theta$, $\bar{\bar{\Theta}} \leq m$, $\varphi \cup (\Delta - \Theta) \vdash A$, show this for any $\Theta$, $\bar{\bar{\Theta}} = m + 1$. Let such $\Theta$ be given then for any $B \in \Theta$, $\varphi \cup$

$(\Delta - \Theta), B \vdash A$. Also by property (*), there exists a $\Vdash$ that agrees with $\vdash$ such that $\varphi, \Delta - \Theta \Vdash \Theta, A$. Now since $\Vdash$ agrees with $\vdash$ we get by Lemma 6 that $\varphi, \Delta - \Theta \Vdash A$, and therefore $\varphi, \Delta - \Theta \vdash A$. This completes the induction step. If we take $\Theta = \Delta$, we get that $\varphi \vdash A$, and thus we see that $\Vdash^+_\vdash$ agrees with $\vdash$.

To show that if $\Vdash$ agrees with $\vdash$ then $\Vdash \subseteq \Vdash^+_\vdash$, assume that $\varphi \Vdash \psi$, then for $\Delta$ empty we get property (*) and so $\varphi \Vdash^+_\vdash \psi$.

**Exercise 14.** Let $\vdash$ be a Tarski consequence relation and let $\Vdash^+_\vdash$ be the maximal Scott consequence relation agreeing with $\vdash$. Let $con(\Delta)$ be $con(\Delta) = \{A \mid \text{for some } \varphi \subseteq \Delta, \varphi \vdash A\}$. Show that:

$$\varphi \Vdash^+_\vdash \psi \quad \text{iff} \quad \bigcap_{B \in \psi} con(\varphi' \cup \{B\}) \subseteq con(\varphi')$$

for all $\varphi' \supseteq \varphi$.

DEFINITION 15. (a) A Hilbert (or axiomatic) system $H$ is a triple $(H_0, H_1, H_2)$ where $H_0$ is a set of wffs called axioms, $H_1$ is a set of rules of the form $A_1, \ldots, A_n/B$, called provability rules and $H_2$ is a set of rules of the form $\varphi/\psi$, with $\bar{\bar{\psi}} = 1$, called consequence rules.

(b) Given a Hilbert system $H$, we define the relation $\vdash_H A$, on wff $A$ as follows: $\vdash_H A$ iff there exists a finite sequence of wff $B_1, \ldots, B_k = A$ such that each $B_i$ of the sequence is either a substitution instance of a member of $H_0$ or for some wffs $A_1, \ldots, A_n$, appearing earlier than $B_i$ in the sequence, we have that $A_1, \ldots, A_n/B_i$ is a rule of $H_1$.

(c) Given a Hilbert system $H$ we define the notion $\varphi \vdash_H \psi$, for $\bar{\bar{\psi}} = 1$ as follows: $\varphi \vdash_H \psi$ iff there exists a sequence of wff $B_1, \ldots, B_n$ such that (i) and (ii) below hold:

(i) For each $i \leq n$ either (1) $B_i \in \varphi$ or (2) $\vdash_H B_i$ ($\vdash_H$ of (b) above) or (3). For some $A_1, \ldots, A_k$ appearing earlier in the sequence $\{A_1, \ldots, A_k\}/\{B_i\}$ is a substitution instance of a rule of $H_2$.

(ii) Either (1) $\psi = \{B\}$ with $B \in \varphi$ or (2) $\{B_1, \ldots, B_n\}/\psi$ is a substitution instance of a rule of $H_2$.

*Remark.* We use the abbreviations of Definition 2 for Hilbert systems as well.

THEOREM 16. *Let H be a Hilbert system, then* $\vdash_H$ *is a Tarski system.*

*Proof.* Let us check the cut rule. Assume $\varphi, C \vdash_H \psi$ and $\varphi \vdash_H C$ we must show that $\varphi \vdash_H \psi$. Let $B_1, \ldots, B_n$ be a proof of $C$ from $\varphi$, and let $A_1, \ldots, A_k$ be a proof of $\psi$ from $\varphi \cup \{C\}$. Then the following is a proof of $\psi$ from $\varphi$: $B_1, \ldots, B_n, C, A_1, \ldots, A_k$. It is easy to verify that $\vdash_H$ is closed under substitution.

THEOREM 17. *Let $\vdash$ be a Tarski system, then there exists a Hilbert system $H$ such that $\vdash = \vdash_H$.*

*Proof.* Let $H$ be the Hilbert system $(H_0, H_1, H_2)$ with $H_0 = \{B \mid \emptyset \vdash B\}$, $H_1 = \emptyset$, $H_2 = \{\varphi/\psi \mid \varphi \vdash \psi\}$; clearly $\vdash \subseteq \vdash_H$. We want to show that $\vdash_H \subseteq \vdash$. Assume $\varphi \vdash_H \psi$. Let $B_1, \ldots, B_n$ be a proof of $\psi$ from $\varphi$. We show by induction on $i$ that $\varphi \vdash B_i$. If $B_i \in \varphi$ this is clear. If $B_i$ is obtained from some $A_1, \ldots, A_k$ appearing previously in the sequence then by the induction hypothesis $\varphi \vdash A_j$ and also by the definition of $H_2$, $A_1, \ldots, A_k \vdash B_i$ therefore by Lemma 10, $\varphi \vdash B_i$. Now $\psi$ is obtained by clause (15c3ii) i.e. either $\psi = \{B\} \subseteq \varphi$, in which case $\varphi \vdash B$ or $\{B_1, \ldots, B_n\}/\psi$ is a rule of $H_2$, i.e. $B_1, \ldots, B_n \vdash \psi$, so again $\varphi \vdash \psi$ by Lemma 10.

DEFINITION 18. Let $\vdash$ be a Tarski consequence relation and let $\Vdash^-_\vdash$ and $\Vdash^+_\vdash$ be the minimal and maximal Scott consequence relations that agree with $\vdash$. Define $S_\vdash$, called *the slash* of $\vdash$ by:

$$S_\vdash = \{\varphi \mid \text{for all } \psi, \quad \varphi \Vdash^+_\vdash \psi \text{ implies } \varphi \Vdash^-_\vdash \psi.\}$$

*Remark.* If $\emptyset \in S_\vdash$, then $\vdash$ has the 'disjunction property' in a certain sense. We shall return to this notion later.

## 2. Scott semantics 1

DEFINITION 1. Let $\Vdash$ be a Scott consequence relation.

(a) By *a theory* we mean a pair $(\Delta, \Theta)$ of sets of wffs.

(b) $(\Delta, \Theta)$ is said to be $\Vdash$-*consistent* iff $\Delta \nVdash \Theta$.

(c) $(\Delta, \Theta)$ is said to be *complete* iff $\Delta \cup \Theta$ is the set of all wffs.

(d) $(\Delta', \Theta')$ is said to *extend* $(\Delta, \Theta)$ iff $\Delta \subseteq \Delta'$, $\Theta \subseteq \Theta'$.

DEFINITION 2. (a) By a *model* we mean a function $\mathbf{t}$ assigning a value in $\{0, 1\}$ to each wff of $L$.

(b) A *semantics* $\mathbf{T}$ is a set of models.

DEFINITION 3. Let **T** be a semantics.

Define: $\varphi \Vdash_{\mathbf{T}} \psi$ iff for all $\mathbf{t} \in \mathbf{T}$ the following holds: If for all $A \in \varphi$, $\mathbf{t}(A) = 1$, then for some $B \in \psi$, $\mathbf{t}(B) = 1$.

LEMMA 4. $\Vdash_{\mathbf{T}}$ *is a Scott consequence relation.*

*Proof.* Exercise.

LEMMA 5. *Let* $(\Delta, \Theta)$ *be* $\Vdash$ *consistent. Then there exists a* $\Vdash$ *consistent and complete extension* $(\Delta', \Theta')$ *of* $(\Delta, \Theta)$.

*Proof.* Let $A_1, A_2, A_3, \ldots$ be an enumeration of all the wffs of $L$. Define by induction a sequence $(\Delta_n, \Theta_n)$ of $\Vdash$-consistent theories such that for all $n$, $\Delta \subseteq \Delta_n \subseteq \Delta_{n+1}$, $\Theta \subseteq \Theta_n \subseteq \Theta_{n+1}$. Let $\Delta_0 = \Delta$, $\Theta_0 = \Theta$. Assume $(\Delta_n, \Theta_n)$ has been defined. Regard $A_n$, if $\Delta_n, A_n \nVdash \Theta_n$, let $\Delta_{n+1} = \Delta_n \cup \{A_n\}$, $\Theta_{n+1} = \Theta_n$. If $\Delta_n, A_n \Vdash \Theta_n$, then $\Delta_n \nVdash \Theta_n, A_n$, since otherwise we can get $\Delta_n \Vdash \Theta_n$, contrary to the inductive hypotheses. So let $\Delta_{n+1} = \Delta_n$, $\Theta_{n+1} = \Theta_n \cup \{A_n\}$. Thus $(\Delta_{n+1}, \Theta_{n+1})$ is defined and is $\Vdash$ consistent in either case.

Now let $\Delta' = \bigcup_n \Delta_n$, $\Theta' = \bigcup_n \Theta_n$.

It is easy to show that $(\Delta', \Theta')$ is the desired extension.

DEFINITION 6. (a) Let $(\Delta, \Theta)$ be a $\Vdash$ consistent and complete theory. Let $\mathbf{t}_{(\Delta,\Theta)}$ be the model with $\mathbf{t}_{(\Delta,\Theta)}(A) = 1$ iff $A \in \Delta$.

(b) Let $\mathbf{T}^*_{\Vdash}$ be the semantics with

$$\mathbf{T}^*_{\Vdash} = \{\mathbf{t}_{(\Delta,\Theta)} \mid (\Delta, \Theta) \Vdash \text{complete and consistent}\}.$$

THEOREM 7. (*Scott completeness theorem*). $\Vdash \, = \, \Vdash_{\mathbf{T}^*_{\Vdash}}$.

*Proof.* $\varphi \Vdash \psi$ iff (by Lemma 5) no $\Vdash$ complete and consistent theory $(\Delta, \Theta)$ extends $(\varphi, \psi)$ iff (by definition) $\varphi \Vdash_{\mathbf{T}^*_{\Vdash}} \psi$.

**Exercise 8.** Let $\Vdash$ be Scott consequence relation, and let $(\Delta, \Theta)$ be a $\Vdash$-consistent theory. Define $\Vdash^*$ by $\varphi \Vdash^* \psi$ iff $\Delta, \varphi \Vdash \Theta, \psi$. Show that $\Vdash^*$ is a Scott consequence relation.

## 3. What is a Classical Connective?

DEFINITION 1. Let $\Vdash$ be a consequence relation and let $f \in 2^{2^n}$. Let $\#$ be an $n$-place connective in the language of $\Vdash$.

Consider the following set of conditions (denoted by $R_f$) on $\Vdash$.

For each $\bar{a} \in \{0, 1\}^n$ take $\#\bar{a}$:

$$\#\bar{a}: \varphi_{\bar{a}} \Vdash \psi_{\bar{a}}$$

Where $\varphi_{\bar{a}}, \psi_{\bar{a}} \subseteq \{A_0, \ldots, A_{n-1}, \#(A_0, \ldots, A_{n-1})\}$ have the property that

$$\bar{a}(i) = 1 \quad \text{iff} \quad A_i \in \varphi_{\bar{a}}$$

$$\bar{a}(i) = 0 \quad \text{iff} \quad A_i \in \psi_{\bar{a}}$$

$$f(\bar{a}) = 1 \quad \text{iff} \quad \#(A_0, \ldots, A_{n-1}) \in \psi_{\bar{a}}$$

$$f(\bar{a}) = 0 \quad \text{iff} \quad \#(A_0, \ldots, A_{n-1}) \in \varphi_{\bar{a}}$$

DEFINITION 2. Let $\Vdash$ be a Scott consequence relation for a language with the $n$-ary connective $\#$. We say the $\#$ is classical in $\Vdash$ with truth table $f$ iff all the conditions of $R_f$ hold for $\Vdash$, for any $A_i$.

When we turn to Tarski consequence relations $\vdash$, the problem of which connectives are classical is more difficult. One may give the following definition.

DEFINITION 3. Let $\vdash$ be a Tarski consequence relation for a language with the connective $\#$. We say that $\#$ is strongly classical in $\vdash$ with truth table $f$ iff for every Scott consequence relation $\Vdash$ agreeing with $\vdash$ we have that $\#$ is classical in $\Vdash$ with truth table $f$.

DEFINITION 4. Let $\vdash$ be a Tarski consequence relation for a language $L$ and $\#$ be a connective of $L$. Then $\#$ is said to be weakly classical with truth table $f$ iff there exists a Scott consequence relation $\Vdash$ agreeing with $\vdash$ in which $\#$ is classical with truth table $f$.

THEOREM 5. *Let $\vdash$ be a consistent Tarski system in a language with the n-ary connective $\#$. Then $\#$ is strongly classical in $\vdash$ iff either* (a) $\emptyset \vdash \#$ *or* (b) *for some $B_1, \ldots, B_k \in \{A_1, \ldots, A_n\}$, $B_1, \ldots, B_k \vdash \#(A_1, \ldots, A_n)$ and $\#(A_1, \ldots, A_n) \vdash B_i$, for each $1 \leq i \leq k$, where $A_1, \ldots, A_n$ are arbitrary atomic wffs.*

*Remark.* Theorem 5 says that essentially only conjunctions can be strongly classical in $\vdash$.

*Proof.* Since $\#$ is strongly classical in $\vdash$, it is classical in $\Vdash^{-}_{\vdash}$, the minimal consequence relation agreeing with $\vdash$. Let $f$ be a truth table, with regard to which $\#$ is classical in $\Vdash^{-}_{\vdash}$. Let $T \subseteq \{1, \dots, n\}$. Let $\bar{a}_T$ be such that $\bar{a}_T(j) = 1$ iff $j \in T$. We proceed to show that the theorem holds. Let $A_1, \dots, A_n$ be atomic, and ask what is the value $f(\bar{a}_\emptyset)$? If the value is 1 then $\emptyset \Vdash^{-}_{\vdash} A_1, \dots, A_n, \#(A_1, \dots, A_n)$, and since $A_i$ are atomic and $\vdash$ consistent, we must have $\emptyset \vdash \#(A_1, \dots, A_n)$, which is case (a) of the theorem.

Otherwise, $f(\bar{a}_\emptyset) = 0$, and so $\#(A_1, \dots, A_n) \vdash A_1, \dots, A_n$ and therefore for some $j \in \{1, \dots, n\}$, $\#(A_1, \dots, A_n) \vdash A_j$. Regard $\bar{a}_{\{j\}}$, if $f(\bar{a}_{\{j\}}) = 1$, then $A_j \Vdash^{-}_{\vdash} \{A_i \mid i \neq j\}, \#(A_1, \dots, A_n)$ and again since $A_i$ are atomic, $A_j \vdash \#(A_1, \dots, A_n)$ and thus case (b) of the theorem holds.

Otherwise, $f(\bar{a}_{\{j\}}) = 0$ and so $A_j, \#(A_1, \dots, A_2) \Vdash^{-}_{\vdash} \{A_i \mid i \neq j\}$. Since $\#(A_1, \dots, A_n) \vdash A_j$, we get that $\#(A_1, \dots, A_n) \Vdash^{-}_{\vdash} \{A_i \mid i \neq j\}$.

Assume by induction on $1 \leqslant i \leqslant n$, that there exists a $T \subseteq \{1, \dots, n\}$, $\bar{\bar{T}} = i$ with the property that either (1) For some $B_{j_1}, \dots, B_{j_k}$ $j_1, \dots, j_k \in T$ case (b) of the theorem holds or (2) $\#(A_1, \dots, A_n) \vdash B_j$ for each $j \in T$.

We find such a $T'$ with $\bar{\bar{T}}' = i + 1$. If case (1) holds, any element can be added to $T$ to form $T'$. If case (2) holds, consider $\bar{a}_T$. If $f(\bar{a}_T) = 1$, then $\{A_j \mid j \in T\} \Vdash^{-}_{\vdash} \{A_j \mid j \notin T\}, \#(A_1, \dots, A_n)$ and since $A_i$ are atomic $\{A_j \mid j \in T\} \vdash \#(A_1, \dots, A_n)$ which yields case (1) for $T$. If $f(\bar{a}_T) = 0$, we get $\{A_j \mid j \in T\}, \#(A_1, \dots, A_n) \Vdash^{-}_{\vdash} \{A_j \mid j \notin T\}$, let $j_0 \notin T$ be such that $\{A_j \mid j \in T\}, \#(A_1, \dots, A_n) \vdash A_{j_0}$. Since $\#(A_1, \dots, A_n) \vdash A_j$ for $j \in T$ we get by Lemma 1.10 that $\#(A_1, \dots, A_n) \vdash A_{j_0}$. This yields case (2) for $T' = T \cup \{j_0\}$.

Now consider the case of $i = n$. If the case (1) holds, then case (b) of the theorem is valid. If case (2) holds, then $\# \vdash A_i$, $1 \leqslant i \leqslant n$. Consider $\bar{a}_T$ for $T = \{1, \dots, n\}$. If $f(a_T) = 1$ we get that $A_1, \dots, A_n \vdash \#(A_1, \dots, A_n)$ which yields case (b) of the theorem. If $f(a_T) = 0$, we get $A_1, \dots, A_n, \#(A_1, \dots, A_n) \Vdash^{-}_{\vdash} \emptyset$ and since $\# \vdash A_i$, $1 \leqslant i \leqslant n$, we get $\# \Vdash^{-}_{\vdash} \emptyset$ which is impossible by the definition of $\Vdash^{-}_{\vdash}$. Thus theorem 5 is proved.

**Exercise 6.** Let $\Vdash$ be a Scott consequence relation and let $\#(A_1, \dots, A_n)$ be an $n$-ary connective. Show that $\#$ is classical in $\Vdash$ with truth table $f$ iff for all $\mathbf{t} \in T_{\Vdash}$, $\mathbf{t}(\#(A_1, \dots, A_n) = f(\mathbf{t}(A_1), \dots, \mathbf{t}(A_n))$.

**Exercise 7.** Let $\vdash$ be a Tarski consequence relation and $\Vdash_{\vdash}^{-}$, $\Vdash_{\vdash}^{+}$ be the maximal and minimal Scott consequence relations that agree with it. Show that:

(a) If $\vdash$ is a Tarski system then $\Vdash_{\vdash}^{-}$, $\Vdash_{\vdash}^{+}$ are Scott systems.

(b) If the connective # is weakly classical in $\vdash$ (with truth table $f$) then it is classical in $\Vdash_{\vdash}^{+}$ (with truth table $f$).

(c) If the connective # is strongly classical in $\vdash$ (with truth table $f$) then it is classical in $\Vdash_{\vdash}^{-}$ (with truth table $f$).

COROLLARY. *If $\#_i$ are weakly classical in $\vdash$, for $1 \leqslant i \leqslant n$, then for some $\Vdash$ that agrees with $\vdash$, $\#_i$, $1 \leqslant i \leqslant n$ are all classical in $\Vdash$.*

**Exercise 8.** Let $\Vdash$ be a Scott consequence relation in a language with some or all of the following connectives:

| | |
|---|---|
| $t, f$ | zero place |
| $\sim$ | one place |
| $\wedge, \vee, \rightarrow$ | two place |

Show that these connectives have their respective classical table in $\Vdash$ iff the following holds (respectively), for all $A$, $B$.

(1) $\emptyset \Vdash t$

(2) $f \Vdash \emptyset$

(3) $A \wedge B \Vdash A$; $A \wedge B \Vdash B$; $A, B \Vdash A \wedge B$

(4) $A \Vdash A \vee B$; $B \Vdash A \vee B$; $A \vee B \Vdash A, B$

(5) $A, \sim A \Vdash \emptyset$; $\emptyset \Vdash A, \sim A$

(6) $A, A \rightarrow B \Vdash B$; $\emptyset \Vdash A, A \rightarrow B$.

DEFINITION 9. Let $\Vdash$ be a Scott consequence relation and let $Q$ be a unary quantifier of the language. We say that $Q$ is the classical universal quantifier in $\Vdash$ iff the following always holds for all $A$, $\varphi$, $\psi$.

(a) $(Qx)A(x) \Vdash A(y)$

(b) $\varphi \Vdash A(x), \psi$ iff $\varphi \Vdash (Qx)A(x), \psi$

where $x$ does not appear free in any wff $\varphi \cup \psi$.

We say that $Q$ is the classical existential quantifier in $\Vdash$ iff the following holds:

(a) $A(y) \Vdash (Qx)A(x)$

(b) $\varphi, A(x) \Vdash \psi$ iff $\varphi, (Qx)A(x) \Vdash \psi$

where $x$ does not appear free in any wff of $\varphi \cup \psi$.

DEFINITION 10. Let $\vdash$ be a Tarski consequence relation and let $Q$ be a unary quantifier. We say that $Q$ is strongly [weakly] the classical universal (existential) quantifier in $\vdash$ iff for every [some] Scott consequence $\Vdash$ that agrees with $\vdash$, $Q$ is the classical universal (existential) quantifier in $\Vdash$ (respectively), [respectively].

**Example 11.** Consider a language with disjunction $\vee$ and the quantifier $\forall$. Consider the smallest Tarski system satisfying:

(a) $A \vdash A \vee B$

$B \vdash A \vee B$

(b) $A \vdash C$ and $A' \vdash C$ imply $A \vee A' \vdash C$

(c) $\varphi \vdash A(x)$ implies $\varphi \vdash \forall x A(x)$, $x$ not free in any member of $\varphi$.

(d) $\forall x A(x) \vdash A(y)$.

Show that $\vee$ is weakly classical in $\vdash$ and $\forall$ is weakly classical in $\vdash$ (as the classical disjunction and classical universal quantifier resp.) but that there cannot exist a $\Vdash$ agreeing with $\vdash$ with the property that both $\vee$ and $\forall$ are classical in $\Vdash$.

**Exercise 12.** Let $\vdash_i$ be Tarski consequence relations and let $\Vdash_i^+$, $\Vdash_i^-$ be the respective maximal and minimal Scott consequence relations agreeing with $\vdash_i$. Is the following true?

(a) $\vdash_1 \subseteq \vdash_2$ implies $\Vdash_1^- \subseteq \Vdash_2^-$

(b) $\vdash_1 \subseteq \vdash_2$ implies $\Vdash_2^+ \subseteq \Vdash_2^+$.

*Dual Connectives*

The reader examining the definition of a Scott consequence relation $\Vdash$ given in Section 1 will notice that conditions (a), (b), (c) of (1.1.) are completely symmetrical in $\Vdash$ or $\dashV$ ($\dashV$ is the inverse relation to $\Vdash$). This leads us to the following notions.

DEFINITION 13. Let $\Vdash$ be a consequence relation. Let $\Vdash^*$ be defined as $\varphi \Vdash^* \psi$ iff $\psi \Vdash \varphi$. Then $\Vdash^*$ is called the dual consequence relation of $\Vdash$.

THEOREM 14. (a) $\Vdash^*$ *of Definition* 13 *is a consequence relation.*

(b) *If the connective # is classical in* $\Vdash$ *with truth table f then # is classical in* $\Vdash^*$ *with truth table* $f^*$, *where* $f^*$ *is the dual truth table, i.e.,*

$$f(x_1, \dots, x_n) = y \Leftrightarrow f^*(1 - x_1, \dots, 1 - x_n) = 1 - y.$$

*Proof.* Exercise.

DEFINITION 15. Let $\Vdash$ be a consequence relation in a language with a certain class of connectives and quantifiers. Let $x$ be a connective (or quantifier) of $\Vdash$ and let $x^*$ be another connective (or resp. quantifier) of $\Vdash$. For each wff $A$, let $A^*$ be the wff obtained from $A$ by interchanging $x$ by $x^*$ and $x^*$ by $x$. Thus $A^{**} = A$. For any $\varphi$ let $\varphi^* = \{A^* \mid A \in \varphi\}$. We say that $x$ and $x^*$ are *dual* in $\Vdash$ whenever the following holds:

$$\varphi \Vdash \psi \quad \text{iff} \quad \psi^* \Vdash \varphi^*$$

Where $\varphi$, $\psi$ are built from wffs $A$ containing only atoms and $x$ and $x^*$ and variables.

DEFINITION 16. $\Vdash$ is said to be *dually complete* iff there exists a pairing function * from connectives to connectives and from quantifiers to quantifiers such that for all $a$, $a$ and $a^*$ are dual in $\Vdash$.

DEFINITION 17. Given a Scott system $\Vdash$ in a certain language, we define the *dual completion* $\Vdash^D$ of $\Vdash$. With each connective or quantifier $x$ of $\Vdash$ associate a new connective or quantifier $x^*$ resp. with the same number of places. For each $A$ in the new language let

$A^*$ be obtained by interchanging all $x$ with $x^*$ and $x^*$ with $x$. Let $\Vdash^D$ be the smallest Scott consequence system such that

(1) $\varphi \Vdash^D \psi$ iff $\psi^* \Vdash^D \phi^*$

(2) The restriction of $\Vdash^D$ to the old language extends $\Vdash$.

THEOREM 18. *The dual completion of* $\Vdash$ *exists but is not necessarily a conservative extension of* $\Vdash$.

*Proof.* Exercise.

**Exercise 19.** Show that:

(a) $\wedge$, $\vee$ are dual.

(b) $(x)$, $(\exists x)$ are dual.

(c) $t$, $f$ are dual.

## 4. Scott semantics 2

In this section the connectives $t$, $f$, $\wedge$, $\vee$, $\rightarrow$, $\sim$ have their classical meaning whenever they appear, i.e., we assume in this section that any $\Vdash$ in a language containing these connectives satisfies the conditions of (3.8). Similarly we assume $(\exists x)$, $(x)$ denote the classical existential and universal quantifiers in this section, respectively, i.e., the conditions of (3.9) hold.

DEFINITION 1. (a) A theory $(\Delta, \Theta)$ is said to be $\Vdash$-saturated in a set of variables $U$ iff it is $\Vdash$ consistent and has the following properties, whenever the respective classical connective is in $\Vdash$.

(1) If $\Delta \Vdash A$ then $A \in \Delta$.

(2) If $A \vee B \in \Delta$ then $A \in \Delta$ or $B \in \Delta$.

(3) If $(\exists x)A(x) \in \Delta$ then for some variable $u$ in $U$, $A(u) \in \Delta$.

(4) If $(\Delta, \Theta \cup \{(x)A(x)\})$ is $\Vdash$-consistent then for some variable $u$ in $U$, $(\Delta, \Theta \cup \{A(u)\})$ is $\Vdash$-consistent.

(5) If $A \vee B \in \Theta$ then $A \in \Theta$ and $B \in \Theta$.

(6) If $A \wedge B \in \Theta$ then $A \in \Theta$ or $B \in \Theta$.

(b) $(\Delta, \Theta)$ is said to be *complete* in $U$ iff for any $A$ in the language with variables $U$, $A \in \Delta$, or $A \in \Theta$.

(c) We say $(\Delta, \Theta)$ is saturated iff it is saturated in the set of variables of the language. We say $(\Delta, \Theta)$ is complete iff for all $A$ of the language, $A \in \Delta \cup \Theta$.

LEMMA 2. *If* $(\Delta, \Theta)$ *is a complete* $\Vdash$*-saturated theory, define* s *by* $s(A) = 1$ *iff* $A \in \Delta$, *then* s *is a model.*

*Proof.* Exercise.

LEMMA 3. *Let* $(\Delta', \Theta')$ *be a* $\Vdash$*-consistent theory such that there exist an infinite number of variables not appearing in* $\Delta \cup \Theta$. *Then* $(\Delta', \Theta')$ *can be extended to a* $\Vdash$*-saturated and complete theory* $(\Delta, \Theta)$.[1]

*Proof.* Let $B_0, B_1, B_2, \ldots$ be an enumeration of all the wff of L. We define by induction a sequence of $\Vdash$-consistent theories $(\Delta_n, \Theta_n)$ with the properties $(\Delta_0, \Theta_0) = (\Delta', \Theta')$ and $\Delta_{n+1} \supseteq \Delta_n$, $\Theta_{n+1} \supseteq \Theta_n$ and $(\bigcup \Delta_n, \bigcup \Theta_n)$ is complete, and that there are infinitely many variables not appearing in $\Delta_n \cup \Theta_n$.

Assume that $(\Delta_n, \Theta_n)$ has been defined, we define $(\Delta_{n+1}, \Theta_{n+1})$, $n \geq 0$.

*Case* 1. $(\Delta_n, \Theta_n \cup \{B_n\})$ is $\Vdash$-consistent. If $B_n$ is not of the form $(x)$ $A(x)$, let $\Delta_{n+1} = \Delta_n$, $\Theta_{n+1} = \Theta_n \cup \{B_n\}$. If $B_n = (x)\ A(x)$, let $\Delta_{n+1} = \Delta_n$, $\Theta_{n+1} = \Theta_n \cup \{B_n, A(u)\}$, where $u$ is the first variable not appearing in $\Delta_n \cup \Theta_n \cup \{B_n\}$. To show that $(\Delta_{n+1}, \Theta_{n+1})$ is $\Vdash$-consistent, assume that for $\varphi \subseteq \Delta_{n+1}$, $\psi \subseteq \Delta_{n+1}$ we have $\varphi \Vdash \psi \cup \{B_n, A(u)\}$. By 3.9 we get $\varphi \Vdash \psi \cup \{B_n\}$, contradicting the assumption of Case 1.

*Case* 2. $(\Delta_n, \Theta_n \cup \{B_n\})$ is not consistent.

In this case we claim that $(\Delta_n \cup \{B_n\}, \Theta_n)$ is consistent. For otherwise for some $\varphi_1, \varphi_2 \subseteq \Delta_n$, $\psi_1, \psi_2 \subseteq \Theta_n$ we have $\varphi_1, B_n \Vdash \psi$, but also for our case, $\varphi_2 \Vdash B_n, \psi_2$. Thus $\varphi_1 \cup \varphi_2, B_n \Vdash \psi_1 \cup \psi_2$ and $\varphi_1 \cup \varphi_2 \Vdash B_n, \psi_1 \cup \psi_2$ and therefore $\varphi_1 \cup \varphi_2 \Vdash \psi_1 \cup \psi_2$, a contradiction to the consistency of $(\Delta_n, \Theta_n)$. Thus $(\Delta_n \cup \{B_n\}, \Theta_n)$ is $\Vdash$-consistent. If $B_n$ is not of the form $(\exists x)\ A(x)$, then let $\Delta_{n+1} = \Delta_n \cup \{B_n\}$, $\Theta_{n+1} = \Theta_n$. If $B_n = (\exists x)\ A(x)$, let $\Delta_{n+1} = \Delta_n \cup \{(\exists x)\ A(x),\ A(u)\}$, $\Theta_{n+1} = \Theta_n$, where $u$ is the first variable not in $\Delta_n \cup \Theta_n \cup \{B_n\}$. To show that $(\Delta_{n+1}, \Theta_{n+1})$ is consistent in the latter case, assume that for some $\varphi$, $\psi$; $\varphi, A(u) \Vdash \psi$, then by (3.9), $\varphi, (\exists x)\ A(x) \Vdash \psi$ which is impossible.

[1] The connectives, $f$, $\wedge$, $\vee$, $\rightarrow$, $\exists$, ( ) are considered classical with their usual meaning whenever they are mentioned in the proof.

Now let $\Delta = \bigcup_n \Delta_n$, $\Theta = \bigcup_n \Theta_n$, then $(\Delta, \Theta)$ is a complete $\Vdash$-saturated extension of $(\Delta', \Theta')$.

DEFINITION 4. Let $\Vdash$ be a consequence relation for a language $L$. Let $\tau = (\Delta, \Theta)$ be a $\Vdash$-saturated and complete theory. Define a model $\mathbf{s}(\tau)$ by:

$$\mathbf{s}(\tau)(A) = 1 \quad \text{iff} \quad A \in \Delta, \qquad \text{for all } A.$$

Let $\mathbf{T}_{\Vdash}$ be $\{\mathbf{s}(\tau) \mid \tau$ is a $\Vdash$-saturated and complete theory$\}$. Then $\mathbf{T}_{\Vdash}$ is called the *canonical semantics for* $\Vdash$.

LEMMA 5. *$T_{\Vdash}$ is a semantics.*

*Proof.* Exercise.

THEOREM 6. (*The Completeness Theorem*). *Let $\Vdash$ be a consequence relation, then*: $\Vdash = \Vdash_{\mathbf{T}_{\Vdash}}$ (*that is, $\Vdash$ is complete for its cononical semantics*).

*Proof.* It is clear that $\Vdash \subseteq \Vdash_{\mathbf{T}_{\Vdash}}$. For the other direction, let $\varphi$, $\psi$ be such that $\varphi \nVdash \psi$. Then $(\varphi, \psi)$ is a $\Vdash$-consistent theory fulfilling the assumptions of (1.4). So there exists a $\Vdash$-complete and saturated extension $\tau$ of $(\varphi, \psi)$. Look at $\mathbf{s}(\tau)$. $\mathbf{s}(\tau) \in \mathbf{T}$ and for all $A \in \varphi$, $\mathbf{s}(\tau)(A) = 1$ and for all $A \in \psi$, $\mathbf{s}(\tau)(A) = 0$. Therefore $\varphi \nVdash_{\mathbf{T}_{\Vdash}}$. Thus $\Vdash_{\mathbf{T}_{\Vdash}} \subseteq \Vdash$.

**Exercise 7.** Let $\mathbf{T}$ be a set of models for the language and let $\Vdash$ be a consequence relation for the language, we say that $\Vdash$ is complete for $\mathbf{T}$ iff $\Vdash = \Vdash_{\mathbf{T}}$. We say that $\Vdash$ is strongly complete for $\mathbf{T}$ iff $\Vdash = \Vdash_{\mathbf{T}}$ and every $\Vdash$-consistent theory has a model in $\mathbf{T}$.

DEFINITION 8. Let $\mathbf{T}$ be a semantics. $\mathbf{T}$ is said to be *compact* iff $\mathbf{T}$ equals the canonical semantics of $\Vdash_{\mathbf{T}}$.

Show the following:

LEMMA. $\mathbf{T}$ *is compact iff* $\Vdash_{\mathbf{T}}$ *is strongly complete for* $\mathbf{T}$.

## NOTES

The central notion, that of a consequence relation is due to D. Scott 1974. Using Scott's notion as basis, the author answered the question of what is a classical connective in a Hilbert system. The material on dual connectives follows naturally from Scott's point of view and was included to accomodate materials of C. Rauszer.

CHAPTER 2

# INTRODUCING HPC

We introduce first Heyting's implication as the weakest implication for which the deduction theorem holds. The full Heyting (intuitionistic) predicate calculus is then introduced and some of its basic properties are studied. Some neighbouring systems are also considered.

## 1. Heyting's implication

In this section we consider a propositional language with only one binary connective $\Rightarrow$ called implication. We are interested in studying consequence relations and axiomatic systems in this language that fulfill various versions of the deduction theorem, *DT*, (i.e., the rule $\varphi$, $A \Vdash B$ iff $\varphi \Vdash A \Rightarrow B$). We shall see that Heyting's implication is the weakest possible implication fulfilling *DT*.

DEFINITION 1. Let $Y$ be the axiomatic system defined by the following axiom schema and provability rule:

*Axiom schema*: Any wff of the form $A_1 \Rightarrow (A_2 \Rightarrow \cdots (A_n \Rightarrow A) \cdots)$, where $A$ is equal to $A_i$ for some $1 \leqslant i \leqslant n$.

*Provability rule*: Modus ponens for $\Rightarrow$, i.e., from $A$ and $A \Rightarrow B$, infer $B$.

We write $Y \vdash B$ iff there exists a sequence of wff $B_0, \ldots, B_m = B$ with the property that each $B_i$ of the sequence is either an axiom of $Y$ or is obtained from two previous members of the sequence of using modus ponens.

Let $\Vdash$ denote the smallest Scott consequence relation for our language that is closed under the condition

$$D: \qquad \text{If } \varphi \cup \{A\} \Vdash \{B\} \quad \text{then} \quad \varphi \Vdash \{A \Rightarrow B\}.$$

THEOREM 2. (*W. Pogorzelski*) *For any* $A$, $Y \vdash A$ *iff* $\emptyset \Vdash \{A\}$.

*Proof.* We first show that $Y$ has the property:

LEMMA 3. *If* $Y \vdash A$ *then* $A$ *is an axiom.*

*Proof.* The proof is by induction on the length of the sequence $A_0, \ldots, A_n = A$ that proves $A$ in $Y$.

(a) The case that $A$ is an axiom is clear.

(b) Suppose $A$ is obtained from $A_k$ and $A_j = A_k \Rightarrow A$ by modus ponens.

Then by the induction hypothesis, both $A_k$ and $A_j$ are axioms. That is

$$A_k = B_1 \Rightarrow (\cdots (B_m \Rightarrow B) \cdots)$$

$$A_j = A_k \Rightarrow (D_1 \Rightarrow \cdots \Rightarrow (D_r \Rightarrow D) \cdots).$$

Where $A = D_1 \Rightarrow (\cdots (D_r \Rightarrow D) \cdots)$, and $D \in \{A_k, D_1, \ldots, D_r\}$, $B \in \{B_1, \ldots, B_m\}$. If $D = A_k$, then certainly $A$ is an instance of an axiom. If $D = D_i$, for some $i$, then again $A$ is an axiom. Thus the lemma is proved.

LEMMA 4. *$Y \vdash A$ implies $\emptyset \Vdash \{A\}$.*

*Proof.* Assume $Y \vdash A$, then by Lemma 1, $A = D_1 \Rightarrow (\cdots \Rightarrow (D_r \Rightarrow D) \cdots)$ with $D \in \varphi = \{D_1, \ldots, D_r\}$. We also have, by definition of $\Vdash$, that $\varphi \Vdash \{D\}$, and so by repeated application of rule $D$ we get $\emptyset \Vdash \{A\}$.

LEMMA 5. *Let $\Vdash^*$ be the binary relation defined below, then $\Vdash^*$ is a consequence relation and rule D holds for $\Vdash^*$: $\varphi \Vdash^* \psi$ iff some $A \in \psi$ can be presented in the form $A = B_1 \Rightarrow (\cdots \Rightarrow (B_n \Rightarrow B) \cdots)$ with $B \in \varphi$ or $B \in \{B_1, \ldots, B_n\}$.*

*Proof.* First we show that this is a Scott consequence relation. We check the cut rule. Assume that $\varphi \cup \{C\} \Vdash^* \psi$, and $\varphi \Vdash^* \psi \cup \{C\}$, we want to show that $\varphi \Vdash^* \psi$. Since $\varphi \Vdash^* \psi \cup \{C\}$, for some $A \in \psi \cup \{C\}$ we have $\varphi \Vdash^* \{A\}$. If $A \neq C$ we are finished. Assume therefore that $\varphi \Vdash^* \{C\}$. We also have that for some $A' \in \psi$, $\varphi \cup \{C\} \Vdash^* \{A'\}$. Let $A' = B_1 \Rightarrow \cdots \Rightarrow (B_n \Rightarrow B) \cdots)$ be such that $B \in \varphi \cup \{C\} \cup \{B_1, \ldots, B_n\}$. If $B \neq C$ we are finished. We therefore assume $B = C$. Now since $\varphi \Vdash^* \{C\}$ let $C = E_1 \Rightarrow \cdots \Rightarrow (E_m \Rightarrow E) \cdots)$ be a presentation of $C$ with $E \in \varphi \cup \{E_1, \ldots, E_m\}$. Thus, since $B = C$, $A' = B_1 \Rightarrow \cdots \Rightarrow (B_n \Rightarrow (E_1 \Rightarrow \cdots \Rightarrow (E_m \Rightarrow E) \cdots)$ is a presentation of $A' \in \psi$ such that $E \in \varphi \cup \{B_1, \ldots, B_n, E_1, \ldots, E_m\}$. Thus $\varphi \Vdash^* \psi$.

It is simple to verify the other rules. Thus $\Vdash^*$ is a Scott consequence relation.

To verify that $D$ holds for $\Vdash^*$ assume that $\varphi \cup \{A\} \Vdash^* B$. Let

$B = B_1 \Rightarrow \cdots \Rightarrow (B_n \Rightarrow C) \cdots)$ be a presentation of $B$ such that $C \in \{B_1, \ldots, B_n\} \cup \varphi \cup \{A\}$. Then clearly $A \Rightarrow (B_1 \Rightarrow \cdots (B_n \Rightarrow C) \cdots)$ is presentation of $A \Rightarrow B$ such that $C \in \{A, B_1, \ldots, B_n\} \cup \varphi$.

LEMMA 6. *If* $\emptyset \Vdash \{A\}$ *then* $Y \vdash A$.

*Proof.* In Lemma 4 we found that $\Vdash^*$ is a consequence relation fulfilling rule *D*. Since $\Vdash$ is the smallest such relation, $\Vdash \subseteq \Vdash^*$. Thus $\emptyset \Vdash^* \{A\}$, but by definition this implies $Y \vdash A$.

Thus the theorem follows.

**Exercise 7.** (a) Let $\Vdash$ be the smallest Scott consequence relation with the property that $\varphi \Vdash A \Rightarrow B$ implies $\varphi, A \Vdash B$. Let $Y$ be the Hilbert system with only one consequence rule $\{A, A \Rightarrow B\} | \{B\}$ and no axioms or provability rules. Show that for all $\varphi$ and A, $\varphi \Vdash A$ iff $\varphi \vdash_Y A$.

(b) *Weak version of the deduction theorem.* Let $\Vdash_n$ be the smallest Scott consequence relation fulfilling:

$D_n$: If $\varphi \cup \{A\} \Vdash_n \{B\}$ then $\varphi \Vdash_n \{A \Rightarrow B\}$,

for any $\emptyset$ with not more than $n$-elements.

*Problem.* Give an axiom system $Y_n$ such that for all $A$, $Y_n \vdash A$ iff $\emptyset \Vdash_n \{A\}$.

DEFINITION 8. Heyting's *Implication*: Let L be a propositional language with just one connective, the non-classical binary connective $\Rightarrow$. Let $\Vdash$ be the smallest consequence relation on L closed under the condition.

*DT*: $\varphi \cup \{A\} \Vdash \{B\}$ iff $\varphi \Vdash \{A \Rightarrow B\}$.

$\Vdash$ is the smallest Scott consequence relation with the property that the full deduction theorem holds (compare with 1.1).

Let $H^{++}$ be the axiom system for $\Rightarrow$ with the rule and axioms below: The rule is taken as both a provability rule and a consequence rule.

*Rule*: From $A$ and $A \Rightarrow B$ deduce $B$.

*Axiom schemas*:

(a) $A \Rightarrow (B \Rightarrow A)$

(b) $[A \Rightarrow (B \Rightarrow C)] \Rightarrow [(A \Rightarrow B) \Rightarrow (A \Rightarrow C)]$.

THEOREM 9. $H^{++} \vdash A$ *iff* $\emptyset \Vdash \{A\}$.

*Remark.* $H^{++} \vdash A$ means that there exists a sequence $B_0, \ldots, B_n = A$ such that each $B_i$ is either an axiom or is obtained from some $B_j$, $B_k$, $j, k < i$ by the rule (modus ponens).

Note also that $H^{++}$ is what is known in the literature as the implication fragment of Heyting's propositional calculus, as we shall see later on.

*Proof.* We now prove the theorem

LEMMA 10. $\varphi \Vdash \{A\}$ *and* $\varphi \Vdash \{A \Rightarrow B\}$ *imply* $\varphi \Vdash \{B\}$.

*Proof.* From the assumptions we get that $\varphi \Vdash \{A, B\}$ and $\varphi \cup \{A\} \Vdash \{B\}$ (using $DT$) and therefore $\varphi \Vdash \{B\}$.

LEMMA 11. $\varphi = \{A \Rightarrow (B \Rightarrow C), A \Rightarrow B, A\} \Vdash \{C\}$.

*Proof.* Since $\varphi \Vdash \{A \Rightarrow B\}$ and $\varphi \Vdash \{A\}$ we get $\varphi \Vdash \{B\}$. Similarly we get $\varphi \Vdash \{B \Rightarrow C\}$ and therefore $\varphi \Vdash \{C\}$.

LEMMA 12. *For any A, if* $H^{++} \vdash A$ *then* $\varphi \Vdash \{A\}$.

*Proof.* We show first that $\varphi \Vdash$ axioms.

(a) $\{A, B\} \Vdash \{A\}$ therefore $\{A\} \Vdash \{B \Rightarrow A\}$ by $DT$ and therefore

$$\varphi \Vdash \{A \Rightarrow (B \Rightarrow A)\}.$$

(b) By Lemma 17,

$$\{A \Rightarrow (B \Rightarrow C), A \Rightarrow B, A\} \Vdash \{C\}$$

and so by repeated use of $DT$ we get

$$\varphi \Vdash \{[A \Rightarrow (B \Rightarrow C)] \Rightarrow [(A \Rightarrow B) \Rightarrow (A \Rightarrow C)]\}.$$

(c) Modus ponens is $\varphi \Vdash \{A\}$ and $\varphi \Vdash \{A \Rightarrow B\}$ implies $\varphi \Vdash \{C\}$ which follows from Lemma 10.

To continue with the proof of the other direction of the theorem, we need a definition. Let $\varphi \vdash_{H^{++}} A$ mean that there exists a sequence of wff $B_0, \ldots, B_n = A$ such that each $B_i$ in the sequence is either an axiom of $H^{++}$ or is a member of $\varphi$ or is obtained from two previous wff of the sequence by using modus ponens. This sequence is called a proof of $A$ from $\varphi$. Note that $\varphi \vdash A$, iff $\varphi \vdash_H \{A\}$ in the sense of definition (1.1.15).

LEMMA 13. $\varphi \cup \{A\} \vdash_{H^{++}} B$ iff $\varphi \vdash_{H^{++}} A \Rightarrow B$.

*Proof.* If $\varphi \vdash_{H^{++}} (A \Rightarrow B)$, let $C_0, \ldots, C_n$ be a proof of $(A \Rightarrow B)$ from $\varphi$. Then $C_0, \ldots, C_n, B$ is a proof of $B$ from $\varphi \cup \{A\}$. This proves one direction of the lemma. To obtain the other direction we use induction on the length of the proof of $B$ from $\varphi \cup \{A\}$.

*Case a.* Length less than 2.

Then $B$ must be either an axiom or a member of $\varphi$.

(a1) If $B \neq A$, then the following is a proof of $A \Rightarrow B$ from $\varphi$:

$$B \Rightarrow (A \Rightarrow B), \qquad B, A \Rightarrow B.$$

(a2) If $A = B$, then the following is a proof of $A \Rightarrow A$ from $\varphi$:

$$B_0 = A \Rightarrow ((A \Rightarrow A) \Rightarrow A) \Rightarrow ((A \Rightarrow (A \Rightarrow A)) \Rightarrow (A \Rightarrow A))$$

$$B_1 = A \Rightarrow ((A \Rightarrow A) \Rightarrow A)$$

$$B_2 = (A \Rightarrow (A \Rightarrow A)) \Rightarrow (A \Rightarrow A)$$

$$B_3 = A \Rightarrow (A \Rightarrow A)$$

$$B_4 = A \Rightarrow A.$$

*Case b.* Length of 2 or more.

(b1) If $B$ is an axiom or a member of $\varphi \cup \{A\}$, we proceed as in Cases (a1) and (a2) above.

(b2) Suppose the sequence is $C_0, \ldots, C_n = B$ and $B$ is obtained by modus ponens. That is for some $i, k$, $C_k = C_i \Rightarrow B$. By the induction hypothesis let $E_0, \ldots, E_m$ be a proof $A \Rightarrow C_i$ from $\varphi$ and let $F_0, \ldots, F_r$ be a proof $A \Rightarrow C_k$ from $\varphi$. The following is a proof of $(A \Rightarrow B)$ from $\varphi$:

$$E_0, \ldots, E_m, F_0, \ldots, F_r, (A \Rightarrow (C_i \Rightarrow B))$$
$$\Rightarrow ((A \Rightarrow C_i) \Rightarrow (A \Rightarrow B)), (A \Rightarrow C_i) \Rightarrow (A \Rightarrow B),$$
$$(A \Rightarrow B).$$

LEMMA 14. *Let $\varphi \vdash_{H^{++}} \psi$ be defined to mean that $\varphi \vdash_{H^{++}} A$ for some $A \in \psi$, then $\vdash_{H^{++}}$ is a consequence relation fulfilling on the rule DT.*

*Proof.* We show that $\vdash_{H^{++}}$ is a Scott consequence relation. Clearly (1.1.1a) and (1.1.1b) are valid. For (1.1.1a) recall that $H^{++} \vdash A \Rightarrow A$ and therefore $\varphi \vdash_{H^{++}} A$ if $A \in \varphi$. To check (1.1.1c) assume that $\varphi \cup \{A\} \vdash_{H^{++}} \psi$ and $\varphi \vdash_{H^{++}} \psi \cup \{A\}$. Then for some $B \in \psi \cup \{A\}$, $\varphi \vdash_{H^{++}} B$. If $B \neq A$ we are finished, as $\varphi \vdash_{H^{++}} \psi$ immediately follows. If $B = A$, notice that $\varphi \cup \{A\} \vdash_{H^{++}} \psi$, and therefore for some $C \in \psi$, $\varphi \cup$

$\{A\} \vdash_{H^{++}} C$, but since $\varphi \vdash_{H^{++}} A$ we conclude $\varphi \vdash_{H^{++}} C$ and therefore $\varphi \vdash_{H^{++}} \psi$.

Lemma 13 shows that $DT$ is valid for $\vdash_{H^{++}}$.

LEMMA 15. *For any* $A$, $\emptyset \Vdash \{A\}$ *implies* $H^{++} \vdash A$.

*Proof.* Look at the consequence relation $\vdash_{H^{++}}$ (of Lemma 14). Since $\Vdash$ is the smallest consequence relation fulfilling $DT$ we obtain that $\Vdash \subseteq \vdash_{H^{++}}$. Now assume that $\emptyset \Vdash \{A\}$, then certainly $\emptyset \vdash_{H^{++}} A$, i.e., $H^{++} \vdash A$.

The theorem follows from Lemmas 12 and 15.

**Exercise 16.** Show that the condition $DT$ is equivalent to the following two conditions.

(1) The condition $D$ of deduction:

$$\varphi, A \Vdash B \quad \text{implies} \quad \varphi \Vdash (A \Rightarrow B).$$

(2) The condition of modus ponens under assumptions:

$$\text{If} \quad \varphi \Vdash (A) \quad \text{and} \quad \varphi \Vdash (A \Rightarrow B) \quad \text{then} \quad \varphi \Vdash (B).$$

**Exercise 17.** Find two consequence relations $\Vdash_1$, $\Vdash_2$ on some language L such that $\Vdash_1 \neq \Vdash_2$ but for all $A$, $\emptyset \Vdash_1 A$ iff $\emptyset \Vdash_2 A$. This example shows that the notion of a Scott consequence relation is stronger than that of an axiomatic logical system.

**Exercise 18.** *Kripke's interpretation for Heyting's implication.*
Let $\Vdash$ be the Scott system of definition 8. Let $\mathbf{T}_{\Vdash}$ be its canonical semantics. For $\mathbf{t}, \mathbf{s} \in \mathbf{T}_{\Vdash}$ let $\mathbf{t} \subseteq \mathbf{s}$ iff for all wff $A$, $\mathbf{t}(A) = 1$ implies $\mathbf{s}(A) = 1$. Show that $\mathbf{t}(A \Rightarrow B) = 0$ iff for some $\mathbf{s}$ such that $\mathbf{t} \subseteq \mathbf{s}$, $\mathbf{s}(A) = 1$ and $\mathbf{s}(B) = 0$.

**Exercise 19.** *The dual completion for Heyting's implication,* (C. Rauzer).
Let $\Vdash^D$ be a Scott consequence system in a language with two binary connectives $\Rightarrow$ and $\Leftarrow$ such that $\Vdash^D$ is the smallest consequence system satisfying:

(D1) $\quad \varphi, A \Vdash^D B \quad \text{iff} \quad \varphi \Vdash^D A \Rightarrow B$

(D2) $\quad B \Vdash^D \varphi, A \quad \text{iff} \quad A \Leftarrow B \Vdash^D \varphi.$

(a) Show that $\Rightarrow$ and $\Leftarrow$ are dual in $\Vdash^D$ (see (1.3.15) for this notion).

(b) Present a finite list of axioms and rules, i.e., a Hilbert system $\vdash^D$ such that $\vdash^D$ and $\Vdash^D$ agree.

(c) Show that in $\mathbf{T}_{\Vdash^D}$,

(1) $\mathbf{t}(A \Rightarrow B) = 0$ iff for some $\mathbf{s}, \mathbf{t} \subseteq \mathbf{s}$ and $\mathbf{s}(A) = 1$ and $\mathbf{s}(B) = 0$.

(2) $\mathbf{t}(A \Leftarrow B) = 1$ iff for some $\mathbf{s} \subseteq \mathbf{t}$, $\mathbf{s}(B) = 1$ and $\mathbf{s}(A) = 0$.

See the previous exercise for the definition of $\subseteq$ and for method of solution.

## 2. Heyting's Predicate Logic (HPC)

The predicate language of HPC contains the quantifiers $\forall$, $\exists$ and also the connectives $\wedge$, $\vee$, $f$, $\rightarrow$.[1] HPC was presented by Heyting as a Hilbert system. We take the following definition.

DEFINITION 1. Let $H$ (sometimes called HPC) be the Hilbert system with the following axioms and rules:

*Axiom schemas*

(a) $A \rightarrow (B \rightarrow A)$

(b) $((A \rightarrow (B \rightarrow C)) \rightarrow ((A \rightarrow B) \rightarrow (A \rightarrow C))$

(c) $(A \wedge B) \rightarrow A; \quad (A \wedge B) \rightarrow B$

(d) $A \rightarrow (B \rightarrow (A \wedge B))$

(e) $A \rightarrow (A \vee B); \quad B \rightarrow (A \vee B)$

(*f*) $(A \rightarrow C) \rightarrow ((B \rightarrow C) \rightarrow ((A \vee B) \rightarrow C))$

(*g*) $f \rightarrow A$

(h) $A(x) \rightarrow \exists y A(y)$

(k) $\forall y A(y) \rightarrow A(x)$.

*Provability rules*

(1) From $A, A \rightarrow B$ deduce $B$.

(2) The following two rules, with $x$ not free in $B$

$$\frac{A(x) \rightarrow B}{\exists x A(x) \rightarrow B} \qquad \frac{B \rightarrow A(x)}{B \rightarrow \forall x A(x)}.$$

[1] $\wedge$ will turn out to be strongly classical in $H$, but we are not *a priori* committed to it.

*Consequence rule*

From $A, A \to B$ deduce $B$.

**Exercise 1.** Let $\vdash_H$ be the Tarski system obtained from $H$. Show that the following holds:

(a) If $x$ is not free in $A$ then

(1) $\vdash_H \forall x(A \to B(x)) \to (A \to \forall x B(x))$

(2) $\vdash_H A \vee \exists x B(x) \to \exists x(A \vee B(x))$

(3) $\vdash_H \forall x(B(x) \to A) \to (\exists x B(x) \to A)$

(b) $\vdash_H A \wedge B \to B \wedge A$

$\vdash_H A \wedge (B \wedge C) \to (A \wedge B) \wedge C$

$\vdash_H A \vee B \to B \vee A$

$\vdash_H (A \vee B) \vee C \to (A \vee B) \vee C.$

(c) From the previous chapter we know that there are two Scott systems $\Vdash_H^+$ and $\Vdash_H^-$, the maximal and minimal Scott systems (respectively) that agree with $\vdash_H$. Show that the following hold for $\Vdash_H^+$:

$$\varphi \Vdash_H^+ A, B \to A \vee C \quad \text{iff} \quad \varphi, B \Vdash_H^+ A, C.$$

(d) Let $A(P)$ be a wff with atomic $P$. Let $Q_1, Q_2$ be such that $\vdash_H (Q_1 \to Q_2) \wedge (Q_2 \to Q_1)$ then $\vdash_H A(Q_1) \to A(Q_2)$.

We now turn to define two Scott systems that will turn to be the maximal and minimal Scott systems that agree with $\vdash_H$. In Definition 2 below note that Exercise 1(b) above makes the definition unambiguous.

DEFINITION 2. (a) Let $\varphi \Vdash_K \psi$ iff $\varphi \vdash_H \vee \psi$ (if $\psi = \emptyset$ replace $\vee \psi$ by $f$) (i.e., $\vee \emptyset = f$).

(b) $\varphi \Vdash_B \psi$ iff for some $\psi' \subseteq \psi$, $\bar{\bar{\psi}}' \leq 1$, $\varphi \vdash_H \psi'$.

*Remark* 3. $\Vdash_K$ is the Kripke consequence relation, and $\Vdash_B$ is the Beth consequence relation. We shall see later that these consequence relations are connected with the Kripke and Beth interpretations for HPC.

THEOREM 4. *$\Vdash_K$ and $\Vdash_B$ are both Scott consequence relations that agree with $\vdash_H$.*

*Proof.* The proof that $\Vdash_B$ is a consequence relation that agrees with $\vdash_H$ is similar to the proof of (1.1.11). (Note that $H \vdash A \to A$).

To show that $\Vdash_K$ is a consequence relation we check the cut rule. Assume that $\varphi \Vdash_K \psi, A$ and $\varphi, A \Vdash_K \psi$ we must show that $\varphi \Vdash \psi$. So we are given that $\varphi \vdash_H \vee \psi \vee A$ and $\varphi, A \vdash_H \vee \psi$. (If $\psi = \emptyset$ let $\vee \psi$ be $f$). We must show that $\varphi \vdash \vee \psi$.

LEMMA 5. *The deduction theorem holds for $\vdash_H$*, i.e.,

$$\varphi, B \vdash_H C \quad \text{iff} \quad \varphi \vdash_H B \to C.$$

*Proof.* Same as that of (1.13).

To continue the proof of (2.4) distinguish two cases.

*Case a.* $\psi = \emptyset$. Then given $\varphi \vdash_H A$ and $\varphi \cup \{A\} \vdash_H \emptyset$ then certainly $\varphi \vdash_H \emptyset$, since $\vdash_H$ is an axiomatic system.

*Case b.* $\psi \neq \emptyset$. Then by the deduction theorem we are given $\varphi \vdash_H \vee \psi \vee A$ and $\varphi \vdash_H A \to \vee \psi$. Since $\varphi \vdash_H (\vee \psi \to \vee \psi) \to ((A \to \vee \psi) \to (\vee \psi \vee A \to \vee \psi))$ we can get, using what is given and *MP*, that $\varphi \vdash_H \vee \psi$.

Thus in either case $\varphi \Vdash_K \psi$ holds.

THEOREM 6. *Let $\Vdash$ be any consequence relation that agrees with $\vdash_H$. Then $\Vdash_B \subseteq \Vdash \subseteq \Vdash_K$.*

*Proof.* (a) Assume $\varphi \Vdash_B \psi$, let $\psi' \subseteq \psi$ with $\bar{\bar{\psi}}' \leqq 1$ and $\varphi \vdash_H \psi'$ and so $\varphi \Vdash \psi'$ and so $\varphi \Vdash \psi$.

(b) Assume $\varphi \Vdash \psi$. If $\bar{\bar{\psi}} \leqq 1$ then $\varphi \vdash_H \psi$ and so $\varphi \Vdash_K \psi$. For $\bar{\bar{\psi}} \geqslant 2$ we proceed by induction. Let $\psi = \{A, B\} \cup \psi_0$. Given $\varphi \Vdash A, B, \psi_0$, $A \neq B$. We want to show that $\varphi \Vdash_K A, B, \psi_0$. Since $A \vdash_H A \vee B$ we get that $A, \varphi \Vdash A \vee B, B, \psi_0$. Also certainly $\varphi \Vdash A \vee B, A, B, \psi_0$ and therefore $\varphi \Vdash A \vee B, B, \psi_0$. But since $\varphi, B \Vdash A \vee B, \psi_0$ we get that $\varphi \Vdash A \vee B, \psi_0$. By the induction hypothesis $\varphi \Vdash_K A \vee B, \psi_0$ therefore $\varphi \Vdash A \vee B, \psi$. But also $\varphi, A \vee B \Vdash_K \psi$ and therefore finally $\varphi \Vdash_K \psi$.

The last theorem shows that $\Vdash_B$ is the smallest consequence relation which agrees with $\vdash_H$ and $\Vdash_K$ is the largest. We are interested in giving a different characterization of $\Vdash_K$.

DEFINITION 7. Consider a language with the connectives $\wedge$, $\vee$, $\rightarrow$, $f$ and quantifiers $\forall$, $\exists$; assume that $\wedge$ is taken as classical conjunction, $\vee$ as classical disjunction, and $f$ as classical falsity and $\exists$ as classical existential quantifier. Let $\Vdash_0$ be the smallest consequence relation in this language (with the above mentioned understanding, see (1.3.8-9)) that is closed under the rules:

*rules for* $\forall$:

(1) $\qquad \varphi \Vdash A(x)$ iff $\varphi \Vdash \forall x A(x)$, $\qquad x$ not free in $\varphi$.

(2) $\qquad \forall x A(x) \Vdash A(y)$.

*rule DT*:

$$\varphi, A \Vdash B \quad \text{iff} \quad \varphi \Vdash A \rightarrow B.$$

PROPOSITION 8. $\Vdash_0 = \Vdash_K$.

*Proof.* We first show that $\Vdash_K \subseteq \Vdash_0$. To show this we prove that if $\emptyset \vdash_H A$ then $\emptyset \Vdash_0 A$. This is shown by induction on the length of the proof of $A$ in $H$. First we check for all the axioms (of definition 2.1). The case of axioms (a) and (b) (of 2.1) follow as in the proof of (1.12). The cases of axioms (c), (e) and (g), (h) follow from $DT$ and the fact that $\wedge$, $\vee$, $f$, $\exists$ are considered classical. The case of axiom (d) follows immediately from the deduction theorem. For the case of axiom $f$ note that since $\vee$ is considered classical, we have $A \vee B \Vdash_0 A, B$. Thus

$$A \vee B, A \rightarrow C, B \rightarrow C \Vdash_0 C, A, B$$
$$A \vee B, A \rightarrow C, B \rightarrow C, A \Vdash_0 C, B$$

and therefore $A \vee B$, $A \rightarrow C$, $B \rightarrow C \Vdash_0 C, B$ but $A \vee B$, $A \rightarrow C$, $B \rightarrow C$, $B \Vdash_0 C$ and therefore $A \vee B$, $A \rightarrow C$, $B \rightarrow C \Vdash_0 C$; using $DT$ we get $\emptyset \Vdash_0$ axiom $f$.

The rule $MP$ and the rule for $\exists$ are clearly valid (as $DT$ is valid). The axiom and rule for $\forall$ follow from the fact that we took the rules of $\forall$ for $\Vdash_0$. Therefore we conclude that if $\emptyset \Vdash_H A$ then $\emptyset \Vdash_0 A$.

Now assume $\varphi \Vdash_K \psi$, then by definition $\varphi \vdash_H \vee \psi$ and so if $\varphi = \{A_1, \ldots, A_n\}$ we get by the deduction theorem $\emptyset \vdash_H A_1 \rightarrow (A_2 \rightarrow \cdots \rightarrow \vee \psi) \cdots)$, and so $\emptyset \Vdash_0 A_1 \rightarrow (A_2 \rightarrow \cdots \rightarrow \vee \psi) \cdots)$ and so again by $DT$, $\varphi \Vdash_0 \vee \psi$ and therefore since $\wedge$, $\vee$, $f$ are classical in $\Vdash_0$, we get $\varphi \Vdash_0 \psi$. Thus $\Vdash_K \subseteq \Vdash_0$.

To show that $\Vdash_K \supseteq \Vdash_0$ we show that $\wedge$, $\vee$, $\exists$, $f$ are classical in $\Vdash_K$, and the rules for $\forall$ are valid. Indeed it is easy to show, using the axioms of (2.1) that the rules of (1.3.8-9) hold for $\Vdash_K$. We also know that $DT$ holds for $\Vdash_K$, and since $\Vdash_0$ is the smallest Scott consequence relation in which $DT$ and the rules for $\forall$ holds and in which $\wedge$, $\vee$, $\exists$, $f$ are classical, we get that $\Vdash_0 \subseteq \Vdash_K$.

PROPOSITION 9. *In H, $\wedge$ is strongly classical and $\vee$ and $f$ are weakly classical but not strongly classical.*

*Proof.* $A \wedge B \vdash_H A$; $A \wedge B \Vdash_H B$ follow directly from the axioms (2.1). $A, B \vdash_H A \wedge B$ follows from axiom (d). Thus it follows that $\wedge$ is strongly classical, since any consequence relation that agrees with $\vdash_H$ must validate the above rules.

$\vee, f$ are weakly classical since they are classical in $\Vdash_K = \Vdash_0$ and they are not strongly classical since they are not classical since they are not classical in $\Vdash_B$. $A \vee B \nVdash_B A, B$ and $f \nVdash_B \emptyset$.

THEOREM 10. *In $\Vdash_B$, the universal quantifier $\forall$ is classical, and so is the existential quantifier $\exists$ and the conjunction $\wedge$.*

*Proof.* For $\forall$ we have to check the rule: if $\varphi \Vdash_B \psi, A(x)$ then $\varphi \Vdash_B \psi, \forall x A(x)$ where $x$ is not free in $\varphi$ or $\psi$.

If $\varphi \Vdash_B \psi, A(x)$ then for some $C \in \psi \cup \{A(x)\}$, $\varphi \vdash_H C$. If $C \in \psi$ then certainly $\varphi \Vdash_B \psi, \forall x A(x)$. If $C = A(x)$ then by the properties of $\vdash_H$, $\varphi \vdash_H \forall x A(x)$ and therefore $\varphi \Vdash_B \psi, \forall x A(x)$.

Similarly, we show for the case of $\exists$, $\wedge$.

COROLLARY 10. *H is a conservative extension of $H^{++}$ (i.e., for any wff A of the language of H, with $\rightarrow$ only, $\vdash_H A$ iff $\vdash_{H^{++}} A$.*

*Proof.* Since $\vdash_{H^{++}}$ agrees with $\Vdash$ of Definition 1.8 and $\vdash_H$ agrees with $\Vdash_0$ of Definition 2.7 and since $\Vdash$ is the smallest Scott consequence relation for which $DT$ is valid in the language of $H^{++}$ which is their common language.

## 3. The disjunction and existence properties

In the previous section we found two consequence relations $\Vdash_B$ and $\Vdash_K$ that agree with $\vdash_H$. $\Vdash_B$ is the smallest one that agrees with $\vdash_H$ and $\Vdash_K$ is the largest one. We are interested in conditions on $\varphi$ such that $\varphi \Vdash_K$ and $\varphi \Vdash_B$ coincide, (i.e., we want to characterise $S_{\vdash_H}$).

DEFINITION 1. (a) Let $\Phi, \Psi$ be any sets of wffs. We write $\Phi \Vdash \Psi$ iff for some finite $\varphi \subseteq \Phi$, $\psi \subseteq \Psi$, $\varphi \Vdash \psi$.

(b) $\Phi$ is said to have the Disjunction property iff for all $\Psi$

$$\Phi \Vdash_K \Psi \quad \text{iff} \quad \Phi \Vdash_B \Psi.$$

(c) $\Phi$ is said to have the existence property iff $\Phi \vdash_H \exists x A(x)$ implies that for some $u$, $\Phi \vdash_H A(u)$.

We are looking for necessary and sufficient conditions on $\Phi$ to have the disjunction property and existence property.

DEFINITION 2. (S. Kleene). (1) We define the notion $\Phi \mid A$ by induction on $A$.

(a) $\quad \Phi \mid A \quad \text{iff} \quad \Phi \vdash_H A, \qquad \text{for } A \text{ atomic or } A = f.$

(b) $\quad \Phi \mid A \wedge B \quad \text{iff} \quad \Phi \mid A \quad \text{and} \quad \Phi \mid B.$

(c) $\quad \Phi \mid A \vee B \quad \text{iff} \quad \Phi \mid\!\vdash_H A \quad \text{or} \quad \Phi \mid\!\vdash_H B.$

(d) $\quad \Phi \mid A \to B \quad \text{iff} \quad \Phi \mid\!\vdash_H A \quad \text{implies} \quad \Phi \mid B.$

(e) $\quad \Phi \mid \exists x A(x) \quad \text{iff} \quad \text{for some } u, \quad \Phi \vdash_H A(u).$

(f) $\quad \Phi \mid \forall x A(x) \quad \text{iff} \quad \text{for all } u, \quad \Phi \mid A(u).$

Where $\Phi \mid\!\vdash_H A$ abbreviates: $\Phi \mid A$ and $\Phi \vdash_H A$
we write $\Phi \mid \Psi$ iff $\Phi \mid A$ for all $A \in \Psi$.

THEOREM 3. If $\emptyset \vdash_H A$ *then for all* $\Phi$ *and all substitution instances* $A'$ *of* $A$ *with other variables,* $\Phi \mid A'$.

*Proof.* We use induction on the length of the proof of $A$. First we show that the theorem holds for all the axioms and then we check that the provability rules preserve this property.

We check the axioms in Definition 2.1 one by one.

(a) Assume $\Phi \mid\!\vdash_H A$ we want to show that $\Phi \mid \mathrm{B} \to \mathrm{A}$ but this is clear.

(b) Assume that $\Phi \mid\!\vdash_H A \to (B \to C)$ we must show that $\Phi \mid (\mathrm{A} \to \mathrm{B}) \to (\mathrm{A} \to \mathrm{C})$. So assume $\Phi \mid\!\vdash_H A \to B$ and show that $\Phi \mid A \to C$. To show the latter assume $\Phi \mid\!\vdash_H A$ and show that $\Phi \mid C$. From what is assumed we get that $\Phi \mid B$ and also $\Phi \vdash_H B$ which together with the assumption that $\Phi \mid B \to C$ yields $\Phi \mid C$.

The cases of axioms (c) and (d) and (e) are easy.

For the case of axiom (f) assume that $\Phi \Vdash_H A \to B$ and show that $\Phi \mid (A \to C) \to ((A \vee B) \to C)$. So assume $\Phi \Vdash_H A \to C$ and that $\Phi \Vdash_H A \vee B$ and show that $\Phi \mid C$. Since $\Phi \Vdash_H A \vee B$ we have $\Phi \Vdash_H A$ or $\Phi \Vdash_H B$. Because of Symmetry in what is given we can assume that $\Phi \Vdash_H A$ and so we immediately get $\Phi \mid C$.

For the case of axiom (h) assume $\Phi \Vdash_H A(x)$ then by definition $\Phi \mid \exists x A(x)$. Axiom (k) is as easy.

To check the provability rules, let $C = A(x) \to B$ and assume that for all substitution instances $C'$ of $C$, $\Phi \Vdash_H C'$. We show that $\Phi \Vdash_H \exists x A'(x) \to B'$. Clearly $\Phi \vdash_H \exists x A'(x) \to B$. Assume $\Phi \Vdash_H \exists x A'(x)$ then for some $u$, $\Phi \Vdash_H A'(u)$. Since $x$ is not in $B$, $A'(u) \to B'$ is a substitution instance of $A(x) \to B$ and so by what is given $\Phi \mid B'$.

Now let $C = B \to A(x)$ and assume that $\Phi \Vdash_H C'$ for every substitution instance $C'$ of $C$. To show that $\Phi \Vdash_H B' \to \forall x A'(x)$ for all substitution instances $B', A'$, assume $\Phi \Vdash_H B'$. Then by what is given $\Phi \mid A'(x)$ and since $x$ is not free in $B$, for all $x$, $B' \to A'(x)$ is a substitution instance of $C$ and therefore for all $x$, $\Phi \mid A'(x)$, i.e., $\Phi \mid \forall x A(x)$.

The case of the rule *MP* is simple.

For the case of axiom (g) assume that $\Phi \Vdash_H f$. We want to show that $\Phi \mid$ A. For this we prove by induction that if $\Phi \vdash_H f$ then $\Phi \mid A$ for all $A$. Clearly $\Phi \mid A$ for $A$ atomic, since $f \to A$ is an axiom of $H$ and thus $\Phi \vdash_H A$ holds. The other induction steps are easy and follow from the definition of $\mid$.

THEOREM 4. *Let $\Phi$ be such that $\Phi \nVdash_H f$ then $\Phi$ has the disjunction and existence properties iff $\Phi \mid \Phi$.*

*Proof.* (a) Assume $\Phi \mid \Phi$ and show that $\Phi \Vdash_K \Psi$ iff $\Phi \Vdash_B \Psi$, $\Psi$ arbitrary. Assume $\Phi \Vdash_K \Psi$ then for some $\varphi \subseteq \Phi$, $\psi \subseteq \Psi$, $\varphi \vdash_H \vee \psi$ (we can assume $\psi \neq \emptyset$ for otherwise take $\vee \psi = f$).

*Case* 1. $\varphi$ is empty. Then $\emptyset \vdash_H \vee \psi$ and by (3.3) $\emptyset \mid \vee \psi$ and so for some $A \in \psi$, $\emptyset \Vdash_H A$ but this implies that $\emptyset \Vdash_B \psi$, and so $\Phi \Vdash_B \Psi$.

*Case* 2. $\varphi = \{A_1, \ldots, A_n\}$. Then by the deduction theorem $\emptyset \vdash_H A_1 \to (A_2 \to \cdots \to \vee \psi) \cdots)$ and so $\varphi \mid A_1 \to (A_2 \to \cdots \vee \psi) \cdots)$, and since $\varphi \Vdash_H A_i$ for all $i$, we get that $\varphi \mid \vee \psi$ and so for some $A \in \psi$, $\varphi \Vdash_H A$ and therefore $\varphi \Vdash_B \psi$ and therefore $\Phi \Vdash_B \Psi$.

Now assume $\Phi \vdash_H \exists x A(x)$ then for some $\varphi \subseteq \Phi$, $\varphi \vdash_H \exists x A(x)$. Since $\Phi \mid \varphi$ we get that $\Phi \mid \exists x A(x)$ and so by definition of $\mid$ for some $u$, $\Phi \Vdash_H A(u)$.

(b) Assume that $\Phi$ has the disjunction and existence property. We show that $\Phi \mid \Phi$. We prove that for such a $\Phi$, $\Phi \vdash_H A$ implies $\Phi \mid A$, by induction on $A$. For $A$ atomic this is the definition. The case of $\wedge$ and the case of $f$ present no difficulty.

If $\Phi \vdash_H A \vee B$ then (since $\Phi$ has the disjunction property) $\Phi \vdash_H A$ or $\Phi \vdash_H B$ and so (by the induction hypothesis) $\Phi \Vdash_H A$ or $\Phi \Vdash_H B$ hence $\Phi \mid A \vee B$.

If $\Phi \vdash_H A \to B$ then if $\Phi \Vdash_H A$ then $\Phi \vdash_H B$ by *MP* and so by the induction hypothesis $\Phi \mid B$.

Assume $\Phi \vdash_H \exists x A(x)$ then by the existence property, for some $u$, $\Phi \vdash A(u)$ and by the induction hypothesis $\Phi \mid A(u)$ and so $\Phi \mid \exists x A(x)$.

Assume $\Phi \vdash \forall x A(x)$ then for all $u$, $\Phi \vdash A(u)$ and so by the induction hypothesis and the definition of $\mid$, $\Phi \mid \forall x A(x)$. Now since $\Phi \vdash_H A$ for all $A \in \Phi$ we get $\Phi \mid \Phi$.

COROLLARY 5. (a) $\vdash_H A \vee B$ *iff* $\vdash_H A$ *or* $\vdash_H B$.

(b) $\vdash_H \exists x A(x)$ *implies* $\vdash_H A(u)$ *for some* $u$.

*Proof.* (a) $\emptyset \vdash_H A \vee B$ iff $\emptyset \Vdash_K A, B$.

Since $\emptyset \nvdash_H f$ (as can be seen by using classical semantics showing that all $H$-theorems are predicate tautologies) we get $\emptyset \Vdash_K A, B$ iff $\emptyset \Vdash_B A, B$ iff $\emptyset \vdash_H A$ or $\emptyset \vdash_H B$.

(b) Since $\{f \to f\} \mid f \to f$ then if $\vdash_H \exists x A(x)$ then $\{f \to f\} \vdash_H \exists x A(x)$ and so $\{f \to f\} \mid \exists x A(x)$ and so for some $u$, $f \to f \vdash_H A(u)$, i.e. $\vdash_H A(u)$.

**Exercise 6.** *Kripke Interpretation.*
Consider the following definition of $\|$

(a) $\Phi \| A$ iff $\Phi \vdash_H A$ for $A$ atomic or $A = f$

(b) $\Phi \| A \wedge B$ iff $\Phi \| A$ and $\Phi \| B$

(c) $\Phi \| A \vee B$ iff $\Phi \| A$ or $\Phi \| B$

(d) $\Phi \| \exists x A(x)$ iff for some $u$, $\Phi \| A(u)$

(e) $\Phi \| \mathrm{A} \to \mathrm{B}$ iff for all $\Phi' \supseteq \Phi$ if $\Phi' \| A$ and $\Phi' \mid \Phi'$ then $\Phi' \| B$.

(f) $\Phi \| \forall x A(x)$ iff for all $u$ appearing in $\Phi$ and all $\Phi' \supseteq \Phi$ if $\Phi' \mid \Phi'$ then $\Phi' \| A(u)$.

Show that $\Phi \| A$ iff $\Phi \vdash_H A$ for all $A$ and all $\Phi$ such that $\Phi \mid \Phi$ and there is an infinite number of variables not in $\Phi$.

**Exercise 7.** *Separability of H (A. Horn).*
Consider the system $H$ of Definition 2.1. For each set of connectives $\mathbb{C} \subseteq \{\wedge, \vee, \rightarrow, f\}$ that contains $\rightarrow$ (i.e. $\rightarrow \in \mathbb{C}$) let $H^{\mathbb{C}}$ be the axiomatic system obtained from $H$ by deleting all axioms mentioning any connective not in $\mathbb{C}$. So e.g. $H^{\{\rightarrow\}}$ is really $H^{++}$ with quantifiers. Show that for each such $\mathbb{C}$, $H$ is a conservative extension of $H^{\mathbb{C}}$, i.e. for any $A$ in the language of $H^{\mathbb{C}}$, $\vdash_H A$ iff $\vdash_{H^{\mathbb{C}}} A$.

*Hint.* Notice that we showed that the consequence relation $\Vdash_0$ (which is the smallest consequence relation with $\wedge, \vee, f, \exists$ classical for which $DT$ and the rules for $\forall$ are valid) agrees with $\vdash_H$.

Define the consequence relation $\Vdash_{\mathbb{C}}$ as the smallest consequence relation in the language of $\mathbb{C}$ which regards the connectives of $\mathbb{C} - \{\rightarrow\}$ as classical and for which $DT$ and the quantifier rules hold and show that it agrees with $\vdash_{H^{\mathbb{C}}}$. To show this use the same proof given for $\Vdash_0$ and $\vdash_H$. In the course of the proofs we treated each connective separately with independent proofs for each connective different from $\rightarrow$.

**Exercise 8.** *Harrop's proof of the disjunction property for propositional H.*
Consider the propositional fragment of $H$ obtained by taking $MP$ and axioms (2.1.a–g). Let $I = \{A \mid \vdash_H A\}$. Show that $A \vee B \in I$ implies $A \in I$ or $B \in I$. To prove that, well order $I$ in such a way that for any $A$, the subformulas of $A$ precede $A$ in the well ordering. Now use this well ordering to define by induction a family of subsets of $I$, of the form $J_A$, for $A \in I$.

(1) Let $J_A = I$ if $A$ is the first element of the well ordering.
(2) If $A$ is a limit point let $J_A = \bigcap_{B<A} J_B$ ($<$ is the well ordering).
(3) If $A$ is the successor of $B$, we define $J_A$ by induction on $B$.

(a) $B = B_1 \wedge B_2$.

Let $J_A = J_B - \{B\}$ if $B_1 \notin J_B$ or $B_2 \notin J_B$ and otherwise let $J_A = J_B$

(b) $B = B_1 \vee B_2$.

Let $J_A = J_B - \{B\}$ if $B_1 \notin J_B$ and $B_2 \notin J_B$ and otherwise let $J_A = J_B$.

(c) $B = B_1 \rightarrow B_2$.

Let $J_A = J_B - \{B\}$ if $B_1 \in J_B$ and $B_2 \notin J_B$ and otherwise let $J_A = J_B$.

(d) No atomic wff are in $I$ and so the atomic case does not arise.

The reader can verify that if $A < B$ then $J_A \supseteq J_B$ and if $A \in J_A$ then $A \in J_B$ for all $B$.

LEMMA. $\bigcap_{A \in I} J_A = I$.

*Proof.* The proof is by induction on the length of the proof of wff in $I$. We first show that for every axiom $A$, $A \in J_A$ and then show that the property '$A \in J_A$' is closed under modus ponens.

## 4. Heyting's Negation

DEFINITION 1. Consider a language with the quantifiers $\forall$, $\exists$ and the connectives $\wedge$, $\vee$, $\rightarrow$ are the additional unary connective $\neg$. Let $H^*$ be the Hilbert system for this language with all axioms and rules of $H$ (of Definition 2.1) with the exception of axiom (g), together with the following axioms (g*) which we take to replace (g):

(g*1) $\quad (A \rightarrow \neg B) \rightarrow (B \rightarrow \neg A)$

(g*2) $\quad A \rightarrow (\neg A \rightarrow B)$.

Actually, $H$ and $H^*$ are essentially the same system in different languages. To see this, we interpret the systems in one another.

DEFINITION 2. In the language of $H$, let $\neg A$ abbreviate $A \rightarrow f$. Thus each $A$ in the language of $H^*$ can be regarded as a wff of $H$ (through the above understanding).

PROPOSITION 3. *With the understanding of Definition* 2,

$$\vdash_{H^*} A \quad \text{implies} \quad \vdash_H A.$$

*Proof.* Show that (g*1) and (g*2) are provable in $H$. First show that $\vdash_H (A \rightarrow (B \rightarrow C)) \rightarrow (B \rightarrow (A \rightarrow C))$ and then substitute $C = f$. (g*2) is really modus ponens.

DEFINITION 4. Let $p$ be any atomic proposition. Regard $f$ in the language of $H^*$ as $p \wedge \neg p$. Thus any wff of $H$ can be considered as a wff of $H^*$.

**Exercise 5.** $\vdash_{H^*} A \wedge \neg A \rightarrow B \wedge \neg B$.

PROPOSITION 6. *With the understanding of Definition* 4 *the following holds*:

$$\vdash_H A \quad \text{implies} \quad \vdash_{H^*} A.$$

*Proof.* We prove axiom (g) ($\vdash f \to B$) for all $B$. For this purpose we show that $\vdash_{H^*} (A \wedge B \to C) \to (A \to (B \to C))$.

COROLLARY 7. *For $A$ without $f$ and without $\neg$, we have $\vdash_H A$ iff $\vdash_{H^*} A$.*

**Exercise 8.** Let $f$ be a fixed atomic propositional constant in the language of $H^{++}$, Section 1. Define for each wff $A$, $\neg A = \text{def } A \Rightarrow f$. Show that

(1) $\qquad H^{++} \vdash (A \Rightarrow \neg B) \Rightarrow (B \Rightarrow \neg A)$

(2) $\qquad H^{++} \vdash A \Rightarrow \neg\neg A.$

*Hint.* Show that

$$H^{++} \vdash (A \Rightarrow (B \Rightarrow C)) \Rightarrow (B \Rightarrow (A \Rightarrow C)).$$

**Exercise 9.** Show that in $H^*$

$$\vdash_{H^*} \neg\neg(A \vee B) \leftrightarrow \neg\neg A \vee \neg\neg B$$

$$\vdash_{H^*} \neg\neg(A \to B) \to (\neg\neg A \to \neg\neg B)$$

$$\vdash_{H^*} A \to \neg\neg A.$$

**Exercise 10.** Let $C^*$ be the extension of $H^*$ with the axiom schema $A \vee \neg A$. Show that $\vdash_{C^*} ((A \to B) \to A) \to A$.

**Exercise 11.** (a) In $H$, show that for all $A, B, C$ if $\vdash_H \neg A \to B \vee C$ then $\vdash_H (\neg A \to B) \vee (\neg A \to C)$.

(b) Show that for some $A, B, C$, $\neg A \to B \vee C \nvdash_H (\neg A \to B) \vee (\neg A \to C)$.

**Exercise 12.** Let $\Phi$ be in the language of HPC without $\vee, \exists$. Show that $\Phi \mid \Phi$ holds. Deduce from that that in the system $\vdash_{H^{++}}$, there exists only one consequence relation that agrees with $\vdash_{H^{++}}$.

**Exercise 13.** In the language of HPC, let

$$\Phi = \{(\neg A \to (B \vee C)) \to ((\neg A \to B) \vee (\neg A \to C)) \quad A \quad \text{any wff}\}.$$

Show that $\Phi \mid \Phi$.

**Exercise 14.** Is $\forall$ strongly classical in $\vdash_H$?
Is $\exists$ strongly classical in $\vdash_H$?

## 5. Classical Predicate Calculus (CPC)

DEFINITION 1. Let $C$ be the Hilbert system obtained from $H$ by adding the axiom schema

(j) $\quad ((A \to B) \to A) \to A.$

$C$ has all the other axioms and rules of $H$ (of 2.1).

THEOREM 2. *The deduction theorem holds for C.*
*Proof.* Same as for $H$.

LEMMA 3.

(a) $\quad (A \to B) \to B \vdash_C A \vee B$

(b) $\quad A \vee B \vdash_C (A \to B) \to B.$

*Proof.* (a) Consider the following proof:

$\vdash_C \quad B \to A \vee B$

$\vdash_C \quad (B \to (A \to B)) \to (C' \to (B \to A \vee B)))$

$\vdash_C \quad C' \to (B \to (A \vee B))$

$\vdash_C \quad (C' \to (B \to (A \vee B))) \to ((C' \to B) \to (C' \to A \vee B))$

$\vdash_C \quad (C' \to B) \to (C' \to (A \vee B)).$

Let $C' = (A \to B) \to B)$, then $\vdash_C (A \to B) \to B) \to B.$
Thus $\vdash_C ((A \to B) \to B) \to (A \vee B)$

(b) $\quad \vdash_C \quad B \to ((A \to B) \to B).$

Also since $A, A \to B \vdash_C B$

$$\vdash_C \quad A \to ((A \to B) \to B).$$

Thus since $\vdash_C (A \to C') \to (B \to C')$

$$\vdash_C \quad ((A \vee B) \to C'))$$

for $C' = (A \to B) \to B$ we get $\vdash_C (A \vee B) \to ((A \to B) \to B)$.

COROLLARY 4. $\vdash_C (A \to B) \vee A$.

THEOREM 5. *Let* $\Vdash_C^+$ *be the maximal consequence relation that agrees with* $\vdash_C$, *then the following hold:*

(a) $\varphi, A \Vdash_C^+ \psi, B$ iff $\varphi \Vdash_C^+ \psi, A \to B$

(b) $\varphi \Vdash_C^+ \psi$ iff $\varphi \vdash_C \vee \psi$ $(\vee \emptyset = f)$.

*Proof.* Let the consequence relation $\Vdash$ be defined by $\varphi \Vdash \psi$ iff $\varphi \vdash_C \vee \psi$. $\Vdash$ is indeed a consequence relation, as we can follow the proof of (2.4) for this case. Note that since $\vdash_C A \vee (A \to B)$ we get that $\varphi \Vdash A, A \to B$. We want to show that rule (a) holds for $\Vdash$.

Assume $\varphi, A \Vdash \psi, B$.

Then $\phi, A \Vdash \psi, B, A \to B$

and since $B \Vdash A \to B$

$\varphi, A, B \Vdash \psi, A \to B$

and hence $\varphi, A \Vdash \psi, A \to B$

and since $\varphi \Vdash A, A \to B$

$\varphi \Vdash \psi, A, A \to B$

and hence $\varphi \Vdash \psi, A \to B$

now assume $\varphi \Vdash \psi, A \to B$

so $\varphi, A \Vdash \psi, A \to B, B$

also $\varphi, A, A \to B \Vdash \psi, B$

hence $\varphi, A \Vdash \psi, B$.

We now want to show that $\Vdash$ is the maximal consequence relation that agrees with $\vdash_C$. We can use the same proof given in (2.6) part (b) of the proof of (2.6) since $C$ is an extension of $H$.

*Remark* 6. In $\Vdash_C^+$ all the connectives and quantifiers have their classical meaning. It is clear that $\wedge$, f, $\vee$, $\exists$, $\rightarrow$ are classical. To show that the rules of (1.3.9) hold for $\forall$ assume $\varphi \Vdash_C^+ \psi, A(x)$, $x$ not appearing in $\varphi, \psi$ (we can assume $\varphi \neq \emptyset$) then by using 3 and 5a we get $\wedge \varphi, \vee \psi \rightarrow f \Vdash_C^+ A(x)$ and therefore $\wedge \varphi, \vee \psi \rightarrow f \Vdash_C^+ \forall x A(x)$ and therefore $\varphi \Vdash_C^+ \psi, \forall x A(x)$. The other direction is easy since $\forall x A(x) \Vdash_C^+ A(x)$.

*Remark* 7. $\Vdash_C^-$ is different from $\Vdash_C^+$, i.e., the minimal consequence relation agreeing with $\vdash_C$ is different from the maximal one. Thus $\vee$ is not strongly classical in $C$. We shall see later that $\Vdash_C^+$ is the consequence relation arising from the classical interpretation for $C$ and $\Vdash_C^-$ is the consequence relation arising from the interpretation of finite Beth structures.

**Exercise 8.** Let $C^*$ be the extension of $H^*$ with the schema $A \vee \neg A$. Show that for any $A$ of $C^*$,

$$\vdash_C A \quad \text{iff} \quad \vdash_{C^*} A.$$

(Where $\neg A$ is understood as $A \rightarrow f$ in $C$).

**Exercise 9.** *Interpretation theorems.*
Let $h$ and $c$ be the propositional fragments of $H$ and $C$ respectively. Show that

$$\vdash_C A \quad \text{iff} \quad \vdash_L (A \rightarrow f) \rightarrow f.$$

**Exercise 10.** Is $\forall$ ($\exists, \wedge, \vee, f, \rightarrow$ respectively) strongly classical in $\vdash_C$?

**Exercise 11.** Is $C$ the smallest Hilbert system extending $H$ for which $A \vee B$ is equivalent to $(A \rightarrow B) \rightarrow B$?

If not, is it true for the propositional fragments?

## 6. The system $CD$

DEFINITION 1. Let $CD$ be the axiomatic system obtained by taking all the axioms and rules of $H$ (of 2.1) together with the additional axiom (cd) where

$$\text{(cd)} \qquad (\forall x)(A \vee B(x)) \rightarrow A \vee (\forall x)B(x) \qquad x \text{ not free in } A.$$

DEFINITION 2. Let $\vdash_{CD}$ be the Tarski system obtained from $CD$. Define the relation:

$$\varphi \Vdash^+ \psi \quad \text{iff} \quad \varphi \vdash_{CD} \vee \psi.$$

THEOREM 3. *$\Vdash^+_{CD}$ of Definition 2 is a Scott system that agrees with $\vdash_{CD}$.*

*Proof.* Same as the proof of Theorem 2.4, for $\Vdash_K$ of (2.4).

THEOREM 4. *Let $\Vdash$ be the smallest Scott consequence relation for the language of CD in which $\wedge$, $\vee$, $f$, $\exists$, $\forall$ are classical (see 1.3.8-9) and in which DT is valid where:*

$$DT \qquad \varphi, A \Vdash B \quad \textit{iff} \quad \varphi \Vdash A \rightarrow B.$$

*Then $\Vdash^+ = \Vdash$.*

*Proof.* First show that $\Vdash^+ \subseteq \Vdash$. To show this prove first that $\emptyset \vdash_{CD} A$ implies $\emptyset \Vdash A$. This is shown by induction on the length of the proof of $A$ in $CD$. We check the axioms and rules of $H$ as we did in the proof of (2.8). The case of the additional axiom $CD$ follows from $DT$ and the fact that $\forall$ is classical in $\Vdash$.

Again, using the argument in the proof of (2.8), we continue and conclude that $\Vdash^+ \subseteq \Vdash$.

To show that $\Vdash \subseteq \Vdash^+$ we show that $\wedge$, $\vee$, $f$, $\exists$, $\forall$ are classical in $\Vdash^+$, and that $DT$ is valid for $\Vdash^+$. The conditions of (1.3.8-9) are easy to verify, (the fact that $\forall$ is classical follows from the additional axiom (cd) and the other conditions of $\forall$) and the condition $DT$ holds for $\rightarrow$ since the proof of (1.13) goes for $CD$ as well.

Thus $\Vdash \subseteq \Vdash^+$, since $\Vdash$ is the smallest Scott consequence relation for which $DT$ is valid and in which $\wedge$, $\vee$, $f$, $\exists$, $\forall$ are classical.

**Exercise 5.** Show that $CD$ is a proper extension of $H$.

**Exercise 6.** Show that $\vdash_{CD} A \vee B$ iff $\vdash_{CD} A$ or $\vdash_{CD} B$.

**Exercise 7.** Is $\forall$ ($\exists$, $\wedge$, $\vee$, $f$, $\neg$, respectively) strongly classical in $\vdash_{CD}$?

## 7. The system $MH^*$

DEFINITION 1. Let $MH^*$ be the axiomatic system obtained by taking all the axioms and rules of $H^*$ of Section 4 together with the axiom scheme (mh) below:

$$\text{(mh)} \qquad \forall x \neg\neg A(x) \rightarrow \neg\neg \forall x A(x).$$

THEOREM 2. $\vdash_{MH^*} \neg\neg\forall x(A(x) \vee \neg A(x))$.

*Proof.* Exercise.

THEOREM 3. *$MH^*$ is the smallest extension $X$ of $H^*$ with the property that*

$$\vdash_{C^*} A \quad \textit{iff} \quad \vdash_X \neg\neg A.$$

*Proof.* First we show that for $C^*$, of (5.8) $\vdash_{C^*} A$ implies $\vdash_{MH^*} \neg\neg A$.

We proceed by induction on the length of the proof of $A$ in $C^*$.

(1) If $A$ is an axiom of $C^*$, then if it is also an axiom of $H^*$, clearly $\vdash_{H^*} \neg\neg A$, since in $H^* \vdash_{H^*} A \rightarrow \neg\neg A$, (see Exercise 4.9).

To check for axiom $A \vee \neg A$ of $C^*$, notice that indeed, $\vdash_{MH^*} \neg\neg(A \vee \neg A)$.

(2) We now check the inference rules.

(a) Modus ponens. If $\vdash_{MH^*} \neg\neg A$ and $\vdash_{MH^*} \neg\neg(A \rightarrow B)$ then certainly $\vdash_{MH^*} \neg\neg B$, since $\vdash_{H^*} \neg\neg(A \rightarrow B) \rightarrow (\neg\neg A \rightarrow \neg\neg B)$.

(b) If $\vdash_{MH^*} \neg\neg(A \rightarrow B(x))$, $x$ not free in $A$, then $\vdash_{MH^*} \forall x \neg\neg(A \rightarrow B(x))$ and hence $\vdash_{MH^*} \neg\neg\forall x(A \rightarrow B(x))$ and therefore $\vdash_{MH^*} \neg\neg(A \rightarrow \forall x B(x))$ by Exercise 2.1.

(c) If $\vdash_{MH^*} \neg\neg(B(x) \rightarrow A)$, $x$ not free in $A$, then $\vdash_{MH^*} \forall x \neg\neg(B(x) \rightarrow A)$ so $\vdash_{MH^*} \neg\neg\forall x(B(x) \rightarrow A)$ and hence by Exercise 2.1 $\vdash_{MH^*} \neg\neg(\exists x B(x) \rightarrow A)$.

Thus we have shown that $\vdash_{C^*} A$ implies $\vdash_{MH^*} \neg\neg A$. Clearly $MH^*$ is the smallest extension of $H^*$ with this property.

COROLLARY 4. *If we understand $\neg A$ as $A \rightarrow f$, in $H$ and $C$ then $\vdash_C A$ iff $\vdash_{MH} \neg\neg A$ where $MH$ is the extension of $H$ with axiom* (cd).

## NOTES

First attempts to study the formal properties of the deduction theorem were made by W. Pogorzelski, 1968. The material is included in Section 1. Characterisation of the intuitionistic $\rightarrow$ as the smallest implication for which the deduction theorem holds follows from work of Scott Grzegorzyk, 1972 and the author. The dual connective to $\rightarrow$ is due to C. Rauszer, 1974. The rest of the material in this chapter is well known, and we present it, from our point of view. We include Kleene slash, 1962 Harrops method 1956 and Godel and Glivenko interpretation theorems. The various forms of negation are beautifully discussed in Curry's book, 1963.

See also Löb's paper.

CHAPTER 3

# THE KRIPKE, BETH AND TOPOLOGICAL INTERPRETATIONS FOR HPC

This chapter gives the basic Kripke, Beth and topological interpretations for a language without equality or function symbols. A later chapter presents the modified Kripke interpretation which is more suitable for languages with equality and function symbols.

## 1. Kripke structures

DEFINITION 1. Kripke structures have the form $(N, S, R, 0, D, U)$ where $N \subseteq S$, $R \subseteq S^2$, $0 \in S$ and the following hold:

(a) $R$ is a reflexive and transitive relation on $S$.

(b) $0Rx$ for all $x \in S$.

(c) $t \in N$ and $tRs$ imply $s \in N$.

(d) $U$ is a function associating with each $t \in S$ a nonempty set $U_t$ and if $sRs'$ then $U_s \subseteq U_{s'}$.

(e) $D$ is a function such that for each $n$-place atomic $A$, and each $t$, $D(t, A) \subseteq U_t^n$. $D$ has the property that if $tRs$ then $D(t, A) \subseteq D(s, A)$, for all atomic $A$. If $t \in N$ then $D(t, A) = U_t^n$.

DEFINITION 2.

(a) Let **K** be the class of all Kripke structures.

(b) Let $\mathbf{K}_p$ be the class of all Kripke structures with the property that $(S, R, 0)$ is a partially ordered set.

(c) Let $\mathbf{K}_0$ be the class of all Kripke structures with $N = \emptyset$.

(d) Let $\mathbf{K}_a$ be the class of all Kripke structures such that $\sim \exists x \exists y \exists z (\sim xRy \wedge \sim yRx \wedge xRz \wedge yRz)$.

(e) Let $\omega^*$ denote the set of all finite sequences of natural numbers and $\Lambda$ be the empty sequence. Let $\leqslant \subseteq \omega^{*2}$ be the relation $x \leqslant y$ iff $x = y$ or $x$ is an initial segment of $y$. A

Kripke structure $(N, S, R, 0, D, U)$ is called a tree-Kripke structure iff $0 = \Lambda$, $S \subseteq \omega^*$ and $R = \leqslant \upharpoonright S$. Let $\mathbf{K}_T$ be the class of all tree Kripke structures, and $\mathbf{K}_{T0}$ be the class of all tree Kripke structures with $N = \emptyset$.

(f) Let $\mathbf{K}_{TS}$ be the class of all tree Kripke structures with the property that

$U_s \cap U_t = U_{s \cap t}$

Where $s \cap t$ is the highest point $x$ in the tree such that $xRs \wedge xRt$ holds.

(g) Let $\mathbf{K}_{TS0} = \mathbf{K}_{TS} \cap \mathbf{K}_0$.

(h) Let $\mathbf{K}_S$ be the class of all Kripke structures such that for all $x, y \in S$ there exists a $z$ such that $zRx \wedge zRy$ and $U_z = U_x \cap U_y$.

DEFINITION 3. Given a Kripke structure and an assignment function $g : V \to U_t$, $t \in S$ ($V$ the set of variables of $HPC$) we define the truth value $[A]^g_t$ by induction on $A$ as follows:

(a) $[A(x_1, \ldots, x_n)]^g_t = 1$ iff $(g(x_1), \ldots, g(x_n)) \in D(t, A)$ if $A$ atomic with $x_1, \ldots, x_n$ free in $A$.

(b) $[A \wedge B]^g_t = 1$ iff $[A]^g_t = 1$ and $[B]^g_t = 1$.

(c) $[A \vee B]^g_t = 1$ iff $[A]^g_t = 1$ or $[B]^g_t = 1$.

(d) $[f]^g_t = 1$ iff $f \in N$.

(e) $[A \to B]^g_t = 1$ iff for all $s$, if $tRs$ and $[A]^g_s = 1$ then $[B]^g_s = 1$.

(f) $[\exists x A(x)]^g_t = 1$ iff for some $g' =_x g$, $[A(x)]^{g'}_t = 1$ where $g' =_x g$ means that for all $y \neq x$, $g(y) = g'(y)$.

(g) $[\forall x A(x)]^g_t = 1$ iff for all $s$, $g'$ if $tRs$ and $g' =_x g$ then $[A(x)]^{g'}_s = 1$.

(h) $A$ is said to hold in the structure under $g$ iff $[A]^g_0 = 1$.

DEFINITION 4. Let $\mathbf{M}$ be a class of Kripke structures. We write $\varphi \Vdash_{\mathbf{M}} \psi$ iff for all Kripke structures in $\mathbf{M}$ and all assignments $g$, if all wffs of $\varphi$ are valid in the structure under $g$ then at least one of the wff of $\psi$ is valid in the structure under $g$. Given an axiomatic system $X$ we say that $X$ is complete for $\mathbf{M}$ if $\vdash_X$ agrees with $\Vdash_{\mathbf{M}}$.

LEMMA 5. (a) *For any* $A$, $[A]^g_t = 1$ *and* $tRs$ *imply* $[A]^g_s = 1$.
(b) *For any* $A$ *and* $t \in N$ $[A]^g_t = 1$.
*Proof.* By induction on $A$.

THEOREM 6. *Let* $\Vdash$ *be the maximal Scott consequence relation agreeing with* $\vdash_H$ (called $\Vdash_K$ in (2.2.2a)) *then* $\Vdash = \Vdash_{\mathbf{K}_{T0}}$.
*Proof.* Let $V$ be the set of variables of the language and decompose $V$ into $\aleph_0$ disjoint infinite sets of variables $V = \bigcup_{n=0} U_n$. Let $V_0 = U_0$ and $V_{n+1} = V_n \bigcup U_{n+1}$, for $n = 0, 1, 2, \ldots.$

We first show that $\Vdash_{\mathbf{K}_{T0}} \subseteq \Vdash$. We assume $\varphi \nVdash \psi$ and show the existence of a structure in $\mathbf{K}_{T0}$ and an assignment $g$ into the structure which validates all members of $\varphi$ and none of the members of $\psi$. We can assume without loss of generality that $\varphi$, $\psi$ are in the language with variables $V_0$.

Let $A_0, A_1, A_2, \ldots$ be an enumeration of the wff of the language with variables $V$. For each wff $B$ let $\bar{B}$ be the place of $B$ in the above enumeration (i.e., $B = A_{\bar{B}}$).

Since we assumed that $\varphi \nVdash \psi$, $(\varphi, \psi)$ can be extended to a saturated theory $(\Delta_\Lambda, \Theta_\Lambda)$ in the language with variables $V_0$ (by Lemma 1.4.3).

We now define a partial function on $\omega^*$, by induction on the length of the sequences of $\omega^*$. For each $x \in \omega^*$ of length $n$, $f(x)$ is a saturated theory $(\Delta_x, \Theta_x)$ in the language with variables $V_n$.

Let $f(\Lambda) = (\Delta_\Lambda, \Theta_\Lambda)$. Assume $f(x) = (\Delta_x, \Theta_x)$ has been defined for $x$ of length $n$. Let $B \in \Theta_x$ be any wff of the form $B = B_1 \to B_2$, then $(\Delta_x \cup \{B_1\}, \{B_2\})$ is a $\Vdash$ consistent theory of the language with variables $V_n$ and can therefore be extended to a saturated theory $(\Delta, \Theta)$, in the language with variables $V_{n+1}$.

Let $f(x * (\bar{B} + 1)) = (\Delta, \Theta)$.

Let $C \in \Theta_x$ be any wff of the form $C = \forall x C_1(x) \in \Theta_x$, then for any $u \in U_{n+1}$, $(\Delta_x, \{C_1(u)\})$ is a $\Vdash$ consistent theory. This theory can be extended to a saturated theory $(\Delta, \Theta)$ in the language with variables $V_{n+1}$. Let $f(x * (\bar{C} + 1)) = (\Delta, \Theta)$, where $(\Delta, \Theta)$ is such a theory for some choice of a $u \in U_{n+1}$.

Consider now the following Kripke structure in $\mathbf{K}_{T0}$, with $S = \{x \in \omega^* \mid f(x) \text{ is defined}\}$, $R = \leqslant \upharpoonright S$, $O = \Lambda$, $U_x = V_{\text{length}(x)}$ and for each $m$-ary atomic $A$, $D(x, A) = \{(u_1, \ldots, u_m) \mid A(u_1, \ldots, u_m) \in \Delta_x$, where $f(x) = (\Delta_x, \Theta_x)\}$. Take $N = \emptyset$.

LEMMA 7. *In the Kripke structure defined above, under any assignment g into the language of* $f(x)$,

$$[A(u_1, \ldots, u_m)]^g_x = 1 \quad \textit{iff} \quad A^g = A(g(u_1), \ldots, g(u_m)) \in \Delta_x$$
$$\textit{where} \quad f(x) = (\Delta_x, \Theta_x)$$

*Proof.* By induction on $A$. For atomic $A$, Lemma 7 holds by definition. $\wedge$, $\vee$, $f$, $\exists$ present no difficulties, since they are classical in $\Vdash$ (of 2.2.2a, by Proposition 2.2.8).

We check the case of $C = A \to B$. Assume $A^g \to B^g \in \Delta_x$, then if $x \leqslant y$, and $[A]^g_x = 1$, then by the induction hypothesis $A^g \in \Delta_y$, and since $\Delta_x \subseteq \Delta_y$ by construction, $A^g \to B^g \in \Delta_y$ and therefore $B^g \in \Delta_y$ i.e., $[B]^g_y = 1$. Thus $[A \to B]^g_x = 1$.

Suppose $A^g \to B^g \notin \Delta_x$, then $C^g = A^g \to B^g \in \Theta_x$ (since $(\Delta_x, \Theta_x)$ is saturated in its language), so by construction $A^g \in \Delta_y$, $B^g \in \Theta_y$ for $y = x * (\bar{\bar{C}}^g + 1)$ and thus $[A]^g_y = 1$, $[B]^g_y = 0$, and hence $[A \to B]^g_x = 0$.

To check the case of $C = (\forall u)(u)$;

Assume $C = (\forall u)A(u) \in \Delta_x$, then $C \in \Delta_y$ for all $y$ such that $x \leqslant y$ and therefore for all $u$ in the language of $\Delta_y$, $A(u) \in \Delta_y$ since $\vdash_H \forall u A(u) \to A(v)$, for all $v$. Thus by the induction hypothesis $[A]^{g'}_y = 1$, for all $y$ such that $x \leqslant y$ and all $g'$ such $g' =_u g'$ and hence $[C]^g_x = 1$.

Assume $C \notin \Delta_x$, then $C \in \Theta_x$ and therefore for some $v_0$, $A(v_0) \in \Theta_y$ for $y = x * (\bar{\bar{C}} + 1)$. Let $g'(v) = g(v)$, for $v \neq u$, $g'(u) = v_0$. Then $[A]^{g'}_y = 0$ by definition and thus $[C]^g_x = 0$.

For the case of $f$ notice that since $f(x)$ is always consistent, $f \in \Theta_x$, and hence $[f]^g_x = 0$, for all $x$, $g$.

Thus Lemma 7 is proved, and we see that if $\varphi \nVdash \psi$, we can construct the structure above and get, under the identity assignment, that all the wff of $\varphi$ hold but none of $\psi$ hold.

To show that $\Vdash \subseteq \Vdash_{\mathbf{K}_{TO}}$ we verify that $\Vdash_{\mathbf{K}_{TO}}$ agrees with $\vdash_H$. We verify that all axioms and rules of $H$ are valid in all structures of $\mathbf{K}_{TO}$. Lemma 5 is useful for this verification.

COROLLARY 8. $\Vdash_{\mathbf{K}_{TO}} = \Vdash_{\mathbf{K}} = \Vdash_{\mathbf{K}_p} = \Vdash_{\mathbf{K}_{pO}} = \Vdash_{\mathbf{K}_O} = \Vdash_{\mathbf{K}_T}$.

*Proof.* One can verify that $\Vdash_{\mathbf{K}}$ equals $\Vdash$ of (2.2.2a) using the previous theorem.

COROLLARY 9. *Let* $\mathbf{K}'_T$ *(resp.* $\mathbf{K}'_{TO}$*) be the class of all Kripke structures in* $\mathbf{K}_T$*(resp.* $\mathbf{K}_{TO}$*) with the property that if x and y are in S and are of the same length then* $U_x = U_y$.

*Then* (1) $\Vdash = \Vdash_{\mathbf{K}'_{TO}}$, *where* $\Vdash$ *is the maximal consequence relation agreeing with* $\vdash_H$. (2) $\Vdash_{\mathbf{K}'_T}$ *agrees with* $\vdash_H$.

*Proof.* The structure constructed in the proof of Theorem 6 was in $\mathbf{K}'_{TO}$.

COROLLARY 10. *H is complete for* $\Vdash_{\mathbf{K}_{TSO}}$.

*Proof.* In the construction of the structure in the proof of Theorem 6, we could have used new set of variables each time we extended a theory to a saturated theory. Thus the condition of (2f) could be satisfied.

## 2. Beth interpretation

Beth structures have the form $(N, S, R, O, D, U)$ where $(S, R, O)$ is a partially ordered set with first element $O \in S$, $N \subseteq S$ and $U$ is a nonempty set of elements. $D$ is a function assigning to each $n$-ary atomic predicate $A$ of the language and each $t \in S$ a subset $D(t, A) \subseteq U^n$. We require that if $tRs$ then $D(t, A) \subseteq D(s, A)$, and also $tRs$ and $t \in N$ imply $s \in N$. If $t \in N$ then $D(t, A) = U^n$.

DEFINITION 1. Let $(N, S, R, O, D, U)$ be a Beth structure.

(a) A subset $S_0 \subseteq S$ is called a path in $S$ through $t \in S$ if it is a maximally linearly ordered subset of $\{s \mid tRs\}$.

(b) A subset $T \subseteq S$ is said to *bar* a point $t \in S$ iff every path through $t$ intersects $T$.

DEFINITION 2. Let $(N, S, R, O, D, U)$ be a Beth structure and let $g$ be a function assigning to each variable of the language an element of $U$. We define by induction the notion of the truth value of a wff $A$ at a point $t \in S$, under $g$, denoted by $\|A\|_t^g$.

(a) $\|A(x_1, \ldots, x_n)\|_t^g = 1$ iff the set $\{s \mid (g(x_1), \ldots, g(x_n)) \in D(s, A)\}$ bars $t$, for $A$ atomic.

(b) $\|A \wedge B\|_t^g = 1$ iff $\|A\|_t^g = 1$ and $\|B\|_t^g = 1$.

(c) $\|f\|_t^g = 1$ iff $N$ bars $t$.

(d) $\|\forall x A(x)\|_t^g = 1$ iff for all $g' =_x g$ $\|A(x)\|_t^{g'} = 1$.

(e) $\|A \to B\|_t^g = 1$ iff for all $s$, if $tRs$ and $\|A\|_s^g = 1$ then $\|B\|_s^g = 1$.

(f) $\|A \vee B\|_t^g = 1$ iff $\{s \mid \|A\|_s^g = 1$ or $\|B\|_s^g = 1\}$ bars $t$.

(g) $\|\exists x A(x)\|_t^g = 1$ iff $\{s \mid$there exists a $g' =_x g$ with $\|A(x)\|_s^{g'} = 1\}$ bars $t$.

We say that $A$ holds (or is valid) in the structure under $g$ iff $\|A\|_0^g = 1$.

LEMMA 3. *If* $\|A\|_t^g = 1$ *and* $tRs$ *then* $\|A\|_s^g = 1$.

*Proof.* Exercise, use induction on $A$.

LEMMA 4. *Let* $(N, S, R, O, D, U)$ *be a Kripke structure in* $\mathbf{K}'_{TO}$, *then there exists a Beth structure* $(N^*, S^*, R^*, O^*, D^*, U^*)$ *that validates the same wffs and that can be effectively obtained from the Kripke structure.*

*Proof.* Let $S^*$ be the set of all finite sequences of the form $(o, t_1, \ldots, t_n)$ where $o, t_1, \ldots, t_n \in S$, $n$ arbitrary and $oRt_1, t_1Rt_2, \ldots, t_{n-1}Rt_n$ holds. For $t^*, s^* \in S^*$ let $t^*R^*s^*$ hold iff $t^*$ is an initial segment (or is equal to) $s^*$. For $t^* \in S^*$ let $|t^*|$ be the last element of $t^*$. Thus we have

(#) $t^*R^*s^* \to |t^*|R|s^*|$.

Let $O^* = (O)$ and $U^* = \bigcup_{t \in s} U_t$. For $n$ place atomic predicate $P$ let $D^*(t^*, P) = D(|t^*|, P)$. Let $N^* = \{t^* \mid |t^*| \in N\}$. It is easy to verify that we thus obtain a Beth structure.

Notice that for any $r \in S^*$, $H_r = \{r * (|r|, \overset{n \text{ times}}{\ldots}, |r|) \mid n \in \omega\}$ (where $*$ denotes concatenation of sequences) is a path through $r$ in $S^*$.

LEMMA 5. *Let* $t^* \in S^*$ *and let* $g$ *be any assignment into* $U_{|t^*|}$ *(of the Kripke structure) then for any* $A$

$$\|A\|_{t^*}^g = [A]_{|t^*|}^g$$

*Proof.* By induction on $A$, for atomic $A(x_1, \ldots, x_n)$, $\|A\|_{t^*}^g = 1$ iff $\{s^* \in S^* \mid (g(x_1), \ldots, g(x_n)) \in D^*(s^*, A)\}$ bars $t^*$. Assume $\|A\|_{t^*}^g = 1$, then the path $H_{t^*}$ intersects the above set and so $(g(x_1), \ldots, g(x_n)) \in D^*(x, A)$ for some $x \in H_{t^*}$ and therefore $(g(x_1), \ldots, g(x_n)) \in D(|t^*|, A)$, since $|x| = |t^*|$ on this path, i.e., $[A]_{|t^*|}^g = 1$.

Assume that $[A]^g_{|t^*|} = 1$, then $(g(x_1), \ldots, g(x_n)) \in D^*(t^*, A)$ and therefore $\|A\|^g_{t^*} = 1$.

The cases of $\wedge$ and $f$ present no difficulties.

$\|A \to B\|^g_{t^*} = 0$ iff for some $s^*$, $t^*Rs^*$ and $\|A\|^g_{s^*} = 1$, and $\|B\|^g_{s^*} = 0$ iff (by # and the induction hypothesis) for some $s$ such that $|t^*|Rs$, $[A]^g_s = 1$ and $[B]^g_s = 0$ iff $[A \to B]^g_{|t^*|} = 0$.

Assume $[A \vee B]^g_{|t^*|} = 1$, then either $[A]^g_{|t^*|} = 1$ or $[B]^g_{|t^*|} = 1$ and in either case $\|A \vee B\|^g_{t^*} = 1$.

If $\|A \vee B\|^g_{t^*} = 1$ then $\{s^* \mid t^*R^*s^*$ and either $\|A\|^g_{s^*} = 1$ or $\|B\|^g_{s^*} = 1\}$ bars $t^*$. So for some $x \in H_{t^*}$, either $\|A\|^g_x = 1$ or $\|B\|^g_x = 1$ and hence either $[A]^g_{|x|} = 1$ or $[B]^g_{|x|} = 1$ and since $|x| = |t^*|$, $[A \vee B]^g_{|t^*|} = 1$.

Assume $[\forall x A(x)]_{|t^*|} = 1$, then any $s$ such that $|t^*|Rs$ and any $u \in U_s$ and any $g' =_x g$, $g'(x) = u$ $[A(x)]^{g'}_s = 1$ and so by the induction hypothesis, for any $s^*$ such that $t^*R^*s^*$ and any $u \in U_{|s^*|} \|A(x)\|^{g'}_{s^*} = 1$. Now since the original structure was in $\mathbf{K}'_{TO}$, for any $u \in U^*$ there exists an $s^*$ such that $t^*R^*s^*$ and $u \in U_{|s^*|}$ and therefore we conclude that for any $u \in U^*$ and any $s^*$ such that $t^*R^*s^*$ $\|A(x)\|^{g'}_{s^*} = 1$, i.e., $\|\forall x A\|^g_{t^*} = 1$.

For the other direction, if for any $u \in U^*$ and $s^*$ s.t. $t^*R^*s^*$, $\|A(x)\|^{g'}_{s^*} = 1$, then since if $|t^*|Rs$ then $t^*R^*t^* * (s)$, we conclude that for any $s$ s.t. $tRs$ and any $u \in U_s$, $[A(x)]^{g'}_s = 1$ i.e., $[\forall x A]^g_{|t^*|} = 1$.

Now assume $[\exists x A(x)]^g_{|t^*|} = 1$ then for some $u \in U_{|t^*|}$ $[A(x)]^{g'}_{|t^*|} = 1$, where $g' =_x g$, $g'(x) = u$, and therefore $\|\exists x A(x)\|^g_{t^*} = 1$.

Now assume that $[\exists x A(x)]^g_{|t^*|} = 0$ then in the path $H_{t^*}$, $[\exists x A(x)]^g_{|v|} = 0$ for any $v \in H_{t^*}$. If it were the case that for some $g' =_x g$ and $v \in H_{t^*}$ $\|A(x)\|^{g'}_v = 1$, then by the induction hypothesis, $[A(x)]^{g'}_{|v|} = 1$, and since $|v| = |t^*|$, we get a contradiction. Thus $\|\exists x A(x)\|^g_{t^*} = 0$.

This proves Lemma 4,

THEOREM 5. *Let* **B** *be the class of all tree Beth structures and let* $\Vdash_{\bar{H}}$ *be the minimal consequence relation agreeing with* $\vdash_H$ (see 2.2.2b) *then* $\Vdash_{\bar{H}} = \Vdash_{\mathbf{B}}$.

*Proof.* First show that $\Vdash_{\mathbf{B}}$ agrees with $\vdash$. It is easy to verify that all axioms and rules of $H$ are valid in all structures of **B**. Now suppose $\varphi \nvdash_H A$, then (by 1.8) for some Kripke structure in $\mathbf{K}'_{TO}$, $\varphi$ is valid in the structure and $A$ is not. By Lemma 4 there exists a tree structure in **B** which validates the same wffs and hance $\varphi \nVdash_{\mathbf{B}} A$.

Therefore $\Vdash_{\bar{H}} \subseteq \Vdash_{\mathbf{B}}$. To show that $\varphi \nVdash_{\bar{H}} \psi$ implies $\varphi \nVdash_{\mathbf{B}} \psi$, let $\psi = \{A_1, \ldots, A_n\}$. Since $\varphi \nVdash_{\bar{H}} \psi$, $\varphi \nvdash_H A_i$, $1 \leqslant i \leqslant n$ and so let $\mathbf{A}_i = (N_i, S_i,$

$R_i$, $O_i$, $D^i$, $U^i$) be Beth structures in which $\varphi$ hold but not $A_i$, respectively. We can assume that $U^i = U^j = U$. Construct $(N, S, R, O, D, U)$ by letting

$$S = \{O\} \cup \bigcup_i (S_i \times \{S_i\})$$

$$R = \{O\} \times S \cup \{((x, S_i), (y, S_i)) \mid 1 \leqslant i \leqslant n, (x, y) \in R_i\}$$

$$D(O, A) = \emptyset, \qquad D((x, S_i), A) = D^i(x, A)$$

$$N = \bigcup_i (N_i \times \{S_i\}).$$

LEMMA 6. *For any* $A$, $\|A\|^{\mathfrak{s}}_{\mathfrak{o}} = 1$ *iff for all* $i$, $\|A\|^{\mathfrak{s}}_{\mathfrak{o}_i} = 1$
*Proof*. by induction.
Thus $\varphi$ holds at $O$ but not $\psi$.

## 3. Structures with Constant Domains

DEFINITION 1. Let **CD** be the class of all Kripke structures with constant domains, i.e., all Kripke structures $(N, S, R, O, D, U)$ such that for all $t, s \in S$, $U_t = U_s$.

THEOREM 2. (*S. Görneman*). *Let* $\Vdash^+_{CD}$ *be the consequence relation of Theorem* (2.6.40) *then* $\Vdash^+_{CD} = \Vdash_{\mathbf{CD}}$.

*Proof*. It is easy to verify that $\forall$ is classical in $\Vdash_{\mathbf{CD}}$ and therefore $\Vdash_{\mathbf{CD}} \supseteq \Vdash^+_{CD}$.

To show the other direction, we prove that if $\varphi \nVdash_{CD} A$, then there exists a Kripke structure in **CD** which validates $\varphi$ but does not validate $A$. We need some lemmas.

LEMMA 3. (a) *If* $x$ *is not free in* $C_1$ *then*

(1) $\quad \forall x(C_1 \to C_2(x)) \vdash_H C_1 \to \forall x C_2(x)$

(2) $\quad \forall x(C_2(x) \to C_1) \vdash_H \exists x C_2(x) \to C_1$

(3) $\quad C_1 \vee \exists x C_2(x) \vdash_H \exists x(C_1 \vee C_2(x))$

(b) $\varphi \vdash_H A \vee (B \to A \vee C)$ iff $\varphi, B \vdash_H A \vee C$.
*Proof*. Exercise.

LEMMA 4. *Let* $(\Delta, \Theta)$ *be a* $\Vdash^+_{CD}$ *theory with property* (*) *below, and assume that* $\Theta$ *is finite, then for any* $A_1$, $A_2$, $(\Delta \cup \{A_1\}, \Theta \cup \{A_2\})$ *has*

(*). *Where,*

(*)(a) *If* $\Delta \Vdash \Theta, B(u)$ *for all* $u$ *of the language then* $\Delta \Vdash \Theta, \forall x B(x)$ *(*$x$ *not free in* $\Theta$*)*

(b) *If* $\Delta, B(u) \Vdash \Theta$ *for all* $u$*, then* $\Delta, \exists x B(x) \Vdash \Theta$.

*Proof.* To show (a) assume $\Delta, A, \Vdash^{+}_{CD} \Theta, A_2, B(u)$, for all $u$, then since $\vee$ is classical in $\Vdash^{+}_{CD}$, $\Delta, A_1 \Vdash^{+}_{CD} \vee \Theta \vee A_2 \vee B(u)$ and hence by the deduction theorem for $\rightarrow$, $\Delta \Vdash^{+}_{CD} A_1 \rightarrow (\vee \Theta \vee A_2 \vee B(u))$ for all $u$ and therefore by (*)

$$\Delta \Vdash^{+}_{CD} \Theta, \forall x(A_1 \rightarrow \vee \Theta \vee A_2 \vee B(x))$$

$x$ not free in $\Theta$ or $A_1$.

Therefore

$$\Delta \Vdash^{+}_{CD} \Theta, A_1 \rightarrow \forall x(\vee \Theta \vee A_2 \vee B(x))$$

hence by *cd*

$$\Delta \Vdash^{+}_{CD} \Theta, A_1 \rightarrow \vee \Theta \vee A_2 \vee \forall x B(x)$$

and hence by Exercise 2

$$\Delta, A, \Vdash^{+}_{CD} \Theta, A_2, (\forall x)B(x).$$

To show (b) assume $\Delta, A_1, B(u) \Vdash^{+}_{CD} \Theta, A_2$, for all $u$ and therefore

$$\Delta, A_1 \Vdash^{+}_{CD} B(u) \rightarrow \vee \Theta \vee A_2 \text{ for all } u$$

so

$$\Delta, A_1 \Vdash^{+}_{CD} \Theta, A_2, \forall x(B(x) \rightarrow \vee \Theta \vee A_2)$$

so

$$\Delta, A_1 \Vdash^{+}_{CD} \Theta, A_2, \exists x B(x) \rightarrow \vee \Theta \vee A_2$$

so

$$\Delta, A_1, \exists x B(x) \Vdash^{+}_{CD} \Theta, A_2.$$

LEMMA 5. *If* $(\Delta_0, \Theta_0)$ *has property* (*) *with* $\Theta_0$ *finite then it can be extended to a saturated and complete theory in the same language.*

*Proof.* We follow the construction in the proof of Lemma 1.4.3. Let $B_n$ enumerate all wffs of the language. Define by induction a sequence of theories $(\Delta_n, \Theta_n)$. We assume that each $(\Delta_n, \Theta_n)$ has property $(*)$, with $\Theta_n$ finite.

*Case* 1. If $(\Delta_n, \Theta_n \cup \{B_n\})$ is consistent, then if $B_n$ is not universal, let $\Delta_{n+1} = \Delta_n$, $\Theta_{n+1} = \Theta_n \cup \{B_n\}$. If $B_n = \forall x B(x)$, then for some $u$, $\Delta_n \Vdash\!\!\!/^{+}_{CD} \Theta_n$, $B(u)$ (by property $(*)$) so let $\Delta_{n+1} = \Delta_n$, $\Theta_{n+1} = \Theta_n \cup \{B(u)\} \cup \{\forall x B(x)\}$.

*Case* 2. $\Delta_n \Vdash^{+}_{CD} \Theta_n, B_n$, if $B_n$ is not existential, let $\Delta_{n+1} = \Delta_n \cup \{B_n\}$, $\Theta_{n+1} = \Theta_n$. $(\Delta_{n+1}, \Theta_{n+1})$ is clearly $\Vdash^{+}_{CD}$ consistent, since $(\Delta_n, \Theta_n)$ is. If $B_n = \exists x B(x)$, then by property $(*)$, since $\Delta_n, B_n \Vdash\!\!\!/^{+}_{CD} \Theta_n$, for some $u$, $\Delta_n, B(u) \Vdash\!\!\!/^{+}_{CD} \Theta_n$, so let $\Delta_{n+1} = \{B(u), B_n\} \cup \Delta_n$, $\Theta_{n+1} = \Theta_n$.

Thus $(\Delta_{n+1}, \Theta_{n+1})$ is defined, is consistent and has $(*)$. Let $\Delta = \bigcup_n \Delta_n$, $\Theta = \bigcup_n \Theta_n$, then $(\Delta, \Theta)$ is the desired extension.

LEMMA 6. *Let $(\Delta, \Theta)$ be a $\Vdash^{+}_{CD}$ complete and saturated theory, with $A \to B \in \Theta$, then $(\Delta \cup \{A\}, \{B\})$ has property* $(*)$.

*Proof.* (a) If $\Delta, A \Vdash^{+}_{CD} B \vee C(u)$ for all $u$ then $\Delta \Vdash^{+}_{CD} A \to (B \vee C(u))$ for all $u$, and since $(\Delta, \Theta)$ is saturated and complete, (remember that $\forall$ is classical in $\Vdash^{+}_{CD}$ and hence condition (4) of (1.4.1a) holds for $(\Delta, \Theta)$), $\Delta \Vdash^{+}_{CD} \forall x(A \to B \vee C(x))$ and hence $\Delta, A \Vdash^{+}_{CD} B, \forall x C(x)$, since $x$ is not free in $A$ or $B$.

(b) If for all $u$

$$\Delta, A, C(u) \Vdash^{+}_{CD} B$$

then for all $u$

$$\Delta, C(u) \Vdash^{+}_{CD} A \to B$$

and since $(\Delta, \Theta)$ is saturated and complete

$$\Delta, \exists x C(x) \Vdash^{+}_{CD} A \to B$$

i.e.,

$$\Delta, A, \exists x C(x) \Vdash^{+}_{CD} B.$$

LEMMA 7. *Let $(\Delta, \Theta)$ be a saturated and complete theory with $A \to B \in \Theta$, then there exists a saturated and complete theory $(\Delta', \Theta')$ in the same language with $\Delta \subseteq \Delta'$, $A \in \Delta'$, $B \in \Theta'$.*

*Proof.* From preceding lemma.

We can now continue with the proof of Theorem 2. Assume that $\varphi \Vdash\!\!\!/^{+}_{CD} \psi$, let $(\Delta_\Lambda, \Theta_\Lambda)$ be a complete and $\Vdash^{+}_{CD}$ saturated theory of the language with $\varphi \subseteq \Delta_\Lambda$, $\psi \subseteq \Theta_\Lambda$ (this exists by (1.4.3)). Let $V_n = V$ for all $n$. Proceed to construct the structure as in the proof of Theorem 1.6.

COROLLARY 8. *CD is complete for the class of all Kripke structures with constant domains with $(S, \leqslant)$ a partially ordered set and $N = \emptyset$.*

*Proof.* Follows from the completeness proof.

THEOREM 9. *Let A be a wff of H without* $\vee$ *and without* $\exists$, *then if* $\nvdash_H A$ *then there exists a Kripke structure with constant domains in which A is false.*

*Proof.* For $B$, $B'$ wffs of $H$, let $B \equiv B'$ iff $\vdash_H B \leftrightarrow B'$. Let $S$ be the set of all equivalence classes of wffs modulo $\equiv$. Define $\leqslant$ on $S$ by $B|\equiv \leqslant B'|\equiv$ iff $\vdash_H B \to B'$. Let $O = (f \to f)|\equiv$ and $N = \{f|\equiv\}$ and $U$ = set of variables of the language. For atomic $P(x_1, \ldots, x_n)$, let $[P(x_1, \ldots, x_n)]_{B|\equiv} = 1$ iff $\vdash_H B \to P(x_1, \ldots, x_n)$.

LEMMA 10. *For any A without* $\vee$ *and* $\exists$, $[A]^g_{B|\equiv} = 1$ *iff* $\vdash_H B \to A^g$.

*Proof.* By induction on $A$. For atomic $A$ this is the definition. The case of $\wedge$ follows since $\vdash_H B \to A \wedge A'$ iff $\vdash_H (B \to A) \wedge (B \to A')$. The case of $\to$ follows since $\vdash_H B \to (A \to A')$ iff for any $C$, $\vdash_H C \to B$ and $\vdash_H C \to A$ implies $\vdash_H C \to A'$.

For the case of $\forall$, if $\vdash_H B \to \forall x A(x)$ then certainly $\vdash_H B \to A(y)$, for all $y$. Conversely, if $\vdash_H B \to A(y)$ for all $y$, choose $y$ not appearing in $B$ and so by the rules of $H$ $\vdash_H B \to \forall x A(x)$.

Thus Lemma 10 is proved. If $\nvdash_H A$, then in the structure defined above $[A]_0 = O$, under the identity assignment.

We want now to change a little the notion of the truth value in a Kripke structure and obtain an interpretation for *CD*.

DEFINITION 11. Let $(N, S, R, O, D, U)$ be a Kripke structure (not necessarily of constant domains) and let $g$ be an assignment. We define by induction on wffs $A$ a new truth value $\lceil A \rceil^g_t$ as follows:

(1) For the cases of atomic $A$, $\wedge$, $\vee$, $f$, $\to$ and $\exists$ use the same clause as in Definition 1.3 for $[A]^g_t$.

(2) $\lceil \forall x A(x) \rceil^g_t = 1$ iff for all $g'$ into $U_t$, if $g' =_x g$ then $\lceil A(x) \rceil^{g'}_t = 1$.

$A$ is said to hold in the structure under $g$ and the interpretation iff $\lceil A \rceil^g_0 = 1$.

(Notice that $\forall x A(x)$ holds at $t$ iff for all $a \in U_t$ $A(a)$ holds. In the [ ] definition we had $[\forall x A(x)]_t = 1$ iff for all $s$, $tRs$ and all $a \in U_s$, $A(a)$ holds at $s$.)

*Remark* 12. In a Kripke structure with constant domains $\lceil A \rceil_t^g = [A]_t^g$ for any $g$, $t$.

*Proof.* Trivial.

COROLLARY 13. *CD is complete for the* $\lceil\ \rceil$ *interpretation in Kripke structures.*

*Proof.* It is sufficient to verify that *CD* is valid under the $\lceil\ \rceil$ interpretation.

LEMMA 14. *Let A be any wff without* $\forall$, *then in any Kripke structure*

$$\lceil A \rceil_t^g = [A]_t^g.$$

*Proof.* Immediate by induction.

THEOREM 15. *H. Fitting. Let A be without* $\forall$, *then if* $H \not\Vdash A$ *then* $CD \not\Vdash A$.

*Proof.* If $H \not\Vdash A$, there exists a Kripke structure and an assignment $g$ such that $[A]_0^g = 0$. Consider $\lceil A \rceil_0^g$. By the previous lemmas, $\lceil A \rceil_0^g = 0$ and $\lceil B \rceil_0^g = 1$ for all $B$ such that $CD \vdash B$, thus $CD \nvdash A$.

THEOREM 16. *The fragments of CD and H without* $\forall$ *and the fragments of CD and H without* $\vee$, $\exists$ *are respectively identical.*

*Proof.* Follows from the previous theorems.

We would like now to sharpen the results of Theorem 16 and show that the fragment of *CD* without $\vee$ is identical with the fragment of *H* without $\vee$.

DEFINITION 17. Let $(N, S, R, O, D, U)$ be a Beth structure. Define a new truth value $|[A]|_t^g$, of a wff $A$ at a point $t \in S$, under the assignment $g$, by induction.

(a) The cases of atomic $A$ and the cases of $\wedge$, $\vee$, $\rightarrow$, $f$, $\forall$ are as in Definition 2.2.

(b) $|[\exists x A(x)]|_t^g = 1$ iff for some $g' =_x g$ $|[A(x)]|_t^{g'} = 1$.

This definition gives $\exists$ the Kripke interpretation truth table. We want to prove:

THEOREM 18. *H is complete for the* $\|[\ ]\|$ *interpretation in Beth structures.*

We can get Theorem 19 below as a corollary, since every Beth structure is essentially a Kripke structure with constant domains. For let $(N, S, R, O, D, U)$ be a Beth structure. Define $D^*$ by

$$D^*(t, P) = \{(x_1, \dots, x_n) \in U^n \mid \{s \mid (x_1, \dots, x_n) \in D(s, P)\} \text{ bars } t\}.$$

Then under any assignment $g$ and any wff without $\vee$ we have:

$$(\#) \qquad [A]_t^{g*} = \|[A]\|_t^g$$

where $[A]*$ is the value in the Kripke structure $(N, S, R, O, D^*, U)$ and $\|[A]\|$ is the value in $(N, S, R, O, D, U)$.

THEOREM 19. *For any A without disjunction,*

$$CD \vdash A \quad \text{iff} \quad H \vdash A.$$

*Proof of Theorem* 18. Given a tree Beth structure $M = (N, S, R, O, D, U)$ we want to transform it into a structure $\bar{M} = (N, S, R, O, \bar{D}, \bar{U})$ in such a way that all wffs retain their truth value and in $\bar{M}$ the $\|[\ ]\|$ and $\|\ \|$ truth values coincide.

DEFINITION 20. (a) Let $T \subseteq S$, denote by $T^+ = \{t' \in S \mid t''Rt' \text{ for some } t'' \in T\}$.

(b) $T$ is open iff $T = T^+$.

(c) Let $\bar{U}$ be the set of all functions $f: \operatorname{dom} f \to U$, such that $\operatorname{dom} f \subseteq S$, $\operatorname{dom} f$ is open and $\operatorname{dom} f$ bars $O$ and such that if $tRt'$ $t \in \operatorname{dom} f$ then $f(t) = f(t')$.

(d) Given $f_1, \dots, f_n \in \bar{U}$ and atomic $n$-place $P$ let

$$C = \bigcap_{i=1}^{n} \operatorname{dom} f_i,$$

and define $f_1, \dots, f_n \in \bar{D}(t, P)$ iff $C$ bars $t$, and for all $t' \in C$,

$$(f_1(t'), \dots, f_n(t')) \in D(t', P).$$

LEMMA 21. *For any* $A(x_1, \dots, x_n)$ *and any* $g$ *with* $g(x_i) = f_i$, $\|[A(x_1, \dots, x_n)]\|_t = 1$ *iff for some* $T \subseteq C = \bigcap_i \operatorname{dom} f_i$, *T bars t, and for all* $t' \in T$, $\|A(x_1, \dots, x_n)\|_{t'}^{g_{t'}} = 1$, *where* $g_{t'}(x) = g(x)(t')$.

*Proof.* (a) For atomic $A$ this is the definition.

(b) The case of $\wedge$ is immediate.

(c) If $\|[A \vee B]\|_t = 1$ then for some $T \subseteq C$, $T$ bars $t$ and for all $t' \in T$, $\|[A]\|_{t'} = 1$ or $\|[B]\|_{t'} = 1$. By the induction hypothesis, for each $t' \in T$ there exists a $T_{t'}$ such that $T_{t'}$ bars $t'$ and is open, $T_{t'} \subseteq \bigcap \mathrm{dom}\, f_i$ $i = 1, \ldots, n$ and for each $t'' \in T_{t'}$ $\|A\|_{t''}^{g_{t''}} = 1$ or $\|B\|_{t''}^{g_{t''}} = 1$. Let $T' = \bigcup_{t' \in T} T_{t'}$. Then $T'$ is open, $T' \subseteq \bigcap \mathrm{dom}\, f_i$, $i = 1, \ldots, n$ and satisfies the other conditions. The other direction is simple.

(d) Assume $\|[\forall x A(x)]\|_t^g = 1$. Then for all $g' =_x g$, $\|[A(x)]\|_t^{g'} = 1$. By the induction hypothesis, for some $T_{g'}$ and any $t' \in T_{g'}$, $\|A\|_{t'}^{g_{t'}} = 1$, and $T_{g'} \subseteq \bigcap \mathrm{dom}\, f_i$, $i = 1, \ldots, n$. Let $t''$ be any minimal point of $C$. Then since $T_{g'}$ bars $t''$, we get $\|A\|_t^{g_{t''}} = 1$ for arbitrary $g'$. Since $g' =_x g$ is arbitrary so $g'_{t''}(x)$ can be made to range over all of $U$ and hence $\|\forall x A(x)\|_{t''}^{g_{t''}} = 1$. Thus for $C$ the required condition holds.

Conversely, assume for some $T \subseteq \bigcap \mathrm{dom}\, f_i$, $T$ open, $\|\forall x A(x)\|_{t'}^{g_{t'}} = 1$, for all $t' \in T$. Thus we conclude that $\|\forall x A(x)\|_{t'}^{g_{t'}} = 1$, for all $t' \in C$. Suppose on the contrary that for some $g' =_x g$ $\|[A]\|_t^{g'} = 0$, then for some $T_{g'} \subseteq C$, and some $t'' \in T_{g'}$ $\|A(x)\|_{t''}^{g'_{t''}} = 0$. But this contradicts the above, since $t'' \in C$.

(e) $\|[A \to B]\|_t^g = 0$ iff for some $tRt''$, $\|[A]\|_{t''}^g = 1$ and $\|[B]\|_{t''}^g = 0$ iff for some suitable $T_1 \subseteq S$, $\|A\|_s^{g_s} = 1$ for all $s \in T_1$ and for *no* suitable $T_2$, $\|B\|_s^{g_s} = 1$ for all $s \in T_2$. Thus for some infinite path $P$, $\|B\|_s^{g_s} = 0$ for all $s \in P$, and $P$ passes through $t''$. Thus $\|A\|_{t''}^{g_{t''}} = 1$ and $\|B\|_{t''}^{g_{t''}} = 0$ and hence $\|A \to B\|_t^{g_t} = 0$. Take $T$ as $\{t\}^+$. The other direction is easy.

(f) The case of $\exists$. If $\|[\exists x A(x)]\|_t^g = 1$, then for some $g' =_x g$, $\|[A]\|_t^{g'} = 1$ and by the induction hypothesis for some $T$, $\|A\|_{t'}^{g'_{t'}} = 1$ for all $t' \in T$ and hence $\|\exists x A(x)\|_{t'}^{g_{t'}} = 1$ for all $t' \in T$.

Conversely assume that for some $T$, $\|\exists x A(x)\|_{t'}^{g_{t'}} = 1$ for all $t' \in T$. Then for each $t' \in T$ a $T_{t'}$ barring $t'$ exist such that for any $t'' \in T_{t'}$, an assignment $h_{t''} =_x g_{t''}$ into $U$ exists with $\|A\|_{t''}^{h_{t''}} = 1$.

Since $(N, S, R, O)$ is a tree structure, a $g' =_x g$ into $\bar{U}$ can be found such that for each $t'' \in T_{t'}$ $g'(x)(t'') = h_{t''}(x)$, and hence on $T' = \bigcup_{t' \in T} T_{t'}$ $\|A\|_{t''}^{g'_{t''}} = 1$ for all $t'' \in T'$ and hence by the induction hypothesis $\|[A(x)]\|_t^{g'} = 1$ and hence $\|[\exists x A]\|_t^g = 1$.

This completes the proof of Lemma 21.

From Lemma 21 we get that for any $A$, $g$, $t$ $\|[A]\|_t^g = \|A\|_t^{g_t}$ and hence Theorem 18 follows, since if $H \nVdash A$, then for some tree structure $M$

and some $g'$, $\|A\|_0^{g'} = 0$ and so for a suitable $g_0$ s.t. $g_0 = g'$, $\|[A]\|_0^{g_0} = 0$ in $\bar{M}$.

**Exercise 22.** Let $\Vdash^+$ be the maximal consequence relation agreeing with $\vdash_H$. This is the consequence relation obtained from the Kripke interpretation for $H$. Following definition 1.3.17 let $\Vdash^*$ be the dual closure of $\Vdash^+$. Show that

(a) $\wedge^* = \vee, \quad \vee^* = \wedge, \quad t^* = f, \quad f^* = t, \quad \exists^* = \forall, \quad \forall^* = \exists.$

(a) shows that the dual closure is not conservative since $\exists^* = \forall$, thus $\forall x(A \vee B(x)) \Vdash^* A \vee \forall x B(x)$ $x$ not free in $A$. ($t$ is $f \to f$).

Let $\mathbf{T}_{\Vdash^*}$ be the canonical semantics for $\Vdash^*$. Show that

(a) $\wedge, \vee, \exists, \forall, t, f$ are classical in $\mathbf{T}_{\Vdash^*}$.

(b) denoting the dual $\to^*$ of $\to$ by $\leftarrow$ show that conditions (C1) and (C2) of (2.1.19) hold for our case.

## 4. Kripke interpretation for *MH*

DEFINITION 1. Let **MH** be the class of all Kripke $\mathbf{K}_0$ (see Definition 1.2) structures with the property **(mh)** below:

**(mh)** $\quad (\forall x \in S)(\exists y \in S)(\forall z \in S)(xRy \wedge (yRz \to y = z)).$

THEOREM 2. *For any $\varphi$ and $A$, $\varphi \vdash_{MH} A$ iff $\varphi \Vdash_{\mathbf{MH}} A$ (where $\Vdash_{\mathbf{MH}}$ is as in Definition* 1.4) *and MH is as in Corollary* 2.7.4).

*Proof.* To prove completeness we follow the lines of the proof of Theorem 1.6. Let $V$, $V_n$ be as in the proof of (1.6).

We first assume that $\varphi \not\Vdash_{MH} A$ and show the existence of a structure in **MH** and an assignment $g$ into the structure which validates all members of $\varphi$ but not $A$. We can assume that $\varphi$, $A$ are in a language with variables $V_0$.

We continue as in (1.6) to define a partial function $f$ on $\omega^*$. To describe the procedures, we need a lemma.

LEMMA 3. *Let $(\Delta, \Theta)$ be a consistent $\Vdash^+_{MH}$ theory in the language with variables $V_n$. Then there exists a consistent and saturated $\Vdash^+_{MH}$ theory $(\Delta^0, \Theta^0)$ in a language with variables $V_{n+1}$ such that for all $A$ of the language of $(\Delta^0, \Theta^0)$ $A \in \Delta_0$ or $\neg A \in \Delta^0$ and such that $\Delta \subseteq \Delta^0$,* *where* $\Vdash^+_{MH}$ *is the maximal consequence relation agreeing with* $\vdash_{MH}$, *defined as in* (2.2.2).

*Proof.* We first show that $\Delta \cup \{(uc)(A \vee \neg A) \mid A \text{ wff of } V_n\}$ is $\Vdash^+_{MH}$ consistent, where (uc) is universal closure.

This is clear since otherwise $\Delta \vdash_{MH} \neg \bigwedge_i (uc)\ (A_i \vee \neg A_i)$ but by Theorem 2.7.2 and Corollary 2.7.4 and the fact that

$$\vdash_H \neg\neg(A \wedge B) \leftrightarrow \neg\neg A \wedge \neg\neg B$$

we get that $\Delta \vdash_{MH} \neg\neg \bigwedge_i (uc)(A_i \vee \neg A_i)$ which is a contradiction.

Thus the theory above is consistent and can be extended to a $\Vdash^+_{MH}$ saturated theory $(\Delta^0, \Theta^0)$ in the language with variables $V_{n+1}$. Since $(uc)(A \vee \neg A)$ is in $\Delta^0$ for all $A$ of the language $V_n$, $A \vee \neg A \in \Delta^0$ for all $A$ of the language $V_{n+1}$ and thus $A \in \Delta_0$ or $\neg A \in \Delta_0$ for all $A$ of $V_{n+1}$ and Lemma 3 is proved.

We now continue with the proof of Theorem 2. We define a partial function $f$ on $\omega^*$ as in the proof of Theorem 1.6. Let $x \in \omega^*$ be such that $f(x)$ is defined, and such that $x$ is not of the form $z * (0)$. Define $f(x * (\bar{B} + 1))$ and $f(x * (\bar{C} + 1))$ as done in the proof of (1.6) (we use the same notation as in (1.6)). Besides that, define $f(x * (0))$ as follows: By Lemma 3 above, for the theory $(\Delta_x, \Theta_x) = f(x)$, there exists a $(\Delta^0_x, \Theta^0_x)$ in the language with $V_{n+1}$ having the properties of Lemma 3, let $f(x * (0))$ be $(\Delta^0_x, \Theta^0_x)$.

Now proceed to define the Kripke structure as in (1.6). We want to show that the analogue of Lemma 1.7 holds for our case. The proof is by induction on $A$. We only have to notice that for $y$'s of the form $y = x * (0)$ we have that $\forall z(yRz \rightarrow y = z)$, since for $y$ of this form we did not continue the construction of the domain of $f$, i.e., $y * (n) \notin \operatorname{dom} f$. For such $y$ we know by construction of $(\Delta_y, \Theta_y)$ that $\Delta_y$ is a classical theory and therefore since $\forall z(yRz \rightarrow y = z)$, $[A]_y = 1$ iff $A \in \Delta_y$, for all $A$, since $[A]_y$ coincides with the classical satisfaction at $y$. Thus the analogue of Lemma 1.7 holds for such $y$'s as well.

The rest of the completeness proof follows as in (1.6).

**Exercise 4.** Let $KC$ be the extension of $H$ with the schema $\neg A \vee \neg\neg A$. Show that $KC$ is complete for the class of all Kripke structures satisfying:

$$(\forall t, t' \in S)(\exists s \in S)(tRs \wedge t'Rs).$$

*Hint.* This is a difficult exercise.

## 5. Topological interpretation

DEFINITION 1. (a) A topological structure has the form $(S, F, O, D, U)$, where $(S, F)$ is a topological space with $F$ the family of open sets, $O \in S$, $U \neq \emptyset$ is a set of individuals and for each $n$-place atomic predicate $P$, $D(t, P) \subseteq U^n$. We require that for any $P$ and $u_1, \ldots, u_n$, $\{t \mid (u_1, \ldots, u_n) \in D(t, P)\}$ is open.

(b) For any wff $A(x_1, \ldots, x_n)$ and any assignment $g$ assigning elements of $U$ to the variables of the language and any $t \in S$ we define the notion $|A|_t^g$, called the truth value of $A$ at $t$ under $g$, by induction as follows:

(1) $A(x_1, \ldots, x_n)|_t^g = 1$ iff $(g(x_1), \ldots, g(x_n)) \in D(t, A)$ for $A$ $n$-place atomic.

(2) $|A \wedge B|_t^g = 1$ iff $|A|_t^g = |B|_t^g = 1$.
$|A \vee B|_t^g = 1$ iff $|A|_t^g = 1$ or $|B|_t^g = 1$.

(3) $|f|_t^g = 0$.

(4) $|A \to B|_t^g = 1$ iff for some open set $t \in S' \subseteq S$ we have that for each $s' \in S'$, $|A|_{s'}^g = 0$ or $|B|_{s'}^g = 1$.

(5) $|\exists x A(x)|_t^g = 1$ iff for some $g' =_x g$, $|A(x)|_t^{g'} = 1$.

(6) $|\forall x A(x)|_t^g = 1$ iff there exists an open set $t \in S'$, such that for each $s' \in S'$ and each $g' =_x g$, $|A(x)|_{s'}^{g'} = 1$.

(7) $A$ is said to be valid (resp. strongly valid) in the structure under $g$ iff $|A|_0^g = 1$. (resp. $|A|_t^g = 1$ for all $t \in S$).

(8) For each $A$ and $g$ let $\bar{A}^g = \{t \mid |A|_t^g = 1\}$.

LEMMA 2. *$\bar{A}^g$ is open, for all $A$, $g$.*

*Proof.* By induction on $A$.

The case of atomic $A$ and the cases of $\wedge$, $\vee$, $f$, $\exists$ follow immediately.

$t \in (A \to B)^{-g}$ iff for some $S'_t \in F$, $t \in S'_t \subseteq (S - \bar{A}^g) \cup \bar{B}^g$. Thus $S'_t \subseteq (A \to B)^{-g} \subseteq \bigcup_t S'_t$ and hence $(A \to B)^{-g} = \bigcup_t S'_t$ and is open. In fact $(A \to B)^{-g} = \text{Interior}\,(B^{-g}) \cup (S - A^{-g})$.

For the case of $\forall$, $t \in (\forall x A(x))^{-g}$ iff for some $S'_t \in F$ and every $g' =_x g$, $t \in S'_t \subseteq \bigcap_{g'=_x g} (A(x))^{-g'}$ and therefore $(\forall x A(x))^{-g} = \bigcup S'_t \subseteq \bigcap_{g'=_x g} (A(x))^{-g'}$. In fact $(\forall x A(x))^{-g} = \text{Interior} \bigcap_{g'=_x g} (A(x))^{-g'}$.

LEMMA 3. *If* $H \vdash A$ *then for any* $g$, $\bar{A}^g = S$.
*Proof.* By induction on the length of the proof of $A$ in $H$.

LEMMA 4. *Let* $\Vdash$ *be the maximal consequence relation agreeing with* $H$. *Let* $(\Delta, \Delta')$ *be a* $\Vdash$ *saturated theory, then there exists a topological structure and an assignment* $g$ *such that*

$$\Delta = \{A \mid O \in \bar{A}^g\}.$$

*Proof.* Let $S$ be the set of all $\Vdash$ saturated theories of the language. Let $U$ be the set of terms (variables and constants) of the language and let $g$ be identity.

Define $F$ by letting $F_0 = \{T \subseteq S \mid \text{for some } A, T = T_A = \{(\Delta, \Theta) \mid A \in \Delta\}\}$ and let $F$ be all the unions of sets from $F_0$. Let, for atomic $n$-ary $P$, $D((\Theta, \Theta'), P) = \{(x_1, \ldots, x_n) \mid P(x_1, \ldots, x_n) \in \Theta\}$.

LEMMA 5. *For any* $(\Theta, \Theta')$ *and any* $A(x_1, \ldots, x_n)$, $g'$, $\bar{A}^{g'} = \{(\Theta, \Theta') \mid A(g'(x_1), \ldots, g'(x_n)) \in \Theta\}$.

*Proof.* This holds for $A$ atomic by definition. The cases of $\wedge$, $\vee$, $f$ present no difficulties.

$(\Theta, \Theta') \in (\exists x A(x))^{-g'}$ iff for some $g'' =_x g'$, $|A(x)|^{g''}_{(\Theta,\Theta')} = 1$ iff for some $g'' =_x g'$, $A(g''(x)) \in \Theta$ iff for some $y$, $A(y) \in \Theta$ iff $\exists x A(x) \in \Theta$.

Assume $(A \to B) \in \Theta$ and consider the open set $S_{A \to B} = \{(\Theta^*, \Theta^{**}) \mid (A \to B) \in \Theta^*\}$. Let $(\Theta^*, \Theta^{**})$ be any element of this set, then certainly if $A \in \Theta^*$ then $B \in \Theta^*$.

Assume $(\Theta, \Theta') \in (A \to B)^{-g}$, then for some open set $S'$, $(\Theta, \Theta') \in S'$ and for every $(\Theta^*, \Theta^{**}) \in S'$, either $A \notin \Theta^*$ or $B \in \Theta^*$. Since $S'$ is open it is the union of sets of the form $S_C = \{(\Gamma, \Gamma') \mid C \in \Gamma\}$. Since $(\Theta, \Theta') \in S'$, for some $C$, $(\Theta, \Theta') \in S_C$. Clearly for any $(\Gamma, \Gamma') \in S_C$, $A \in \Gamma$ implies $B \in \Gamma$. We claim $C \vdash_H A \to B$, for otherwise, $C \wedge A \nVdash_H B$ and so for some saturated $(\Gamma, \Gamma')$, $C \wedge A \in \Gamma$, $B \notin \Gamma$, a contradiction. Now since $C \in \Theta$, we conclude $A \to B \in \Theta$ which was to be shown. This completes the case of $\to$.

Assume $\forall x A(x) \in \Theta$, then $S_{\forall x A(x)}$ is open and contains $(\Theta, \Theta')$. Clearly since $\vdash_H \forall x Ax \to A(y)$ for any $(\Gamma, \Gamma') \in S_{\forall x A(x)}$ and any $y$, $A(y) \in \Gamma$ and hence for any $(\Gamma, \Gamma') \in S_{\forall x A(x)}$ and any $g'' =_x g'$, $A(g''(x)) \in \Gamma$, i.e., $(\Gamma, \Gamma') \in \bar{A}^{g''}$.

Assume $(\Theta, \Theta') \in (\forall x A(x))^{-g'}$ then for some open set $S'$, $(\Theta, \Theta') \in S'$

and for any $g'' =_x g'$, $A(g''(x)) \in \Gamma$ for any $(\Gamma, \Gamma') \in S'$. Since $S' = \bigcup_C S_C$ for some wffs $C$, for some $C$, $(\Theta, \Theta') \in S_C$, and for any $(\Gamma, \Gamma') \in S_C$ and any $g'' =_x g'$, $A(g''(x)) \in \Gamma$. We claim $C \vdash_H \forall x A(x)$, for otherwise for some saturated theory $(\Gamma, \Gamma')$, and some $y$, $C \in \Gamma$ and $A(y) \notin \Gamma$ (note that $C$, $A$ contain only a finite number of free variables and hence there is a saturated theory as above). Then $(\Gamma, \Gamma') \in S_C$ and $A(y) \notin \Gamma$, in contradiction to what is known about $S_C$. Thus $\vdash_H C \to \forall x A(x)$ and hence $\forall x A(x) \in \Theta$.

Thus Lemma 5 is proved and we conclude that $(\Delta, \Delta') \in S$ is a point where $(\Delta, \Delta') \in \bar{A}^g$, for all $A \in \Delta$, which proves Lemma 4.

THEOREM 6. *Let $\Vdash_H^+$ be the maximal consequence relation agreeing with $\vdash_H$ and let $\Vdash_{\text{top}}$ be the consequence relation arising from the topological interpretation, then $\Vdash_H^+ = \Vdash_{\text{top}}$.*

*Proof.* From the three previous lemmas.

**Exercise 7.** Show that if $(\Delta, \Delta')$ is a $\Vdash_H^+$ consistent theory then there exists a topological structure and an assignment $g$ such that $\bar{A}^g = S$, for $A \in \Delta$ and $\bar{A}^g = \emptyset$ for $A \in \Delta'$.

DEFINITION 8. (a) Let $(S, \leqslant)$ be a partially ordered set. Define a topology $F_\leqslant$ on $S$ by taking the base for open sets all sets of the form $a^+$, $a \in S$ where $a^+ = \{y \in S \mid a \leqslant y\}$.

(b) Let $(S, F)$ be a topological space, define an order $\leqslant_F$ on $S$ by letting $a \leqslant_F b$ iff $\forall S' \in F\ (a \in S' \to b \in S')$.

THEOREM 9. (a) *If $(S, \leqslant)$ is a partially ordered set then $(S, F_\leqslant)$ is a topological space which is $T_0$ and the topology is closed under arbitrary intersections. We have that $\leqslant_{F_\leqslant} = \leqslant$.*

(b) *If $(S, F)$ is a $T_0$-topological space whose topology is closed under arbitrary intersection then $(S, \leqslant_F)$ is a partially ordered set with $F_{\leqslant_F} = F$.*

*Proof.* (a) (1) To show that the space is $T_0$ notice that if $a \in b^+ \wedge b \in a^+$ then $a = b$, by antisymmetry of $\leqslant$.

(2) To show that $\leqslant_{F_\leqslant} = \leqslant$ notice that $a \leqslant b$ implies that $b \in a^+$ and hence $a \leqslant_{F_\leqslant} b$, since $a^+$ is the smallest open set containing $a$. On the other hand if $a \leqslant_{F_\leqslant} b$ then certainly $b \in a^+$, since $a \in a^+$ and hence $a \leqslant b$.

(3) Clearly $F_\leqslant$ is closed under arbitrary intersections.

(b) (1) $T_0$ of $F$ ensures that $\leqslant_F$ is a partial order.

(2) To show that $F_{\leqslant_F} = F$, we first show that any $S_0 \in F$ is in $F_{\leqslant_F}$. Let $a \in S_0$, then not $a \leqslant_F b$ for any $b \notin S_0$. So $a^+ \subseteq S_0$. Thus $S_0 = \bigcup_{a \in S_0} a^+$ and hence $S_0$ is open in $F_{\leqslant F}$. Assume that $S_0$ is open in $F_{\leqslant_F}$. Since $S_0 = \bigcup a^+$ for some $a$s, it is enough to show that for any $a$, $a^+ \in F$. Let $y \notin a^+$, then for some $S_y \in F$, $a \in S_y$, $y \notin S_y$. Then $a^+ = \bigcap_{y \notin a^+} S_y$, which is open. Thus $a^+ \in F$.

THEOREM 10. *The logic CD is complete for the class of all topological structures that are $T_0$ and whose topology is closed under arbitrary intersections.*

*Proof.* We observe that by the previous theorem any Kripke structure with constant domains with $(S, \leqslant)$ a partially ordered set and $N = \emptyset$ can be regarded as a topological structure in the above mentioned class and vice versa.

We observe the following connection between $\leqslant$ and the topology $F$, (proved by induction), $[A]_t^g = 1$ in the Kripke structure iff $|A|_t^g = 1$ in the topological structure.

## NOTES

We include here the basic Kripke, Beth and topological interpretations for Heyting's logic. Both Kripke and Beth completeness proofs use essentially semantic tableaux. R. Thomason and P. Aczel later used Henkin methods. For topological and algebraic interpretations see H. Rasiowa's book.

The refinements of Kripke and Beth interpretation and completeness proofs are due to the author and are included here. S. Gornemann first proved completeness for *CD*, using algebraic methods. The present proof is original. So is the result for *MH**.

For topological models and sheaf models see the publications of D. Scott, M. Fourman and R. Grayson. As mentioned in the introduction, their work deserves a separate volume.

CHAPTER 4

# HEYTING'S PROPOSITIONAL CALCULUS AND EXTENSIONS

A great deal of research has been done on Heyting's propositional calculus and on intermediate logics, logics between Heyting's propositional calculus, $h$, and the classical propositional calculus, $c$. In this chapter we shall

(a) study the correspondence between axioms added to $h$ and geometrical conditions on the Kripke structures,

(b) introduce special methods, such as selective filtration, to study certain intermediate logics.

The language is propositional and contains the connectives $\wedge$, $\vee$, $\rightarrow$, $f$.

## 1. *h* AND *c*

DEFINITION 1. Let $h$ be the Hilbert system defined by the axiom schemas:

(a) $A \rightarrow (B \rightarrow A)$

(b) $(A \rightarrow (B \rightarrow C)) \rightarrow ((A \rightarrow B) \rightarrow (A \rightarrow C))$

(c) $A \wedge B \rightarrow A$; $A \wedge B \rightarrow B$

(d) $A \rightarrow (B \rightarrow A \wedge B)$

(e) $A \rightarrow A \vee B$; $B \rightarrow A \vee B$

(f) $(A \rightarrow C) \rightarrow ((B \rightarrow C) \rightarrow (A \vee B \rightarrow C))$

(g) $f \rightarrow A$.

The following rule serves as both a consequence and provability rule:

MP From $A, A \rightarrow B$ deduce $B$.

DEFINITION 2. Let $c$ be the Hilbert system obtained by adding to $h$ the axiom

$$((A \rightarrow B) \rightarrow A) \rightarrow A.$$

The reader should note that $h$ and $c$ are propositional fragments of $H$ and $C$ of 2.2.1 and 2.5.1 respectively.

The following theorems can be proved in the same way as in Chapter 2.

THEOREM 3. (*Deduction theorem for h*).

$$A_1, \ldots, A_n, C \vdash_h B \quad \text{iff} \quad A_1, \ldots, A_n \vdash_h C \to B.$$

*Proof.* As in 2.1.13.

THEOREM 4.

$$\vdash_c \quad ((A \to B) \to B) \to A \vee B$$

$$(A \vee B) \to ((A \to B) \to B).$$

*Proof.* As in (2.5.3).

DEFINITION 5. Let $\neg A$ be $A \to f$.

THEOREM 6.

$$\vdash_h \quad (A \to \neg B) \to (B \to \neg A)$$

$$\vdash_h \quad A \to (\neg A \to B).$$

*Proof.* Exercise.

THEOREM 7.

$$\vdash_c \neg\neg A \to A.$$

*Proof.* Exercise.

DEFINITION 8. (a) A propositional Kripke structure has the form $(S, R, O, D)$, where $(S, R, O)$ is a partially ordered set with a first element $O$ and $D$ is a function such that for each $t \in S$ and atomic $q$, $D(t, q) \in \{0, 1\}$. We have that if $tRs$ and $D(t, q) = 1$ then $D(s, q) = 1$.

(b) The truth value of a wff $A$ at a point $t \in S$ denoted by $[A]_t$ is defined by induction as follows:

(1) $\qquad [A]_t = D(t, A), \quad \text{for } A \text{ atomic}; [f]_t = 0.$

(2) $[A \wedge B]_t = 1$ iff $[A]_t = 1$ and $[B]_t = 1$

$[A \vee B]_t = 1$ iff $[A]_t = 1$ or $[B]_t = 1$.

(3) $[A \to B]_t = 1$ iff for all $s$, if $tRs$ and

$[A]_s = 1$ then $[B]_s = 1$.

(c) A structure is said to validate $A$ iff $[A]_O = 1$, in such a case $A$ is also said to hold in the structure; we say that $(S, R, O)$ validates $A$ iff for any $D$, $(S, R, O, D)$ validates $A$.

(d) Let $x$ be an intermediate logical system between $h$ and $c$. Let **M** be a class of propositional Kripke structures. $x$ is said to be complete for **M** iff for all $A$, $\vdash_x A$ iff $A$ holds in all structures of **M**.

(e) Let $x$ be an intermediate logic and let **M** be a class of partially ordered sets. $x$ is said to be complete for **M**, or to be characterised by **M**, iff for all $A$, $\vdash_x A$ iff $A$ holds in any structure of the form $(S, R, O, D)$ with $(S, R, O)$ in **M**.

(f) Let $F$ be a condition on partially ordered sets and let $x$ be an intermediate logic. $x$ is said to be characterised by $F$ iff $x$ is characterised by the class of all partially ordered sets satisfying $F$.

THEOREM 9. *$h$ is characterised by the class of all partially ordered sets of the form* $(S, R, O)$.

This theorem follows from the completeness theorem for $H$ given in Chapter 2. We are going to give a direct proof here.

DEFINITION 10. (a) By a theory $(\Delta, \Theta)$ we mean a pair of sets of wff.

(b) If $x$ is an intermediate logic, $(\Delta, \Theta)$ is said to be $x$ consistent iff for no finite $\varphi \subseteq \Delta$, $\psi \subseteq \Theta$ do we have $\vdash_x \wedge \varphi \to \vee \psi$, where $\vee \emptyset = f$, $\wedge \emptyset = f \to f$.

(c) $(\Delta, \Theta)$ is said to be $x$-saturated iff it is $x$-consistent and:

(1) $\Delta \vdash_x A$ implies $A \in \Delta$

(2) $A \vee B \in \Delta$ iff $A \in \Delta$ or $B \in \Delta$.

(d) $(\Delta', \Theta')$ is said to be an extension of $(\Delta, \Theta)$ iff $\Delta \subseteq \Delta'$, $\Theta \subseteq \Theta'$.

(e) A Kripke structure is said to validate $(\Delta, \Theta)$ iff every member of $\Delta$ holds in the structure and none of the members of $\Theta$ hold in it.

LEMMA 11. *Let* $(\Delta, \Theta)$ *be an* $x$*-consistent theory, then* $(\Delta, \Theta)$ *can be extended to an* $x$*-saturated theory.*

*Proof.* As in Section 4 Chapter 1, note that the relation $\varphi \Vdash \psi$ iff (def) $\vdash_x \wedge \varphi \rightarrow \vee \psi$ is a Scott consequence agreeing with $\vdash_x$. Show this as in the proof of (2.2.4).

LEMMA 12. *Let* $(\Delta, \Theta)$ *be a theory such that* $\Delta \nvdash_x A \rightarrow B$, *then* $(\Delta \cup \{A\}, \{B\})$ *is* $x$*-consistent.*

*Proof.* By the deduction theorem for $x$.

COROLLARY 13. *If* $(\Delta, \Theta)$ *is such that* $\Delta \nvdash_x A \rightarrow B$, *then there exists an* $x$*-saturated theory* $(\Delta_{A\rightarrow B}, \Theta_{A\rightarrow B})$ *such that* $\Delta \cup \{A\} \subseteq \Delta_{A\rightarrow B}$, $B \in \Theta_{A\rightarrow B}$.

DEFINITION 14. Let $x$ be an intermediate logic. Let $S_x$ be the set of all $x$-saturated theories. Define $R_x$ by $(\Delta, \Theta) R_x (\Delta', \Theta')$ iff $\Delta \subseteq \Delta'$. Let $D_x((\Delta, \Theta), q) = 1$ iff $q \in \Delta$.

$(S_x, R_x, D_x)$ is called the canonical structure for $x$.

LEMMA 15. *For any wff* $A$ *and any* $(\Delta, \Theta)$, $[A]_{(\Delta, \Theta)} = 1$ *in the canonical structure for* $x$ *iff* $A \in \Delta$.

*Proof.* By induction. For the case of $\rightarrow$, use Corollary 13.

COROLLARY 16. (a) *Let* $(\Delta, \Theta)$ *be an* $h$*-consistent and saturated theory, then it is validated by some Kripke structure.*

(b) *Let* $(\Delta, \Theta)$ *be a* $c$*-consistent and saturated theory, then it is validated by some Kripke structure with exactly one element.*

*Proof.* Let $S_x^{(\Delta,\Theta)} = \{(\Delta', \Theta') \in S_x \mid \Delta \subseteq \Delta'\}$. By Lemma 15, $(S_x^{(\Delta,\Theta)}, R_x, (\Delta, \Theta), D)$ validates $(\Delta, \Theta)$.

For $x = c$, notice that since $c \vdash A \vee \neg A$, $S_x^{(\Delta,\Theta)} = \{(\Delta, \Theta)\}$.

DEFINITION 17. (a) Let $kc$ be the Hilbert system obtained by adding to $h$ the axiom schema $\neg A \vee \neg\neg A$.

(b) Let $lc$ be the Hilbert system obtained by adding to $h$ the axiom schema $(A \rightarrow B) \vee (B \rightarrow A)$.

**Exercise 18.** (a) Show that in $lc$ $p \vee q$ is definable as $((p \rightarrow q) \rightarrow q) \wedge ((q \rightarrow p) \rightarrow p)$.

(b) Show that $lc$ is the smallest intermediate logic in which $((q \rightarrow p) \rightarrow p) \wedge ((p \rightarrow q) \rightarrow q))$ is $p \vee q$.

*Solution.* (a) Let

$$((q \to p) \to p) \wedge ((p \to q) \to q) = \text{def } p \dot{\vee} q.$$

(1) Clearly $\vdash_{lc} p \to p \dot{\vee} q$
$\vdash_{lc} q \to p \dot{\vee} q$.

(2) Since $\vdash_{lc} (p \to r) \to ((q \to r) \to (p \vee q \to r))$, we get $\vdash_{lc} p \vee q \to p \dot{\vee} q$.

(3) To show that $\vdash_{lc} (p \to r) \to ((q \to r) \to (p \dot{\vee} q \to r))$, assume $p \dot{\vee} q$, $(p \to r)$, $(q \to r)$, since $\vdash_{lc} (p \to q) \vee (q \to p)$ we can assume that as well. Thus we get $r$.

(b) Assume $x$ is the logic with $p \dot{\vee} q \leftrightarrow p \vee q$ as axiom. We show that $\vdash_x (p \to q) \vee (q \to p)$.

Assume $(p \to q) \to (q \to p)$, $q$, thus we get $q \to (p \to q)$ and hence $(p \to q)$ and hence $q \to p$ and hence $p$. Thus $\vdash_x ((p \to q) \to (q \to p)) \to (q \to p)$.

Similarly $\vdash_x ((q \to p) \to (p \to p)) \to (p \to q)$, i.e., $\vdash_x (p \to q) \dot{\vee} (q \to p)$.

THEOREM 19. (a) *kc is characterised by all the partially ordered sets with the property*

$$\exists u \forall t (tRu).$$

(b) *lc is characterised by all the partially ordered sets with*

$$\forall t, s(tRs \vee sRt).$$

*Proof.* It is easy to verify that any theorem of $x$ is validated in any structure of the proposed class, where $x$ is $kc$ and $lc$, resp.

For the other direction let $(\Delta, \Theta)$ be an $x$ saturated theory, $x$ being $kc$ or $lc$, consider $(S_x^{(\Delta,\Theta)}, R_x, (\Delta, \Theta), D)$. It is sufficient to show that $(S_x^{(\Delta,\Theta)}, R_x)$ has the desired property, since $(S_x, R_x, D)$ is the canonical structure for $x$.

*Case of lc.* Assume $(\Delta_i, \Theta_i)$, $i = 1, 2$, are in $S_{lc}^{(\Delta,\Theta)}$. If neither $\Delta_1 \subseteq \Delta_2$, nor $\Delta_2 \subseteq \Delta_1$, then for some $A$, $B$, $A \in \Delta_1$, $A \notin \Delta_2$, $B \in \Delta_2$, $B \notin \Delta_1$. Since $\vdash_{lc} (A \to B) \vee (B \to A)$, one of the disjuncts must be in $\Delta$, say $A \to B$. Since $\Delta \subseteq \Delta_1$, $A \to B \in \Delta_1$, a contradiction.

*Case of kc.* We want to find a $(\Delta', \Theta') \in S_{kc}^{(\Delta,\Theta)}$ such that for any $(\Delta'', \Theta'')$ in $S_{kc}^{(\Delta,\Theta)}$, $\Delta'' \subseteq \Delta'$. We claim that the theory $(\cup\{\Delta'' \mid (\Delta'', \Theta'') \in S_{kc}^{(\Delta,\Theta)}\}, \emptyset)$ is $kc$-consistent. Otherwise for some $(\Delta_i, \Theta_i)$, $A \in \Delta_1$, $B \in \Delta_2$, $\vdash_{kc} A \wedge B \to f$, or $\vdash_{kc} A \to \neg B$. Thus $\neg B \in \Delta_1$. Since $\vdash_{kc} \neg B \vee \neg\neg B$, one of the disjuncts must be in $\Delta$. It must be $\neg B$, since $\neg B \in \Delta_1 \supseteq \Delta$.

Thus $\neg B \in \Delta_2$, but $B \in \Delta_2$, a contradiction. Thus $(\Delta_1 \cup \Delta_2, \emptyset)$ is $kc$-consistent. Thus there exists a saturated theory $(\Delta', \Theta')$ with $\Delta'' \subseteq \Delta'$, for any $(\Delta'', \Theta'') \in S_{kc}^{(\Delta, \Theta)}$.

This completes the proof of Theorem 19.

**Exercise 20.** (a) show that $h$ has the disjunction property.

(b) Show that $\vdash_c A$ iff $\vdash_h \neg\neg A$.

(c) For $A$ with $\neg$ and $\wedge$ only: $\vdash_c A$ iff $\vdash_h A$.

**Exercise 21.** (a) Show that $lc$ is an extension of $kc$.

(b) Show that if we add to $h$ the axiom schema

$$((p \to q) \to r) \wedge ((q \to p) \to r) \to r, \text{ we get } lc.$$

(c) Show that if we add to $h$ the axiom schema

$$(p \to q \vee r) \to ((p \to q) \vee (p \to r)), \text{ we get } lc.$$

DEFINITION 22. A propositional Kripke structure is called a tree structure iff $(S, R, O)$ is a tree, i.e., every path through $O$ is well ordered and for no $x, y, z \in S$ do we have $x \neq y \wedge xRz \wedge yRz$.

THEOREM 23. *h is complete for the class of tree Kripke structures.*

*Proof.* Let $(S, R, O, D)$ be a propositional Kripke structure. Let $S^*$ be the set of all finite $R$-increasing sequences of $S$ with 0 as first element. For $t^* = (0, \ldots, x_n)$ let $|t^*| = x_n$. Let $R^*$ be the relation of 'being equal or an initial segment of'. Let $D^*(t^*, q) = D(|t^*|, q)$, for $t^* \in S^*$.

LEMMA 24. *If* $[A]_{t^*}$ *denotes truth value in* $(S^*, R^*, (0), D^*)$ *then*

$$[A]_{t^*} = [A]_{|t^*|}.$$

*Proof.* By induction on $A$. The interesting case is that of $\to$.

$[A \to B]_{t^*} = 0$ iff for some $s^*$, $t^* R^* s^*$ and $[A]_{s^*} = 1$ and $[B]_{s^*} = 0$ iff for some $s$, $|t^*| Rs$ and $[A]_s = 1$ and $[B]_s = 0$. This holds, for if $s^*$ is given take $s = |s^*|$. If $s$ is given let $s^* = t^* * (s)$ ($*$ is concatenation).

Theorem 23 follows from the Lemma 24.

## 2. Filtration

DEFINITION 1. (a) By a condition $F$ on partially ordered sets $(S, R, O)$ we mean any condition formulated in higher order language with names for $R$, $O$, and $=$.

(b) A condition $F$ is said to be *absolute* iff for every $(S, R, O)$ satisfying $F$, there exists a *finite* $S_0 \subseteq S$ such that for any $S_0 \subseteq S' \subseteq S$, $(S', R \upharpoonright S', O)$ also satisfies condition $F$.

(c) Let $x$ be an intermediate logic. $x$ is said to have the finite model property (f.m.p.) iff whenever $x \nVdash A$, then there exists a propositional Kripke structure $(S, R, O, D)$ with $S$ finite which validates $x$ and in which $A$ does not hold.

THEOREM 2. *Let $x$ be an intermediate logic characterised by an absolute condition $F$, then $x$ is characterised by the class of all* finite, *partially ordered sets satisfying $F$.*

*Proof.* Assume $x \nVdash A$, we show a structure $(S, R, O, D)$ with $S$ finite, satisfying condition $F$ in which $[A]_O = 0$.

There exists a structure $(S', R', O', D')$ satisfying $F$ with $[A]_{O'} = 0$. We want to define a set $S \subseteq S'$ as follows. First let $S_0' \subseteq S'$ be the set promised by the fact that $F$ is absolute. Now let $\Theta$ be the set of all subformulas of $A$ and define a family of subsets $S_n \subseteq S'$ as follows:

$$S_O = S'_O \cup \{O'\}.$$

Suppose $S_n$ has been defined. For each $t \in S_n$ and each subformula $B = B_1 \rightarrow B_2 \in \Theta$ such that $[B_1]_t = 0$, $[B]_t = 0$, there exists an $y(t, B) \in S'$ with $[B_1]_{y(t,B)} = 1$, $[B_2]_{y(t,B)} = 0$, and $tR'y(t, B)$.

Let $S_{n+1} = \{y(t, B) \mid t \in S_n, B \in \Theta\}$.

Let $S = \bigcup S_n$.

We claim $S$ is finite. This is so because if $t \in S_n$ and $y(t, B) \in S_{n+1}$, then at $y(t, B)$ one more subformula of $A$ is true, and since there is only a finite number of those, $S_m$ is empty, for sufficiently large $m$.

Since $F$ is absolute and $S'_O \subseteq S$, $(S, R' \upharpoonright S, O')$ satisfies condition $F$.

Lemma 3. *For each $B \in \Theta$, $t \in S : [B]_t$ in $(S, R' \upharpoonright S, O', D \upharpoonright S) = [B]_t$ in $(S', R', O', D')$.*

*Proof.* By induction. The cases of atomic $B$ and $\wedge$, $\vee$, $f$ present no difficulty.

Assume $[B_1 \rightarrow B_2]_t = 0$ in $S'$, then if $[B_1]_t = 1$ in $S'$, then certainly

$[B_1]_t = 1$ in $S$ and so since $[B_2]_t = 0$ in $S'$ and also in $S$ we get $[B_1 \to B_2]_t = 0$ in $S$. If $[B_1]_t = 0$ in $S'$, then by construction, $y(t, B_1 \to B_2) \in S$ and $B_1$ holds there and $B_2$ doesn't and hence by the induction hypothesis $[B_1 \to B_2]_t = 0$ in $S$.

If $[B_1 \to B_2]_t = 0$ in $S$, then for some $s \in S$, $tR'S$ and $[B_1]_s = 1$, $[B_2]_s = 0$, both in $S$. By the induction hypothesis $[B_1 \to B_2]_t = 0$ in $S'$.

Thus $[A]_0 = 0$ in $S$ and we have found a finite structure satisfying $F$ in which $A$ is false.

THEOREM 4. (a) *h is characterised by the class of all finite trees.*

(b) *lc is characterised by the class of all finite partially ordered sets with* $\forall xy(xRy \vee yRx)$.

(c) *kc is characterised by the class of all finite partially ordered sets with* $\exists x \forall y(yRx)$.

*Proof.* The respective conditions are absolute.

DEFINITION 5. Define the following sequence of Jaskowski trees:

$J_1 = \cdot$ (one point tree)

$J_2 =$ 

$J_3 =$ 

$J_n =$ $J_{n-1}$ $n$ times $J_{n-1}$

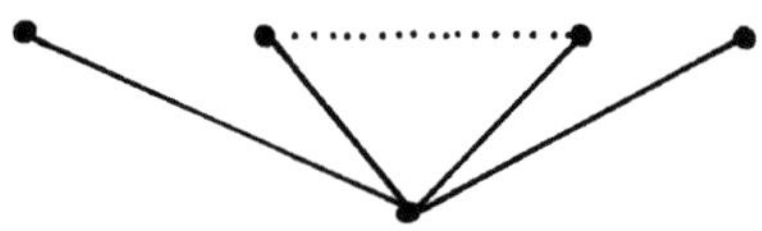

$R_n$ is defined on $J_n$ by going upwards.

DEFINITION 6. Let $\mathbf{A}_i = (S_i, R_i, O_i, D_i)$, $i = 1, 2$ be two propositional Kripke structures. A function $\alpha : S_1 \to S_2$ is said to be a homomorphism from $\mathbf{A}_1$ into $\mathbf{A}_2$ iff the following holds:

(a) $tR_1 s \to \alpha(t) R_2 \alpha(s)$.

(b) $\alpha$ is onto $S_2$.

(c) $\alpha(t) R_2 y \to \exists s(tR_1 s \wedge \alpha(s) = y)$.

(d) $D_1(t, q) = D_2(\alpha(t), q)$ for all atomic $q$.

THEOREM 7. *Let $\alpha$ be a homomorphism from $\mathbf{A}_1$ onto $\mathbf{A}_2$, then for any wff $A$, $[A]^1_t = [A]^2_{\alpha(t)}$.*

*Proof.* By induction. The case of atomic $A$ follows from the definition. The cases of $\wedge$, $\vee$, $f$ present no difficulties.

Suppose $[A \to B]^1_t = 0$, then for some $s \in S_1$, $tR_1s$ and $[A]^1_s = 1$, $[B]^1_s = 0$. By the induction hypothesis $\alpha(t)R_2\alpha(s)$ and $[A]^2_{\alpha(s)} = 1$, $[B]^2_{\alpha(s)} = 0$ and so $[A \to B]^2_{\alpha(t)} = 0$.

Assume $[A \to B]^2_{\alpha(t)} = 0$, then for some $y$, $\alpha(t)R_2y$ and $[A]^2_y = 1$, $[B]^2_y = 0$. There exist $s \in S_1$ such that $\alpha(s) = y$ and $tR_1s$. Therefore $[A]^1_s = 1$, $[B]^1_s = 0$. Hence $[A \to B]^1_t = 0$.

This completes the proof.

THEOREM 8. *Let $J_n$ be the sequence of trees of Definition 5 and let $T$ be any finite tree, then for some $n$, there exists a homomorphism $\alpha$ from $J_n$ onto $T$.*

*Proof.* By induction on the number of points of $T$. First a lemma.

LEMMA 9. *If there exists a homomorphism from $J_n$ onto $T$ then there exists a homomorphism from $J_{n+1}$ onto $T$.*

*Proof.* Let $\alpha : J_n \to T$.

Define $\beta : J_{n+1} \to T$ as in the drawing

image obtained by applying $\alpha$ $J_n$

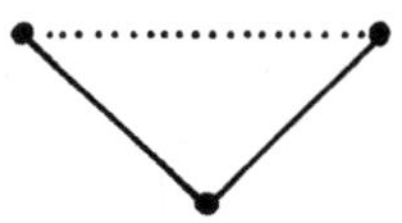

$J_n$ image obtained by applying $\alpha$

image of this point is the first point of $T$.

To continue the proof of Theorem 8, first if $T$ has only one point then it is equal to $J_1$.

Assume $T$ is of the form

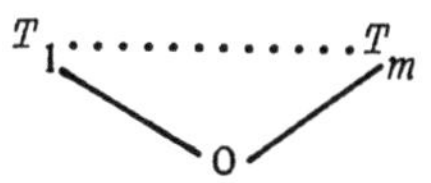


By Lemma 9 and the induction hypothesis, we can assume that there exists an $n \geq m$ and homomorphisms $\alpha_i : J_n \to T_i$, $1 \leq i \leq m$. Consider the following homomorphism drawing for $\alpha : J_{n+1} \to T$:

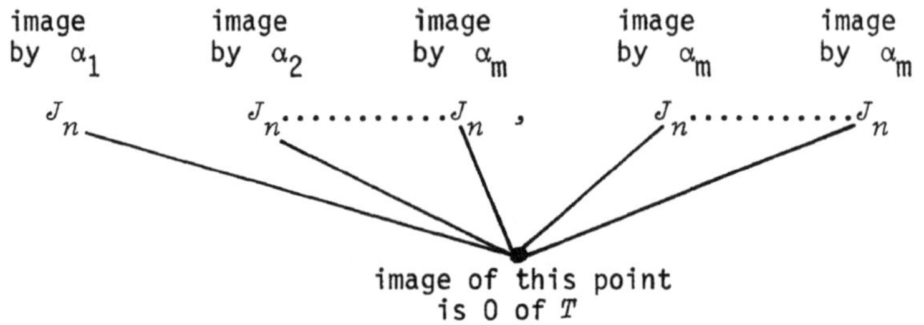


Clearly $\alpha$ is a homomorphism.

COROLLARY 10. *h is complete for the class of trees* $J_n$.
*Proof.* By Theorems (4a), (7), (8).

DEFINITION 11. A tree $T$ is said to be $n$-ary iff every point in the tree has at most $n$ immediate successors.

THEOREM 12. *h is complete for the full infinite binary tree* (i.e., obtained from all the finite sequences of zeros and ones).

*Proof.* We show that this tree is homomorphic onto $J_n$, for any $n$.

The proof is by induction on $n$. Certainly it is homomorphic onto $J_1$.

Suppose $\alpha$ is a homomorphism onto $J_n$. We show a homomorphism onto $J_{n+1}$. Let $m$ be such that $2^m \geqslant n+1$. Consider all sequences of 0's and 1's of length $m$. Denote them by $x_1, \ldots, x_{2^m}$. The binary tree $T$ can be decomposed into the sets $T_i$ = all extensions of $x_i$, $1 \leqslant i \leqslant 2^m$ and $T_*$ = all sequence of length $\leqslant 2^m$. Since $T_i$ is isomorphic to $T$, there exists a homomorphism $\alpha_i$ of $T_i$ onto $J_n$, by the induction hypothesis.

The following is a homomorphism of $T$ onto $J_{n+1}$. The diagram shows what is mapped onto what.

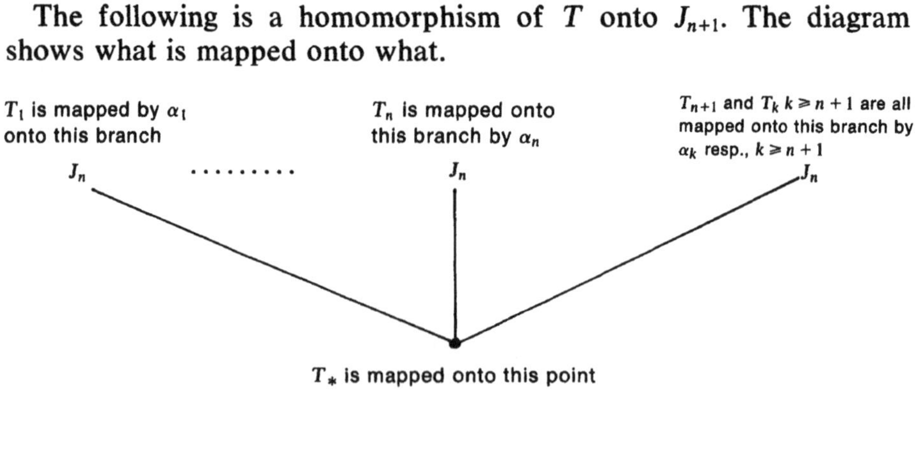

The reader can verify that we get a homomorphism. We can conclude that Theorem 11 holds by (4a), (7–8).

A similar proof can be given for $n$-ary trees.

*Remark* 13. $h$ is *not* complete for the class of all *finite* $n$-ary trees. We get a proper extension $x_n$ of $h$. We shall investigate these logics in the next chapter. However, the fragment without $\vee$, is complete for the class of finite $n$-ary trees.

**Exercise 14.** Let $J_n$ be the tree of Definition 5. Let $S$ be the set of all endpoints of $J_n$. For $t \in J_n$, let $S_t$ be the set of all endpoints above $t$. It is a simple observation to note that if $t \neq s$ then $S_t \neq S_s$. For each $t \in S$ let $P_t$ be an atomic proposition. Let $D$ be an assignment such that $D(s, P_t) = 1$ iff $s = t$. Show that in the propositional Kripke structure thus obtained we have:

$$\Big[\bigvee_{s \in S_t} \neg\neg P_s\Big]_r = 1 \quad \text{iff} \quad t \text{ is below } r \text{ or } t = r.$$

## 3. Incomparable intermediate logics

DEFINITION 1. Let $(S, R, O)$ be a partially ordered set with first element 0. We define a wff $\alpha$, associated with this partially ordered set. For each $t \in S$, let $P_t$ be a distinct atomic variable. Consider the assignment $D$ defined by $D(t, P_s) = 1$ iff $sRt$.

For any $t \in S$, let $A_t = \bigwedge_{sRt} P_s$ and $B_t = \bigvee_{\sim sRt} P_s$.

Let $\Delta = \{P_t \mid t \in S\}$, let $\Theta \subseteq \Delta$, define $C_\Theta = \bigwedge_{P_t \in \Theta} P_t \rightarrow \bigvee_{P_s \notin \Theta} P_s$.

For each $t$, let $\Theta_t = \{s \mid sRt\}$, then $C_{\Theta_t} = A_t \rightarrow B_t$. For $tRs$ let $\alpha_{t,s}$ be $A_t \rightarrow ((A_s \rightarrow B_s) \rightarrow B_t)$. Let $\alpha'$ be $\bigwedge_{\sim \exists t(\Theta = \Theta_t)} C_\Theta \wedge \bigwedge_{tRs} \alpha_{t,s}$ and let $\alpha''$ be $\bigvee_{t \in S} (A_t \rightarrow B_t)$.

LEMMA 2. *In the structure* $(S, R, O, D)$ *of Definition* 1,

$$[\alpha']_O = 1, \qquad [\alpha'']_O = 0.$$

*Proof.* Exercise.

LEMMA 3. *Let* $(S', R', O', D)$ *be any structure such that* $[\alpha']_{O'} = 1$, $[\alpha'']_{O'} = 0$, *then there exists a homomorphism* $h: S' \rightarrow S$ *satisfying conditions* (a), (b), (c) *of Definition* 2.6.

*Proof.* To define $h$, let $x \in S'$ be arbitrary, let $\Theta^x = \{t \mid [P_t]_x = 1\}$. We claim that there exists a unique $s \in S$ such that $\Theta^x = \Theta_s$. Clearly such an $s$ must exist for otherwise $\alpha' \vdash_h C_{\Theta^x}$ and so $[C_{\Theta^x}]_x = 1$, contrary to the definition of $\Theta^x$.

Since for $s \neq s'$, $\Theta_s \neq \Theta_{s'}$, such a unique $s$ exists. Let $h(x)$ be this $s$. We now have to show that we get a homomorphism.

(a) If $xR'y$ then clearly $\Theta^x \subseteq \Theta^y$ and hence $h(x)Rh(y)$.

(b) $h$ is onto, since given a $t \in S$, then $C_{\Theta_t}$ is a disjunct of $\alpha''$ and so for some $x \in S'$, $[A_t]_x = 1$, $[B_t]_x = 0$ and so $\Theta^x = \Theta_t$, i.e., $h(x) = t$.

(c) Assume $h(x) = t$, $tRs$, we are looking for a $y$ such that $xR'y$ and $h(y) = s$. Since $[A_t \to ((A_s \to B_s) \to B_t)]_{O'} = 1$ and at $x$, $[A_t]_x = 1$, $[B_t]_x = 0$, we must have $[A_s \to B_s]_x = 0$ and so for some $y$, $xR'y$, $[A_s]_y = 1$, $[B_s]_y = 0$ and so $h(y) = s$.

DEFINITION 4. Let $\mathbf{A} = (S', R', L', D')$ be a propositional structure and let $x \in S'$, and let $\mathbf{A}^x$ denote the structure $(S'_x, R'_x, x, D'_x)$ where $S'_x = \{y \in S' \mid xR'y\}$, $R'_x = R' \upharpoonright S'_x$, $D'_x = D' \upharpoonright S'_x$. $\mathbf{A}^x$ is called the truncation of $A$ at $x$.

COROLLARY 5. *Let $(S', R', O')$ be a partially ordered set such that no truncation $(S'_x, R'_x, x)$ of it is homomorphic onto $(S, R, O)$. Then $\alpha$ (of $(S, R, O)$) is valid in any structure with $(S', R', O')$.*

*Proof.* Follows from Theorem 3.

DEFINITION 6. We now introduce a sequence of finite partially ordered sets denoted by $K_n$. The definition is as in the diagram, with upward lines denoting the order

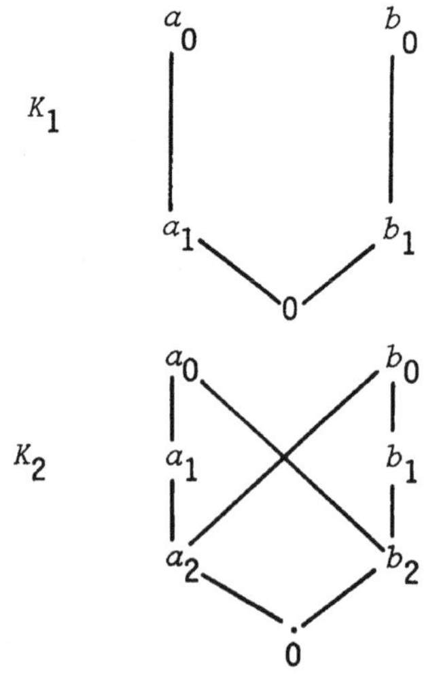

$K_3$

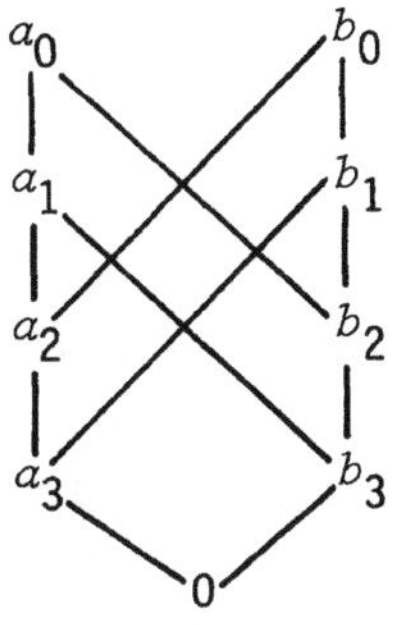


$K_4$

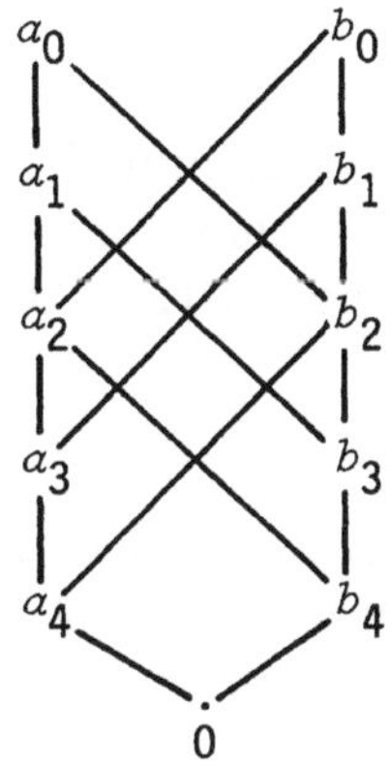


That is $K_n = \{0\} \bigcup \{a_i, b_i\}$, $0 \leqslant i \leqslant n$. 0 is the first point and the order is the reflexive and transitive closure of the conditions $a_{i+1}Ra_i$, $b_{i+1}Rb_i$, $a_{i+2}Rb_i$, $b_{i+2}Ra_i$.

LEMMA 7. *Let $K_n^x$ be a truncation of $K_n$ at $x$, where $x = 0 \vee x = a_m$, $0 \leqslant m \leqslant n$, then there exists no homomorphism, $\alpha$, of $K_n^x$ on $K_r$, $r \neq n$.*

*Proof.* First observe that since $\alpha$ is onto, if $x = a_m$ then $m$ cannot be less than $r+1$, or if $x = 0$, $n \geqslant r$. Since $\alpha$ is a homomorphism, it must take endpoints to endpoints. Thus $\alpha(a_0)$ and $\alpha(b_0)$ are endpoints. We show that $\alpha(a_0) \neq \alpha(b_0)$. Assume that both are $a_0$ (i.e., $\alpha(a_0) = \alpha(b_0) = a_0$, $a_0$ of $K_r$). Since $\alpha$ is onto for some $t \in K_n^{a_m}$, $\alpha(t) = b_0$, but then $t$ is below some endpoint $s$, ($s = a_0$ or $s = b_0$) and hence $\alpha(t)$, which is $b_0$, is below $\alpha(s)$ which is $a_0$ and so $b_0$ is below $a_0$ in $K_r$, which is false. Thus $\alpha(a_0) \neq \alpha(b_0)$ and since $K_r$ is symmetrical in the $a$'s and $b$'s we can assume that $\alpha(a_0) = a_0$, $\alpha(b_0) = b_0$.

We show by induction that $\alpha(a_i) = a_i$ and $\alpha(b_i) = b_i$.

Assume this is true up to $i$, and show $\alpha(a_{i+1}) = a_{i+1}$, $\alpha(b_{i+1}) = b_{i+1}$.

We have

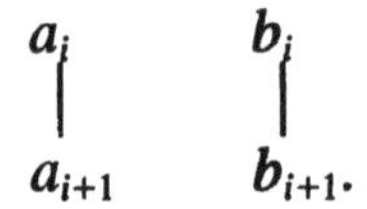

Since $a_{i+1}$ is below $a_i$, $\alpha(a_{i+1})$ is below $\alpha(a_i) = a_i$. Now if $\alpha(a_{i+1}) = a_{i+1}$, we are finished. If not, then $\alpha(a_{i+1})$ must be either $a_{i+2}$ or $b_{i+2}$ or anything below these two points. In any case $\alpha(a_{i+1})$ is below $b_i$ and hence for some $y$ above $a_{i+1}$, $\alpha(y) = b_i$. But the only such $y$ is $b_i$ itself and $b_i$ is not above $a_{i+1}$. Thus $\alpha(a_{i+1}) = a_{i+1}$. From symmetry we get that $\alpha(b_{i+1}) = b_{i+1}$.

Having shown that $\alpha(a_i) = a_i$ and $\alpha(b_i) = b_i$ for all $i$, we get a contradiction and thus prove Lemma 7. Consider $K_n^x$, $n \neq r$. If $x = a_m$, for some $m$, then $m \geqslant r+1$, then what is $\alpha(a_m)$? It should be $a_m$ by what we proved but $a_m$ is not in $K_r$. If $x = 0$, then $n > r$, again what is $\alpha(a_m)$?

Thus (7) is proved.

DEFINITION 8. (a) For each $n$, let $\alpha_n$ be the wff associated with $K_n$.

(b) For a set $M \subseteq \omega_0$, let the logic $y_M$ be defined as the intermediate logic with all the axiom schemas $\alpha_n$, for $n \notin M$.

THEOREM 9. *For* $M \neq N$, $y_M \neq y_N$.

*Proof.* Assume $M \neq N$, let $n \in N - M$. Then $\alpha_n$ is a theorem of $y_M$. We show that $\alpha_n$ is not a theorem of $y_N$. Consider $K_n$. $\alpha_n$ is false in $(K_n, D)$ for some assignment $D$. However, $y_N$ is valid in $(K_n, D)$, for otherwise, for some $m \notin N$ (and hence $m \neq n$) $\alpha_m$ is false in $(K_n, D)$. But by Corollary 5, since no truncation of $K_m$ is homomorphic onto $K_n$, $m \neq n$, $\alpha_m$ cannot be false in $(K_n, D)$. Thus $y_N \nVdash \alpha_n$. Hence $y_M \neq y_N$.

COROLLARY 10 (*Jankov*; *Fine*). *There exists a strictly increasing chain of intermediate logics.*

## 4. The fragment without $\vee$

DEFINITION 1. Let $h^+$ be the Hilbert system with axioms (a), (b), (c), (d), and (g) of Definition 1.1 and with rule MP, serving as both a consequence and provability rule.

Let $c^+$ be the system obtained from $h^+$ by adding the axiom schema $((A \to B) \to A) \to A$.

THEOREM 2. *$h^+$ is the positive fragment of $h$, i.e., for any $A$ without disjunction, $h^+ \vdash A$ iff $h \vdash A$.*

*Proof.* The deduction theorem can be proved for $h^+$ as in (2.1.13), then the completeness theorem can be proved for $h^+$ for the propositional Kripke interpretation as in Section 1 of this chapter.

A direct completeness proof for a special class of propositional Kripke structure will be given below. In the sequel we allow for the possibility that the language contains only a finite number of atomic propositions.

DEFINITION 3. (a) A set $\Delta$ of wffs of $h^+$ is a saturated theory iff $\Delta \vdash_{h^+} A$ implies $A \in \Delta$. (This definition coincides with previous definitions of saturated theories in a Hilbert system since no disjunction is available in our case).

(b) If $\Delta$ is a saturated theory and $A$ is a wff, let $\Delta + A$ denote the theory $\{B \mid \Delta \vdash_{h^+} A \to B\}$ $\Delta + A$ is also saturated.

(c) Let $p$ be a propositional variable, and $\Delta$ be a saturated theory. $\Delta$ is said to be $p$-prime iff $p \notin \Delta$ and for any $A$, $A \in \Delta$ or $A \to p \in \Delta$.

A saturated theory is *prime* iff it is $p$-prime for some $p$.

(d) A prime theory $\Delta$ is *minimal* iff there is no prime theory $\Delta' \subsetneq \Delta$.

(e) A wff is called a $p$-formula, $p$ a propositional variable, iff it has the form $A \to p$.

THEOREM 4. *Let $\Delta$ be a saturated theory in a certain language with possibly a finite number of atoms, and $A = C \to p \notin \Delta$, then there exists a $p$-prime saturated theory (in the language) $\Theta \supseteq \Delta$ such that $C \to p \notin \Theta$.*

*Proof.* Let $B_1, \ldots, B_n \ldots$ enumerate the wffs. We define a sequence of saturated theories $\Delta_n$, $\Delta_0 \subseteq \Delta_1 \subseteq \Delta_2 \ldots$ with the property that $A \notin \Delta_n$, $\Theta = \bigcup \Delta_n$ is $p$-prime and saturated.

(a) Let $\Delta_0 = \Delta$.

(b) Assume $\Delta_n$ has been defined.

*Case b1*: $\Delta_n \nvdash_{h^+} B_{n+1} \to A$.

Let $\Delta_{n+1} = \Delta_n + B_{n+1}$.

*Case b2*: $\Delta_n \vdash_{h^+} B_{n+1} \to A$.

Let $\Delta_{n+1} = \Delta_n + (B_{n+1} \to A) = \Delta_n$.

LEMMA 5. *$A \notin \Delta_n$, for all $n$.*

*Proof.* By induction. $A \notin \Delta_0$ by construction. Assume $A \notin \Delta_n$.

*Case* 1. $\Delta_n \nvdash_{h^+} B_{n+1} \to A$. In this case $\Delta_{n+1} = \Delta_n + B_{n+1}$ and so if $A \in \Delta_{n+1}$, then $\Delta_n \vdash_{h^+} B_{n+1} \to A$, contrary to the case.

*Case* 2. $\Delta_n \vdash_{h^+} B_{n+1} \to A$. In this case $\Delta_{n+1} = \Delta_n$, so if $A \notin \Delta_n$ then $A \notin \Delta_{n+1}$.

Now let $\Theta = \bigcup \Delta_n$. Clearly $A \notin \Theta$. Since $A = C \to p$, $p \notin \Theta$ either, $\Theta$ is clearly saturated. We want to show $\Theta$ is $p$-prime.

LEMMA 6. *$\Theta$ is $p$-prime.*

*Proof.* Notice that by construction for any wff $B$, $B \in \Theta$ or $B \to A \in \Theta$. Thus since $A = C \to p$, and $A \notin \Theta$, $C \in \Theta$. Now let $B$ be arbitrary. If $B \notin \Theta$, then $B \to (C \to p) \in \Theta$ and $C \in \Theta$, we get $(B \to p) \in \Theta$. Thus $\Theta$ is $p$-prime.

Thus Theorem 4 is proved.

LEMMA 7. *If $\Delta$ is a $p$-prime theory and $\Delta \subsetneq \Theta$, then $\Theta \vdash_{h^+} p$.*

*Proof.* If $A \in \Theta - \Delta$, then $A \to p \in \Delta$ and thus $\Theta \vdash_{h^+} p$.

DEFINITION 8. We define the prime canonical structure for $h^+$. Let $S$ be the set of all prime saturated theories of $h^+$, let $D(\Delta, q)$ be defined by $D(\Delta, q) = 1$ iff $q \in \Delta$, for $q$ atomic.

LEMMA 9. *In the prime canonical structure $(S, \subseteq, D)$. for any $A$ and any $\Delta$, $[A]_\Delta = 1$ iff $A \in \Delta$.*

*Proof.* By induction on $A$. For atomic $p$ this holds by definition. $[f]_\Delta = 0$, since all the theories are consistent.

$[A \wedge B]_\Delta = 1$ iff $[A]_\Delta = 1$ and $[B]_\Delta = 1$ iff $A \in \Delta$ and $B \in \Delta$ iff (by the axioms of $h^+$) $A \wedge B \in \Delta$.

$(A \to B) \in \Delta$ implies for any $\Delta' \supseteq \Delta$, if $A \in \Delta'$ then $B \in \Delta'$ and hence by the induction hypothesis $[A \to B]_\Delta = 1$.

Assume $A \to B \notin \Delta$. Then $\Delta \nvdash_{h^+} A \to B$ and so $B \notin \Delta + A$. We are looking for a saturated prime $\Theta$ such that $\Delta + A \subseteq \Theta$, $B \notin \Theta$. If $B$ were a wff of the form $C \to p$, we could have used Theorem 4. To overcome this difficulty we use the following

LEMMA 10. *Any wff $B$ is equivalent in $h^+$ to a conjunction of wffs of the form $B_i \to q_i$, $q_i$ atomic.*

*Proof.* By induction. For atomic $q$ take $(f \to f) \to q$. The case of $\wedge$ presents no difficulties. Assume $B = B' \to B''$. By the induction hypothesis, $B''$ is equivalent to $\wedge(B''_i \to q''_i)$. Thus $B$ is equivalent to $B' \to \bigwedge_i(B''_i \to q''_i)$ which is equivalent to $\bigwedge_i(B' \wedge B''_i \to q''_i)$.

To continue the proof of Lemma 9, we can assume $B$ is of the form $\bigwedge_i(C_i \to p_i)$, $p_i$ atomic. Since $B \notin \Delta + A$, for some $i$, $C_i \to p_i \notin \Delta + A$. There exists therefore a saturated prime theory $\Theta \supseteq \Delta + A$, with $C_i \to p_i \notin \Theta$ and therefore $B \notin \Theta$. Thus by the induction hypothesis $[B]_\Theta = 0$ and therefore $[A \to B]_\Delta = 0$.

This completes the proof of Lemma 9.

THEOREM 11. (a) *Let $\Delta$ be a saturated theory such that $\Delta \nvdash_{h^+} A$, then there exists a propositional Kripke structure $(S', R', O, D')$ such that $\Delta$ holds at $O$, $A$ does not hold at $O$ and for each $t$, $\{A \mid [A]_t = 1\}$ is a prime saturated theory.*

*Proof.* By Lemma 10, we can assume $A$ is of the form $C \to p$, $p$ atomic. By Theorem 4 there exists a prime saturated theory $\Theta$ containing $\Delta$, with $A \notin \Theta$. Let $(S, \subseteq, \Theta, D)$ be the prime canonical structure. By Lemma 9, $\Theta$ is valid in the structure and $A$ is not.

COROLLARY 12. *Let $\Delta$ be a prime saturated theory, then there exists a propositional Kripke structure $(S, R, O, D)$ such that $\Delta = \{A \mid [A]_0 = 1\}$.*

*Proof.* By Lemma 9 for the prime canonical structure.

LEMMA 13. *Let $\Delta$ be a prime saturated theory and let $p \in \Delta$, $p$ atomic. Let $\Delta_p = \{A \in \Delta \mid A$ does not contain $p$ as a subformula$\}$.*

*Then* (a) *$\Delta_p$ is a prime saturated theory in the language without $p$.*

(b) *In the language with $p$, $\Delta = \Delta_p + p$.*

*Proof.* (a) Clear, since $\Delta$ is $q$-prime for some $q \neq p$.

(b) Let $t = f \to f$. Clearly $\vdash_{h^+} p \to (A(t) \to A(p))$. Since $A(p) \in \Delta$ implies $A(t) \in \Delta_p$, we get that if $A(p) \in \Delta$ then $A(p) \in \Delta_p + p$. Thus $\Delta \subseteq \Delta_p + p$.

LEMMA 14. *Let $\Delta$ be prime and saturated. Then $\Delta$ is minimal iff for no atomic $p$ do we have $p \in \Delta$.*

*Proof.* (a) Suppose $p \in \Delta$, for $p$ atomic. We show $\Delta$ is not minimal. By Corollary 12, let $(S, R, O, D)$ be a propositional Kripke structure with $\Delta = \{A \mid [A]_0 = 1\}$. Let $O^*$ be a new point and let $S^* = S \cup \{O^*\}$. $R^* = R \cup \{O^*\} x S^*$, $D^*(t, q) = D(t, q)$, $t \in S$, $q$ atomic and

$D^*(O^*, q) = 0$. Let $\Delta^* = \{A \mid [A]_{0^*} = 1\}$. Clearly $\Delta^* \subseteq \Delta$. We claim $\Delta^*$ is $p$-prime. This is clear since if $[A]_{0^*} = 0$, certainly $[A \to p]_{0^*} = 1$, since $[p]_0 = 1$ and 0 is next to $0^*$.

(b) The other direction follows from Lemma 7.

THEOREM 15. (a) *The set of prime saturated theories in a language with a finite number of atomic propositions is finite.*

(b) (Diego) *The number of* $h^+$ *non-equivalent wffs built up from a fixed finite number of atomic wffs is finite.*

*Proof.* (a) By induction on the set of atoms $V$. The claim is trivial for $V$ empty. Assume (a) true for $\bar{\bar{V}} = n$, show for $\bar{\bar{V}} = n + 1$. By Lemma 14, if $\Delta$ is non-minimal prime saturated theory, then for some $p \in V$, $p \in \Delta$. Hence by Lemma 13, $\Delta = \Delta_p \dot{+} p$. Thus using our inductive assumption there are only a finite number of non-minimal prime saturated theories built up on $V$.

Now let $\Delta$ be a minimal theory built up from $V$. Let $S^\Delta = \{\Theta \supseteq \Delta \mid \Theta$ is prime saturated built up from $V\}$. By Lemma 9, $(S^\Delta, \subseteq, \Delta, D)$, the canonical structure, characterises $\Delta$. The elements of $S^\Delta$ are all non-minimal prime saturated theories. Since there is only a finite number of such theories, there is a finite number of different canonical structures and hence only a finite number of minimal prime saturated theories.

(b) If $\nVdash_{h^+} A \to B$, then for some prime saturated theory in the language of $A, B$, $A \in \Delta$, $B \notin \Delta$ (note that $\{C \mid \vdash_{h^+} C\}$ is saturated). Thus since, on a finite set of atoms there is only a finite number of prime saturated theories, (b) follows.

THEOREM 16. *Let x be any extension of* $h^+$ *(in a language without* $\vee$*) then x has the finite model property.*

*Proof.* Assume $\nVdash_x A$. Let $V$ be the set of atoms appearing in $A$. Let $\Delta_x = \{A'$ in the language $V \mid \vdash_x A'\}$, then $\Delta_x$ is a saturated theory of $h^+$ in the language $V$. Since $A$ is not a theorem of $x$, then $\Delta_x \nVdash_{h^+} A$. By Lemma 10 and Theorem 4 there exists a prime saturated theory $\Theta \supseteq \Delta_x$ such that $A \notin \Theta$, $\Theta$ in the language $V$. Let $(S^\Theta, \subseteq, \Theta, D)$ be the prime canonical structure for $\Theta$ in the language $V$. By Theorem 15, this structure is finite. $x$ is valid in this structure and $A$ is not.

THEOREM 17 (*K. Segerberg*). $h^+$ *is complete for the class of finite binary trees.*

*Proof.* By Theorem 2, if $h^+ \not\Vdash A$ then $h \not\Vdash A$, and so by (2.4) there exists a propositional Kripke structure with a finite tree in which $A$ is false (not valid). It is sufficient for us to show how to transform this structure to another structure with binary tree in such a way that disjunction free wff preserve their truth value.

Let $(S, \leqslant, O, D)$ be a propositional Kripke structure with a finite tree $(S, \leqslant, O)$. A point $z \in S$ is called *bad* iff it has more than two immediate successors. We prove by induction on the number of bad points of $S$ that

(*) There exists an $S' \supseteq S, \leqslant', O, D'$ such that $\leqslant = \leqslant' \restriction S, D = D' \restriction S$ and if $[A]'$ denotes value in $(S', \leqslant', O, D')$ then for any wff $A$ of $h^+$, $t \in S$ $[A]'_t = [A]_t$.

If there are no bad points take $S' = S$.

Let $S$ have $n$ bad points. Let $z_0 \in S$ be point with more than two immediate successors. Let $y_1, y_2$ be two such successors. Take $S' = S \cup \{x\}$, $x$ a new point. Define $R'$ as the transitive and reflexive closure of $R \cup \{(z_0, x), (x, y_1), (x, y_2)\}$. Let $D'(t, q) = D(t, q)$, for $t \in S$ and $D'(x, q) = 1$ iff $D(y_1, q) = D(y_2, q) = 1$.

We claim that for any $A$ of $h^+$, $t \in S$: (a) $[A]'_x = 1$ iff $[A]_{y_1} = [A]_{y_2} = 1$. (b) $[A]'_t = [A]_t$. Show by induction on $A$. For $A$ atomic this holds by definition. $\wedge$ and $f$ present no difficulty.

Assume $[A \to B]'_x = 0$. Then for some $t$, $xR't$, $[A]'_t = 1$ and $[B]'_t = 0$. If $t = x$, by the induction hypothesis, since $[B]'_x = 0$, for some $i$, $i \in \{1, 2\}$, $[B]_{y_i} = 0$ and also $[A]_{y_i} = 1$, thus $[A \to B]_{y_i} = 0$. If $t \neq x$ then for some $i$, $y_iRt$ and again $[A \to B]_{y_i} = 0$.

If $[A \to B]_{y_i} = 0$ for some $i$, then for some $t$, $y_iRt$ and $[A]_t = 1$ and $[B]_t = 0$. By the induction hypothesis of (b), $[A]'_t = 1$ and $[B]'_t = 0$ and hence $[A \to B]'_x = 0$.

This proves (a). To prove (b) let $[A \to B]'_t = 0$, then for some $s$, $tR's$, $[A]'_s = 1$, $[B]'_s = 0$. If $s \neq x$, then by the induction hypothesis $[A]_s = 1$, $[B]_s = 0$ and by the definition of $R'$, $tRs$ holds and hence $[A \to B]_t = 0$. If $s = x$, by part (a) for $A \to B$, for some $i$, $[A]_{y_i} = 1$ and $[B]_{y_i} = 0$. Again by the definition of $R'$, $tRy_i$ and hence $[A \to B]_t = 0$.

Now assume $[A \to B]_t = 0$, then for some $s$, $tRs$ and $[A]_s = 1$, $[B]_s = 0$. By definition of $R'$ and the induction hypothesis, $tR's$ and $[A]'_s = 1$, $[B]'_s = 0$, and hence $[A \to B]'_t = 0$.

This completes the induction step for the proof of (a) and (b). Thus (a), (b) hold and hence the induction step in the proof of (*) is completed.

Theorem 17 follows from (*) and the fact that $h^+$ is complete for the class of finite trees.

**Exercise 18.** (a) Let $lc^+$ be $h^+$ extended by the axiom schema $((p \to q) \to r) \to (((q \to p) \to r) \to r)$. Show that $lc^+$ is the disjunctionless fragment of $lc$.

(b) Let $h^{++}$ be the fragment of $h$ with $\to$ only. Show it can be axiomatized by modus ponens as a provability and consequence rule and axioms (a) and (b) of Definition 1.1.

(c) Show that adding the axiom in (a) of this exercise to $h^{++}$ yields the $\to$ fragment of $lc$.

(d) Can you axiomatize the $\to$ fragment of $kc$? Is it equal to $h^{++}$? How about the fragment of $kc$ without disjunction?

(e) Consider the schema

$$(((p \to q) \to q) \to r) \to (((p \to q) \to r) \to r).$$

Let $x_1$ be the extension of $h^+$ with this schema ($x_1$ is in the language of $h^+$). Show that $x_1 = lc^+$.

Let $x_2$ be the extension of $h^{++}$ with this schema ($x_2$ is in the language with $\to$ only). Show that, however, $x_2 \subsetneq h^{++}$, i.e., the schema of (a) above is not provable in $x_2$.

## NOTES

In writing this Chapter I used profitably C. Smorynski's thesis. The idea of filtration was first introduced in modal logic by E. J. Lemmon and D. Scott, 1966. K. Segerberg introduced it in Kripke structures and developed the method further. The author continued with it, calling the refined form selective filtration.

Incomparable intermediate logics were given independently by Jankov, 1968 and K. Fine. We base our presentation here on Fine's results. The fragment without $\vee$ was studied by A. Diego. We base our presentation on A. Urquhart. We also use results of K. Segerberg, 1974.

## CHAPTER 5

# THREE INTERMEDIATE LOGICS

## 1. THE LOGIC OF FINITE $n$-ARY TREES

We saw in previous chapters the following properties of $h$:

(1) $h$ is complete for the class of all finite trees

(2) $h^+$ is complete for the class of all finite $n$-ary trees, for any $n \geqslant 2$.

(3) $h$ is complete for the infinite full binary tree.

We can therefore ask the following:

(4) Is $h^+$ complete for the class of finite 1-ary trees?

(5) Is $h$ complete for the infinite 1-ary tree?

(6) Is $h$ complete for the class of finite $n$-ary trees?

The exercises of the last chapter indicate that

(7) $lc^+$, the disjunctionless fragment of $lc$ is complete for the class of all 1-ary trees. In fact $lc^+$ can be obtained from $h^+$ by adding the axiom

(8) $((p \to q) \to r) \to (((q \to p) \to r) \to r)$.

*Remark* 9. The logics that extend $lc$ are the logics $s_n$, obtained by adding the axiom $P_n$, where

(9) $P_1 =_{\text{def}} ((A_1 \to A_0) \to A_1) \to A_1$

(10) $P_n =_{\text{def}} ((A_n \to P_{n-1}) \to A_n) \to A_n$.

$s_n^+$ is obtained by adding $P_n$ to $lc^+$. It is complete for a finite linear chain of length $n$.

Turning to Question 6, the answer is negative. This section axiomatizes the logics $D_n$, determined by the class of $n+1$-ary finite trees.

The axiomatization answers another question of Łukasiewicz. We noticed in earlier chapters that $h$ has the disjunction property, i.e.,

(11) $\vdash_h A \vee B$ iff $\vdash_h A$ or $\vdash_h B$.

We ask, can we characterise $h$ by this property? Can we have

(12) $h$ is the smallest intermediate logic that has the disjunction property.

The answer is no, there are extensions of $h$ that have the disjunction property, e.g., the logics $D_n$.

### *The Sequence $D_n$*

Define $D_n$ to be the theory of class of all structures whose tree is finite and has the property that every point has not more than $n+1$ immediate successors (e.g., $D_1$ is the theory of all finite binary trees). It is clear that $D_n \supseteq D_{n+1}$, and (since $h$ has the finite model property) that $\bigcap_{nD_n} = h$. Note that $D_n$ is closed under modus ponens!

LEMMA 14. *Let $\tau$ be a class of trees such that if $T_1$, $T_2 \in \tau$ then $T_3 \in \tau$ where $T_3$ is obtained by*

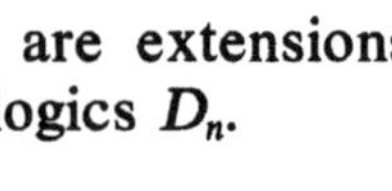


*Then the logic $D_\tau$ defined as the theory of the class of all structures with tree in $\tau$ has the disjunction property.*

*Proof.* Assume that $\nVdash B$ and $\nVdash C$; then $B$ is false and some structure with tree $T_1$ and $C$ in $T_2$. Take $T_3$ and give propositional variables value 0 at the base and the respective old values at $T_1$ and $T_2$; then $B \vee C$ is false at $T_3$.

(15) We immediately conclude that $D_n$ has the disjunction property. Let us show now that $D_n \supsetneq h$.

(16) Consider the following formula $A_n$, with the propositional variables $p_0, \ldots, p_{n+1}$:

$$\bigwedge_{i=0}^{n+1}\left(\left(p_i \to \bigvee_{j\neq i} p_j\right) \to \bigvee_{j\neq i} p_j\right) \to \bigvee_{i=0}^{n+1} p_i.$$

LEMMA 17. $h \nVdash A_n$.

*Proof.* Consider the following structure:

(18)

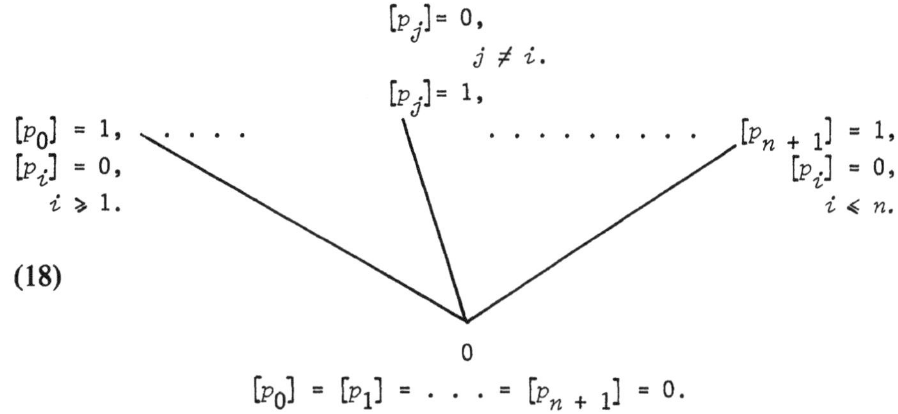


LEMMA 19. $D_n \vdash A_n$.

*Proof.* Let us assume that there exists a model with a finite tree such that no point in the tree can have more than $n+1$ successors and such that $A_n$ is false at the base point. Since the tree is finite we may assume that at any higher point the formula $A_n$ is true. So we have that in our tree the following holds:

(20) $$\left[\bigwedge_i\left(\left(p_i \to \bigvee_{j\neq i} p_j\right) \to \bigvee_{j\neq i} p_j\right)\right]_0 = 1;$$

(21) $$\left[\bigvee_{i=0}^{n+1} p_i\right] = 0.$$

But (21) implies $[\bigvee_{j\neq i} p_j]_0 = 0$ and so by (20) we get that the following diagram holds for points $x_0, \ldots, x_{n+1}$:

(22)

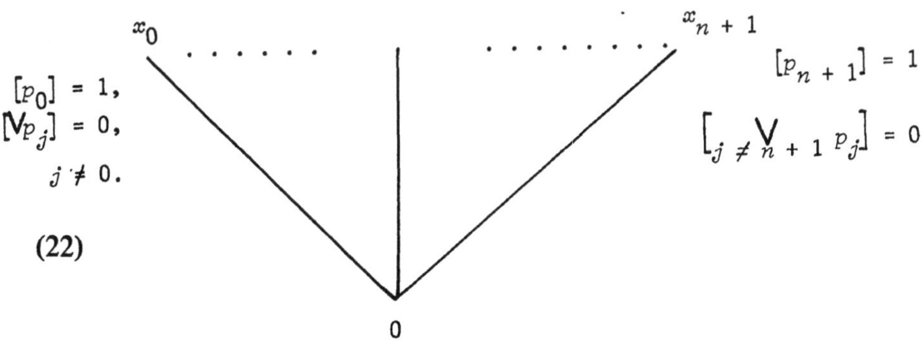

(Here $0Rx_0, \dots, 0Rx_n$ hold, and the truth value of the $p$'s is indicated beside the points in the diagram.) Since our tree fulfills the condition that no point can have $n+2$ successors we have that, for some $i$, and $j$ and $a$, we get the following:

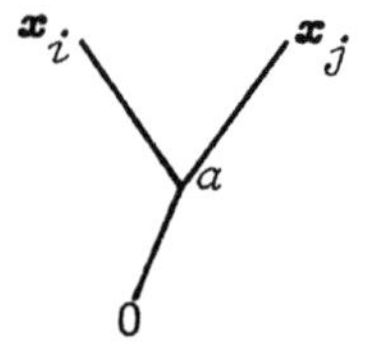


But this means that $[\bigvee_{i=0}^{n+1} p_i]_a = 0$ and so in view of (20) we get that $[A_n]_a = 0$ which is a contradiction since we assumed that 0 is the *only* point with $[A_n]_0 = 0$.

We shall show later that $A_n$ actually axiomatizes $D_n$.

*$D_n$ is Finitely Axiomatizable*

Now we shall present the selective filtration method and prove that the scheme $A_n$ indeed does axiomatize the system $D_n$.

Our goal is to show that

(34) $h + A_n \nVdash C$ implies $C$ is false in a structure with a finite tree each point of which has at most $n+1$ immediate successors.

To prove this we shall need some definitions:

Let $\Gamma_n$ be the set of all substitution instances of the theorems of $h + A_n$. The following is true:

(35) $h + A_n \nVdash C$ iff the theory $(\Gamma_n, \{C\})$ is $h$-consistent iff there exists a structure validating $\Gamma_n$ in which $C$ is false.

Let $C$ be a nonprovable formula and let **M** be a tree structure validating $\Gamma_n$ in which $C$ is false. Let $(T_{\mathbf{M}}, \leqslant_{\mathbf{M}})$ be the tree of the structure **M**. Our strategy is to derive from **M** a structure with tree $(T_{\mathbf{N}}, \leqslant_{\mathbf{N}})$ which is finite and has the property that no point in $T_{\mathbf{N}}$ has more than $n+1$ successors, and at which $C$ is false! Having done this we will have concluded the proof of the completeness theorem.

Before we embark on the details of the proof let us describe briefly the intuitions motivating it and its general outlines.

Suppose $D_n \nVdash C$. We are interested in a structure with a finite

$(n+1)$-ary tree in which $C$ is false. Our idea is to use the filtration method. We first take any structure $\mathbf{M} = (M, \leqslant, O, D)$ validating $D_n$ in which $C$ is false. We notice that the truth value of $C$ depends only on the truth values of the subformulas of $C$. If $\Theta$ is set of all subformulas of $C$, our first step is to try and identify any two points of $\mathbf{M}$ that agree on all the truth values of elements of $\Theta$. Let, for $t, s \in M$, $tRs$ mean that for all $\alpha \in \Theta$, $[\alpha]_t = 1 \Rightarrow [\alpha]_s = 1$. Thus if $t \equiv s =_{\text{def}} (tRs$ and $sRt)$, then we are interested in the set of equivalence classes $M^* = M/\equiv$. $M^*$ is essentially the set of points that matter, as far as $C$ is concerned. Set $0^* = 0/\equiv$; $t/\equiv \; R^* s/\equiv$ iff by definition $tRs$; then $(M^*, R^*, 0^*)$ is a finite partially ordered set. This is the finite part of $M$ relevant to the evaluation of the truth value of $C$.

Let, for atomic $p \in \Theta$, $[p]^*_{t/\equiv} = [p]_t$, then the following lemma shows that indeed, we succeeded in extracting a finite substructure of $\mathbf{M}$ containing all the information relevant to $C$.

LEMMA*. *For* $\alpha \in \Theta$, $[\alpha]^*_{t/\equiv} = [\alpha]_t$.

The proof is by induction.

This line of proof fails, unfortunately, because we want a finite $\mathbf{M}^* = (M^*, R^*, 0^*)$ that is also an $(n+1)$*-ary tree*! One can try to use the fact that $\mathbf{M}$ validates $D_n$ to show that $\mathbf{M}^*$ has the required properties, but things do not work. (Incidentally, the above construction when applied to $h$ shows that $h$ has f.m.p.)

The next idea is to be more careful in the construction. Maybe we identified too many points. Perhaps if we start with a *tree* $(T_M, \leqslant_M, 0)$ and identify points very carefully, we can preserve the tree property and also hopefully show that it is an $(n+1)$-ary tree. This turns out to be correct, in principle, but a careful construction is needed. It seems best, rather than identify points, to choose a finite subtree $(T_N, \leqslant_N, 0) \subseteq (T_M, \leqslant_M, 0)$ that has all the required properties. We hope to choose the points in such a way that we can prove that $(T_N, \leqslant_N)$ is $(n+1)$-ary. This method can be applied to a variety of situations and is called the selective filtration method. (Essentially one can call it a "guided truth tableaux" method; that is, we construct a model using truth tableaux, but whenever we have to make a choice (e.g. $\alpha \vee \beta$ is truth) we look at $\mathbf{M}$, but this need not concern us here.)

To describe the selection process, start by letting $0 \in T_N$. Now suppose you already put $x \in T_N$. Look at $T^x_M = \{s \mid x \leqslant_M s\}$. This is the set of all successors and $x$ in $M$ and we want to select some of them and put them in $T_N$. Of course whatever we select are going to be the *immediate* successors of $x$ in $T_N$ and thus we are interested in selecting as few as possible. $\equiv$ divides $T^x_M$ into equivalence classes. Choose one $t_i(x)$ of each class. Call this set $K_x = \{t_i(x)\}$. We observe that if, for $i \neq j$, $t_iRt_j$) then maybe we can postpone putting $t_j \in T_N$, until we reach the stage of treating $T^{t_i}_M$. Also, is it possible that $x \equiv a_x$, such that $x \leqslant_M a_x$ and $T^{a_x}_M$ has less equivalence classes? Of course in this case we had better select from $T^{a_x}_M$ *instead of* $T^x_M$, because from our point of view, $x$ and $a_x$ are identical, except that $a_x$ gives rise to *less* immediate successors in $T_N$? The correct inductive definitions are given in (41)–(50) below. (41)–(43) give the definition of the first stage and (43)–(46) describe some properties of the selection and (47)–(50) give the induction step, i.e. the choice of $\{t_i(x)\} = K_x \subseteq T_N \cap T^x_M$, for $x$ already chosen.

In practice, we must remember that we must also get a finite tree (51) and so some points $x \in T_N$ may be troublesome. In the construction such points are declared RED and not touched.

Thus we obtain in (51) below a finite $(T_N, \leqslant_N) \subseteq (T_M, \leqslant_M)$ and our next task in (52) below is to show that indeed this is an $(n+1)$-ary tree. The construction is done in such a way that the assumption that $t \in T_N$ has more than $n+1$ immediate successors leads to the contradiction that $D_n$ does not hold in $t$. This is another advantage of the choice of construction and the fact that $t \in T_N \subseteq T_M$.

Lemma 57 is the analogue of Lemma* and shows that for $\alpha \in \Theta$, $[\alpha]^N_t = [\alpha]^M_t$. Let us begin with the construction:

(36) Let $\Theta$ be the set of all subformulas of the formula $C$. Let $x, y \in T_M$.

Define

(37) $xRy$ iff for all $\beta \in \Theta$ we have $[\beta]_x = 1 \Rightarrow [\beta]_y = 1$.

(38) $x \equiv y$ iff $xRy \wedge yRx$.

Clearly $\equiv$ is an equivalence relation and thus $T_M$ is divided into a finite number of equivalence classes $S_0, \ldots, S_m$. We may also assume that $0 \in S_0$ (0 is the base point of $T_M$). Before we continue to construct $(T_N, \leqslant_N)$ observe that

(39) $x \leqslant_M y \Rightarrow xRy$.

(Note that $\leqslant_M$, $\leqslant_N$ are taken as the transitive and reflexive closures of the successorship relation in the respective trees.) Another useful definition is the following:

(40) $T_M^{x_0}$ for $x_0 \in T_M$ is defined as $\{y \mid x_0 \leq_M y\}$.

Let us now turn to the construction of the tree $T_N$. This will be done in stages.

The classes $S_0, \ldots, S_m$ divide the points of $T_M^0 = T_M$ into disjoint classes. $S_0^0 = S_0 \cap T_M^0, \ldots, S_m^0 = S_m \cap T_M^0$. (Here $S_i^0 = S_i$ but this distinction is made for the induction step!)

Our immediate objective is to choose one element out of each class $S_i^0$. Now since we have certain matters in mind, which we would like to be able to prove later, we make our choice carefully. It may happen that for some $a_0 \in T_M^0 \cap S_0 = S_0^0$ (recall that $0 \in S_0$!) has the property that whenever we take an element $b$ above it, i.e., $a_0 \leq_M b$ then $b \notin S_1$ let us say. In this case we are more interested in $a_0$ than in 0 itself. So let us assume that $a_0$ is a point with the following properties:

(41) $a_0 \equiv 0$.

(42) The number of classes from among $S_0^0, \ldots, S_m^0$ that intersect $T_M^{a_0}$ is *minimal* (with respect to other choices of $a_0 \in S_0$).

Let $S_0^{a_0}, \ldots, S_k^{a_0}$, $k \leq m$, be exactly the classes such that $T_M^{a_0} \cap S_j^{a_0} \neq \emptyset$, $j = 0, \ldots, k$. We know that $k$ is minimal with respect to other choices of $a_0$. We now distinguish two possibilities:

*Possibility* 1. $T_M^{a_0} \subseteq S_0$ (i.e., $a_0 \equiv 0$ and $\forall y(a_0 \leq_M y \rightarrow y \equiv 0)$. In this case we declare the point 0 to be a RED point.

*Possibility* 2. We assume that 0 is *not* a RED point. In this case, by the minimality of $k$ we know that for any $a$ such that $a_0 \leq_M a$ and $a \equiv 0$ the following holds:

$$S_j^{a_0} \cap T_M^a \neq \emptyset, \quad 0 \leq j \leq k.$$

There are no other possibilities.

Let us now begin the construction of $T_N$.

*Stage* 0. Put the point 0 in $T_N$.

*Stage* 1. Let us look at the point $a_0$ which has the properties discussed above. And let us consider each of the two possibilities.

(1) If 0 was declared RED we do not put in $T_N$ any successors of 0.

(2) If Possibility 2 holds for the point $a_0$ we begin to put successors to the point $0 \in T_N$.

Let $x_0(0) \in S_0^{a_0} \cap T_M^{a_0}, \ldots, x_k(0) \in S_k^{a_0} \cap T_M^{a_0}$.

The index $i$ in $x_i(0)$ shows where the point comes from and the 0 in $x_i(0)$ shows that we are defining now the successors of 0 in $T_N$. Let $K_0 \subseteq \{x_0(0), \ldots, x_k(0)\}$ be a set with the following properties:

(43)(a) For $x \neq y$, $x, y \in K_0$ we have that $(xRy \vee yRx)$ is *false*! In particular $x \in K_0$ implies $x \not\equiv 0$.

(b) For each $x_i(0)$, $x_i(0) \not\equiv 0$, there exists a $y \in K_0$, $y \not\equiv 0$ such that $yRx_i(0)$ holds.

Such a $K_0$ can be easily constructed since we have only a finite number of points to consider. Notice that in (43)(b) we required $y \not\equiv 0$ and $x_i(0) \not\equiv 0$. This we have to do since $S_0 \cap T_M^{a_0}$ may not be empty and we would like our process to terminate. Notice that although $k$ may equal $m$ and although all $x_i(0) \not\equiv 0$ may be incompatible with respect to $R$, no representative of the class $S_0$ is in $K_0$. The index 0 of $K_0$ shows that we are working above $0 \in T_N$.

Now having defined such a $K_0$ we let each $y \in K_0$ be in $T_N$ and say that they are successors of 0 in $T_N$. So $0 \leqslant_N y$ for $y \in K_0$ holds.

Thus stage 1 of the construction is completed. We have either put $K_0 \subseteq T_N$ or declared 0 to be RED.

LEMMA 44. *Let $x \in K_0$, then $xRa_0$ is false.*

*Proof.* Since $a_0 \equiv 0 \; Rx$ holds (since $a_0 \leqslant_M x$) we cannot have that $xRa_0$ since this means $x \equiv a_0$ which contradicts (43)(a).

LEMMA 45. *For $x \in K_0$ we have $x \leqslant_M y \Rightarrow y \in S_j^{a_0}$ for some $j$.*

*Proof.* Clear.

(45a) Lemma 44 actually says that for any point $b$ such that $a_0 \leqslant_M b$ we have that either $b \equiv a_0$ *or* $xRb$ for some $x \in K_0$.

(46) We shall say that any $x \in K_0$ was put in $T_N$ in stage 1.

*Stage* $n+1$. Let $s \in T_N$ be a point which was put into $T_N$ in stage

$n$. Let $a_s$ be a point above $s(s \leqslant_M a_s)$ with the following properties:

(47) $a_s \equiv s$.

(48) The number of classes from among the class that intersects $T_M^s$ which have nonempty intersection with $T_M^{a_s}$ is *minimal* with respect to other choices of $a_s \equiv s$.

We shall, again, distinguish two possibilities.

*Possibility* 1. $x \in T_M^{a_s} \Rightarrow x \equiv s \equiv a_s$.

In this case we declare $s$ to be a RED point.

*Possibility* 2. $s$ is not a RED point. Let $S_0^{a_s}, \ldots, S_l^{a_s}$ be the classes such that $S_i^{a_s} \cap T_M^{a_s} \neq \emptyset$.

We also assume that for any $a$ such that $a \equiv s$ and $a_s \leqslant_M a$ and any $S_i^{a_s}$ we have $S_i^{a_s} \cap T_M^a \neq \emptyset$.

Now to continue our construction, proceed as follows: If $s$ is a RED point, $s$ will have no successors in $T_N$.

If $s$ is not a RED point, let $x_0(s) \in S_0^{a_s} \cap T_M^{a_s}, \ldots, x_l(s) \in S_l^{a_s} \cap T_M^{a_s}$. Now let $K_s$ be a maximal set of $R$-incomparable points from among $\{x_0(s), \ldots, x_l(s)\}$. We have again that the following holds:

(49a) For every $x_i(s) \not\equiv s$, a $y \in K_s$ exists such that $yRx_i(s)$ holds.

(49b) $x \in K_s \Rightarrow x \not\equiv s$.

We now put all points of $K_s$ into $T_N$ as successors of $s$.

We shall say that each point in $K_s$ was put into $T_N$ in the $(n+1)$th stage. The following is true:

(50) If $a_s \leqslant_M b$ then either $b \equiv a_s$ or for some $y \in K_s$ we have $yRb$.

Having done this for each point $s$ which was put into $T_N$ in the $n$th stage we say that the $(n+1)$th stage is completed.

THEOREM 51. *The construction terminates.*

*Proof*. We shall prove by induction on the stage $n$ that if $s$ was constructed in stage $n$ then the number of classes that have nonempty intersection with $T_M^{a_s}$ does not exceed $m - n$.

*Case* $n = 1$. If $s$ was constructed in stage 1, then $s = x_i(0)$ for some $i$. Now by Lemma 44, no point above $x_i(0)$ can be in the class $S_0$.

*Case* $n = n$. In this case if $s$ was constructed in stage $n$ let $t$ be the predecessor of $s$ in the tree $T_N$, then $t$ was constructed in the stage $n-1$ and so by the induction hypothesis the number of classes intersecting $T_M^{a_t}$ does not exceed $m-n+1$. Now $s$ itself is some $s = x_j(t)$ and since $s \neq t$ we know that the class in which $t$ is a member does not intersect $T_M^s$ by (49) and therefore does not intersect $T_M^{a_s}$ and so the number is diminshed by one more unit.

We therefore conclude that at the stage $m$ all points are declared RED.

Now since the construction terminates we get that $T_N$ is a finite tree with $\leqslant_N$ reflexive and transitive. We also have some information on how the immediate successors of an $x \in T_N$ were chosen. This fact, together with the structure which we shall define on $T_N$ (with $C$ false in **N**) will conclude our completeness proof for $D_n$.

LEMMA 52. *There are at most $n+1$ immediate successors to any point $z \in T_N$.*

*Proof.* Let $z \in T_N$. Assume that $z$ has more than $n+1$ immediate successors and arrive at a contradiction. First note that $z$ cannot be a RED point since such points have no successors. This being the case, we do know that for $z$, Possibility 2 did hold and in fact $K_z$ is the set of immediate successors of $z$. Let us assume that $K_z = \{x_1(z), \ldots, x_{n+1+k}(z)\}$, $k \geqslant 1$, we know that neither $x_i(z)Rx_j(z)$ nor $x_j(z)Rx_j(z)$ holds for $1 \leqslant i, j \leqslant n+1+k$, $i \neq j$.

By definition there exist $\gamma_{i,j} \in \Theta$, $1 \leqslant i \neq j \leqslant n+k+1$, such that

$$[\gamma_{i,j}]_{x_i(z)} = 1, \qquad [\gamma_{i,j}]_{x_j(z)} = 0$$

in $M$.

Define

$$\gamma_{i,*} = \bigwedge_{j \geqslant n+2} \gamma_{i,j} \quad \text{for} \quad i \leqslant n+1,$$

$$\gamma_{*,i} = \bigvee_{j \geqslant n+2} \gamma_{j,i} \quad \text{for} \quad i \leqslant n+2.$$

Then the following holds:

$$(53) \qquad [\gamma_{i,*}]_{x_i(z)} = 1, \qquad [\gamma_{i,*}]_{x_j(z)} = 0, \qquad j \geqslant n+2,$$

$$[\gamma_{*,i}]_{x_i(z)} = 0, \qquad [\gamma_{*,i}]_{x_j(z)} = 1, \qquad j \geqslant n+2.$$

What we did is, in fact, to 'identify' all points $x_j(z)$ for $j \geqslant n+2$ into one point $*$.

Now define

$$p_1 = \gamma_{1,2} \wedge \gamma_{1,3} \wedge \cdots \wedge \gamma_{1,n+1} \wedge \gamma_{1,*},$$
$$p_2 = \gamma_{2,1} \wedge \gamma_{2,3} \wedge \cdots \wedge \gamma_{2,n+1} \wedge \gamma_{2,*},$$
$$\vdots$$
$$p_{n+1} = \gamma_{n+1,1} \wedge \cdots \wedge \gamma_{n+1,n} \wedge \gamma_{n+1,*},$$
$$p_* = \gamma_{*,1} \wedge \cdots \wedge \gamma_{*,n+1}.$$

Let us note that for any sentence $A$ which is formed using conjunctions and disjunctions of sentences in $\Theta$ the following holds:

(54) $\quad xRy \wedge [A]_x = 1 \Rightarrow [A]_y = 1.$

In view of the above remark, and the fact that $a_z \equiv z$ and $a_z \leqslant_M x$ for $x \in K_z$ we get the following diagram:

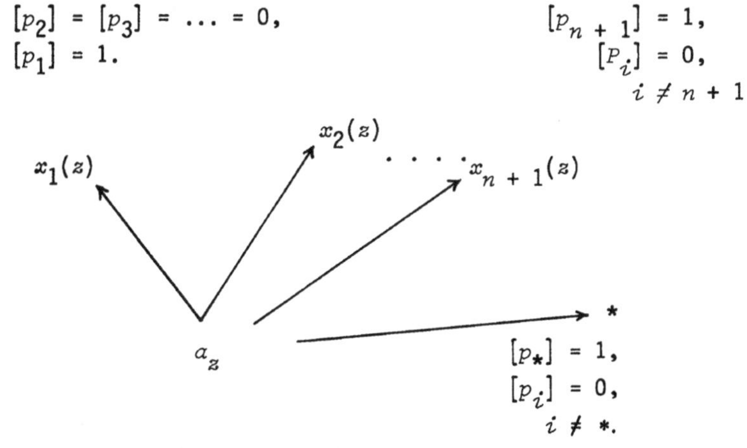


(Arrows indicate relation $R$, and beside each point the truth of the $p_i$'s at the point is indicated.)

(56) We shall now show that at **M** at the point $a_z$ the following is true:

$$\left[\left(\left(p_* \rightarrow \bigvee_{i=1}^{n+1} p_i\right) \rightarrow \bigvee_{i=1}^{n+1} p_i\right)\right]_{a_z} = 1,$$

$$\left[\left(\left(p_j \rightarrow (p_* \vee \bigvee_{i \neq j} p_i\right)\right) \rightarrow \left(p_* \vee \bigvee_{i \neq j} p_i\right)\right)\right]_{a_z} = 1,$$

$i = 1, \ldots, n+1.$

Let us prove for example that $((p_* \to \bigvee p_i) \to \bigvee p_i)$ holds at $a_z$. Assume the contrary, so for some $b_1$ such that $a_z \leqslant_M b_1$ we have $[p_* \to \bigvee p_i]_{b_1} = 1$ and $[\bigvee p_i]_{b_1} = 0$. Now by (50) we have that either $b_1 \equiv a_z$ or for some $1 \leqslant i \leqslant n+k+1$ we have $x_i(z)Rb_1$. Let us assume the latter. Now if $i \geqslant n+2$ then by (54) we have $[p_*]_{b_1} = 1$ and so by our assumption $[\bigvee p_i]_{b_1} = 1$ which contradicts the other part of our same assumption. Now certainly if $i \leqslant n+1$ we get $[p_i]_{b_1} = 1$ which is impossible. So we are forced to conclude that $b_1 \equiv a_z$.

Now if this is the case by Possibility 2 the class of $x_{n+2}(z)$ must have a representative above $b_1$ (for otherwise, for $a_z$, $n+k+1$ is not *minimal*). Let this representative be $b_2$. So $b_1 \leqslant_M b_2$. But in $b_2 \equiv x_{n+2}(z)$ we have $[p_*]_{b_2} = 1$, $[\bigvee p_i]_{b_2} = 0$. But this stands in contradiction with $[p_* \to \bigvee p_i]_{b_1} = 1$.

We have therefore concluded the proof of (56).

We shall proceed to complete the proof of (52).

Clearly, in view of (54), we have that $[p_1 \vee \cdots \vee p_{n+1} \vee p_*]_{a_z} = 0$. But this together with (56) stands in contradiction with our initial assumption that **M** validates $A_n$. Thus, the proof of (52) is concluded.

We shall now proceed to define a structure **N** with tree $(T_N, \leqslant_N)$. Define for a propositional variable $q \in \Theta$. $[q]_x$ in $\mathbf{N} = [q]_x$ in **M**.

LEMMA 57. *If $\delta$ is a subformula of $C$ then for $x \in T_N$ we have $[\delta]_x$ in $\mathbf{N} = [\delta]_x$ in* **M**.

*Proof.* The proof is by induction. For atomic $\delta$ this holds by definition. Conjunctions and disjunctions present no difficulties. Let us examine the case of implication.

*Case* 1. $[\beta \to \gamma]_x = 0$ in **M**. We want to prove that $[\beta \to \gamma]_x = 0$ in **N**. To do this we have to examine under what circumstances $x$ was put in $T_N$. Clearly, by the construction $x$ was put in $T_N$ at some stage $n$. From this stage onwards all the points in $T_N$ above $x$ were put into $T_N$. The immediate successors were put in stage $n+1$, the successors of these successors in stage $n+2$ and so on.

Consider the immediately next stage. There are two possibilities:

(1) $x$ was declared RED, i.e., we had an $a_x$ above $x$ such that $a_x \equiv x$ and all points $y$ above $a_x$ have the property that $a_x \equiv y$. Now, in **M**, $[\beta \to \gamma]_x = 0$; since this sentence is in $\Theta$ and $a_x \equiv x$, we have $[\beta \to \gamma]_{a_x} = 0$ in **M**; so for some $y$ above $a_x (a_x \leqslant_M y)$ we have, in **M**, $[\beta]_y = 1$, $[\gamma]_y = 0$, but $x \equiv a_x \equiv y$; so $[\beta]_x = 1$ in **M**, $[\gamma]_x = 0$ in **M**. So we are finished by the induction hypothesis.

(2) $x$ was not declared RED. So $a_x \equiv x$, and so $[\beta \to \gamma]_{a_x} = 0$ in **M**. So we have a $y$ above $a_x$ such that $[\beta]_y = 1$, $[\gamma]_y = 0$. Thus $y$ is either $\equiv a_x$ or $y$ fulfills that, for some $v \in K_x$, $vRy$ (by (50)). Now if $y \equiv a_x$ we get that $[\beta]_{a_x} = [\beta]_x = 1$ and $[\gamma]_{a_x} = [\gamma]_x = 0$ and we are finished. If $vRy$ then clearly $[\beta \to \gamma]_v = 0$ in **M**.

In this later case we repeat the whole process (since $v \in T_N$) with respect to $v$. And again *either* we get that $[\beta]_v = 1$ and $[\gamma]_v = 0$, in which case by the induction hypothesis and the fact that $x \leqslant_N v$ we get $[\beta \to \gamma]_x = 0$ in **N**, *or* we find another $v_1 \in T_N$, $v \leqslant_N v_1$, with $[\beta \to \gamma]_v = 0$. Now since $T_N$ is finite there must be a stage $k$ where $[\beta]_{v_k} = 1$ and $[\gamma]_{v_k} = 0$.

This concludes one direction of the implication case.

*Case* 2. $[\beta \to \gamma]_x = 1$ in **M**. Let $x \leqslant_N y$. Since by construction $x \leqslant_N y$ implies $xRy$ we get by the definition of $R$ that $[\beta \to \gamma]_y = 1$ in **M**. Now if we assume that $[\beta]_y = 1$ in **N** we get by the induction hypothesis that $[\beta]_y = 1$ in **M** and therefore $[\gamma]_y = 1$ in **M** and so $[\gamma]_y = 1$ in **N**. This shows that indeed $[\beta \to \gamma]_x = 1$ in **N**.

The case of negation follows from the case of implication.

We therefore have completed the proof of Lemma 57 and

COROLLARY 58. $[C]_0 = 0$ *in* **N**.

Now (52) and Corollary 58 yield

CONCLUDING THEOREM 59. *There exists a sequence* $D_1 \supsetneq D_2 \supsetneq \dots$ *of logics such that*

(1) $D_n$ *is a finitely axiomatizable, proper extension of* $h$

(2) $D_n$ *has the disjunction property, and*

(3) $\bigcap_n D_n = h$

## 2. THE KREISEL–PUTNAM SYSTEM

DEFINITION 1. Let $KP$ be the extension of $h$ with the schema:

(2) $$(\neg p \to (q \vee r)) \to ((\neg p \to q) \vee (\neg p \to r)).$$

*Remark* 3. As we shall see later, $KP$ has the disjunction property. It was originally introduced as an example of a proper extension of $h$ having the disjunction property. Another such example is the extension of $h$ with Scott's axiom:

(4) $((\neg\neg p \to p) \to (\neg p \vee p)) \to \neg p \vee \neg\neg p.$

The *KP* axiom (2) is the analogue of the disjunction property itself for negated sentence.

We have seen in the exercise in Chapter 3, that if we take the schema

(5) $(p \to (q \vee r)) \to (p \to q) \vee (p \to r)$

We get the logic *lc*. *lc* does not have the disjunction property as it can also be axiomatised by $(p \to q) \vee (q \to p)$.

*Remark* 6. Note that the rule corresponding to the Axiom (2), that is:

(7) $\vdash \neg p \to q \vee r$ implies $\vdash \neg p \to q$ or $\vdash \neg p \to r$

is valid as a rule of *h* while the rule

(8) $\vdash p \to q \vee r$ implies $\vdash p \to q$ or $\vdash p \to r$ is not valid in *h*.

*Remark* 9. It has been shown that $KP^+$, the positive fragment of *KP* is identical with $h^+$.

We now proceed to describe the class of partially ordered sets which characterise *KP*. We need some definitions.

Let $(T, \leqslant)$ be a partially ordered set with a first element 0. Let $E \subseteq T$. Define:

(10) $E^+ = \{y \in T \mid (\exists x \in E)(x \leqslant y\},$
$E^- = \{y \in T \mid (\exists x \in E)(y \leqslant x\}.$

(11) $T$ is said to have property # iff for every $E \subseteq T$ we have that $T - (E^+)^-$ is *either* empty *or* has a first element.

THEOREM 12. *Let* $(T, \leqslant)$ *be a partially ordered system fulfilling* (13) *below, then any structure with* $(T, \leqslant)$ *as a set of possible worlds validates the Kreisel-Putnam system*:

(13) *For any* $z \in T$ *the partially ordered system* $\{z\}^+$ *has the property #. (In particular* $\{0\}^+ = T$ *has this property.)*

*Proof.* Let **A** be a Kripke structure with $(T, \leqslant)$ as a set of possible worlds and let $(T, \leqslant)$ satisfy (13). We assume that $[\neg p \to (q \vee r)]_z = 1$ for some $z \in T$, we shall now show that either $[\neg p \to q]_z = 1$ or $[\neg p \to r]_z = 1$.

To do this let us consider the partially ordered system $\{z\}^+$. By our assumptions this set fulfills #. Let us consider a particular subset $E = \{y \geq z \mid [p]_y = 1\}$. Clearly $E^+ = E$ and thus $E^{+-} = E^-$. Now let us look at $\{z\}^+ - E^-$. If this set is empty then both $(\neg p \to q)$ and $(\neg p \to r)$ hold at $z$. Otherwise by # this set has a first element $a$, $a \notin E^-$, $a \in \{z\}^+$. Since $z \leq a$ we have that $[\neg p \to (q \vee r)]_a = 1$ and since $a \notin E^-$ implies that $[\neg p]_a = 1$ we get that $[q \vee r]_a = 1$. We may therefore assume that $[q]_a = 1$.

We claim now that $[\neg p \to q]_z = 1$. We shall prove it by contradiction. If $[\neg p \to q]_z = 0$ then, by definition, for some $b \geq z$ we have $[\neg p]_b = 1$ and $[q]_b = 0$. But $[\neg p]_b = 1$ implies that $b \notin E^-$ and hence $a \leq b$ which contradicts $[q]_b = 0$.

We therefore conclude that these partially ordered systems are adequate for our logic. In fact we shall prove the completeness theorem in the following paragraphs.

In what follows we shall use a modification of a method of Segerberg to prove both the completeness and the finite model property for the Kreisel-Putnam system; another modification of this method is used to find a sequence of logics with the disjunction property which are finitely axiomatizable.

THEOREM 14. *The Kreisel-Putnam logic is complete for the class of all Kripke structures with a finite partially ordered set with the property* (13).

To prove this, we need some machinery. Let *KP* be the set of all substitution instances of theorems of the Kreisel-Putnam logic. Then we have that for any $C$ the following holds:

(15) $C$ is not provable in the system $KP \Leftrightarrow$ the theory $(KP, \{C\})$ is $h$-consistent $\Leftrightarrow$ there exists a (general) Kripke structure at which all sentences of *KP* hold and $C$ does not hold.

Our strategy is as follows: Let $C$ be a nonprovable sentence then there exists a structure validating *KP* at which $C$ is false. Let $(T, \leq)$ be the partially ordered set associated with the structure. We shall derive from $(T, \leq)$ a finite partially ordered set $(S, R)$ which fulfills (13) and define a structure on it such as to have $C$ false.

Before we go to the actual construction let us recall Theorem 4.3.15.

THEOREM 16 (*Diego-McKay*). *Let $p_1, \ldots, p_k$ be propositional variables and let $\Theta$ be the set of all sentences built out of $p_1, \ldots, p_k$ using the connectives $\neg, \wedge, \rightarrow$ then there exist only a finite number of nonequivalent sentences in $\Theta$ (over the intuitionistic logic).*

We now continue our construction. Let **A** be a structure validating $KP$ at which $C$ is false. Let $(T, \leqslant)$ be the partially ordered set of the structure.

(17) Let $\Theta_0$ be the set of all subformulas of $C$. Let $\Theta$ be the closure of $\Theta_0$ under the connectives of negation, conjunction and implication.

Define for $x, y \in T$

(18) $xRy$ iff for all $\gamma \in \Theta$ we have $[\gamma]_x = 1 \Rightarrow [\gamma]_y = 1$.

(19) $x \equiv y$ iff $xRy \wedge yRx$.

Clearly $\equiv$ is an equivalence relation. Moreover, in view of (17) and (18) $T$ is divided into a *finite* number of equivalence classes. Let $S$ be the set of all these classes.

Our next step is to show that $(S, R)$ fulfills property (13).

Let $z/\equiv$ be a given equivalence class. We want to show that $\{z/\equiv\}^+$ fulfills $\#$.

Let $E \subseteq \{z/\equiv\}^+$. Let us regard $E^+$. Now since $S$ is finite and $E^+$ has the property that $x \in E^+$ and $xRy$ imply that $y \in E^+$ we can choose points $y_1/\equiv, \ldots, y_k/\equiv$ such that $z/\equiv Ry_i/\equiv$ for $i = 1 \cdots k$ and for all $b/\equiv$ we have that $b/\equiv \in E^+$ iff $y_i/\equiv Rb/\equiv$ for some $1 \leqslant i \leqslant k$. We may also assume that all of $y_1/\equiv \cdots y_k/\equiv$ are $R$ incompatible, for otherwise we can drop the unnecessary $y_i/\equiv$'s.

Now let us regard the set $\{z/\equiv\}^+ - E^{+-}$. This set is either empty, in which case we are done, or is nonempty. In the latter case we have that $\{z/\equiv\}^+ - E^{+-} = (\{z/\equiv\}^+ - E^{+-})^+$ holds and so again we have that this set is *determined* by some $x_1/\equiv \cdots x_m/\equiv$, i.e. $b/\equiv \in \{z/\equiv\}^+ - E^{+-}$ iff $x_j/\equiv Rb/\equiv$ for some $j$. (Of course $z/\equiv Rx_j/\equiv$ holds!) We may again assume that $x_1/\equiv \cdots x_m/\equiv$ are not $R$-compatible.

(20) Clearly for no $v/\equiv$ do we have that both $x_j/\equiv Rv/\equiv$ and $y_i/\equiv Rv/\equiv$ for some $i, j$. In view of this let $x/\equiv$ be such that $x_j/\equiv Rx/\equiv$ for some $j$. So there exists a $\gamma_{i,x} \in \Theta$ such that $[\gamma_{i,x}]_{y_i} = 1$, $[\gamma_{i,x}]_x = 0$.

Let $p_i = \bigwedge_x \gamma_{i,x}$. We have that $p_i \in \Theta$ since $\Theta$ is closed with respect to $\wedge \rightarrow$ and $\neg$.

(21) Now let $p = \bigvee_i p_i$. Since we have that $\vdash_h \neg(A \vee B) \Leftrightarrow \neg A \wedge \neg B$ we get that $\neg p$ is equivalent to a sentence $p_* \in \Theta$.

(22) Similarly, for any class $b/\equiv$ such that we do not have $x_j/\equiv\ Rb/\equiv$ there exists an $\gamma_b \in \Theta$ such that $[\gamma_b]_{x_j} = 1$ and $[\gamma_b]_b = 0$. Let $q_j = \bigwedge_b \gamma_b$.

Clearly $q_j \in \Theta$.

LEMMA 23. $[\neg p \rightarrow \bigvee_j q_j]_z = 1$.
(Where $z \in T$ is a representative of $z/\equiv$.)

*Proof.* If this is not true then there exists a $b$ such that $b \geqslant z$ and $[\neg p]_b = 1$ and $[q_j]_b = 0$ for all $j$. Let us regard $b/\equiv$. Now since $\vdash_h \neg p \leftrightarrow p_*$ and $p_* \in \Theta$ (see (21) above) we have that for no $v/\equiv\ \in E^+$ does $b/\equiv\ Rb/\equiv$ hold, since by (20) we would have $[p]_v = 1$ which contradicts $[p_*]_b = 1$.

We conclude that $b/\equiv\ \notin E^{+-}$ and therefore by the definition of $x_1/\equiv, \ldots, x_m/\equiv$ we have $x_j/\equiv\ Rb/\equiv$ for some $j$. But this implies by (22) that $[q_j]_b = 1$ which is a contradiction.

Now to complete the proof observe that since $[\neg p \rightarrow \bigvee q_j]_z = 1$ we have that $[\bigvee_j(\neg p \rightarrow q_j)]_z = 1$. Therefore we may assume that $[\neg p \rightarrow \bigvee q_j]_z = 1$. Since $\vdash_h \neg p \leftrightarrow p_*$ we get $[p_* \rightarrow q_j]_z = 1$. Now since $(p_* \rightarrow q_j) \in \Theta$ and $z/\equiv\ Rx_j/\equiv$ we have that $[p_* \rightarrow q_j]_{x_j} = 1$. To get a contradiction, observe now that $[\neg p]_{x_j} = 1$ for otherwise for some $v \geqslant x_j$, $[p]_v = 1$ and then $v/\equiv$ fulfills $x_j/\equiv\ Rv/\equiv$ and $[p]_v = 1$ which contradicts the definition of $p$ in (20) and (21).

Thus we finally have that $[\neg p]_{x_j} = 1$, $[q_j]_{x_j} = 0$ and $[p_* \rightarrow q_j]_{x_j} = 1$ which is a contradiction.

We must therefore have that $m = 1$, i.e. there exists only one $x_1/\equiv$.

(25) Define for a propositional variable $\gamma \in \Theta_0$, $[\gamma]_{x/\equiv} = [\gamma]_x$.

LEMMA 26. *For any* $B \in \Theta_0$ *we have* $[B]_{x/\equiv} = [B]_x$.

*Proof.* For propositional $B$ this is the definition (and the definition is legitimate since $B \in \Theta$!).

The cases of $\wedge$, $\vee$ present no difficulties.

Assume $[B \to B']_x = 1$ then since $B \to B' \in \Theta$ we get that for any $b/\equiv$ such that $x/\equiv Rb/\equiv$ we have $[B \to B']_b = 1$. So if $[B]_b = 1$ we would get $[B']_b = 1$ and so by the induction hypothesis, we get that if $[B]_{b/\equiv} = 1$ then $[B']_{b/\equiv} = 1$.

Assume that $[B \to B']_x = 0$. Then for some $b \geqslant x$ we have that $[B]_b = 1$, $[B']_b = 0$ and then by the induction hypothesis $[B]_{b/\equiv} = 1$, $[B']_{b/\equiv} = 0$ and we have that $x/\equiv Rb/\equiv$. The case of $\neg$ is similar to the case of $\to$.

Now since $C \in \Theta_0$ and $[C]_0 = 0$ so is $[C]_{0/\equiv}$. Therefore we found a structure $(S, R)$ – as required – where $C$ is false.

In this section we shall use the interpretation we have provided for the Kreisel-Putnam logic to show that it has the disjunction property. We begin with a construction.

Let $S$ and $S^*$ be two finite partially ordered systems fulfilling (13). We shall construct another partially ordered system $T$, which also fulfills (13). If $B$ is refuted in $S$ and $C$ in $S^*$ then $T$ will refute $B \vee C$.

Let $V = \{v_1, \ldots, v_n\}$ and $U = \{u_1, \ldots, u_m\}$ be the sets of endpoints of $S$ and $S^*$ respectively. Let $T = S \cup S^* \cup K$ where

$$K = \{a \subseteq V \cup U \mid a \cap V \neq \emptyset \text{ and } a \cap U \neq \emptyset\}.$$

We now turn to define a partial ordering, $\leqslant$, on $T$.

(30)(i) For $a, b \in K$ let $a \leqslant b \Leftrightarrow a \supseteq b$.

(ii) If both elements are in $S$ let the order be as of $S$. If both elements are in $S^*$ let the order be as of $S^*$.

(iii) Let $a = A \cup B$, $A \subseteq V$, $B \subseteq U$. We now proceed to define for what elements $t \in S \cup S^*$ do we have $a \leqslant t$

Let $s_a$ and $s^*_a$ be the *first* elements of $S - (V - A)^-$ and $S^* - (U - B)^-$ respectively, then let $a \leqslant t$ hold for all $t$ such that $s_a \leqslant t$ or $s^*_a \leqslant t$.

Note that the ordering just defined is transitive since

(31) $\quad a \supseteq b \Rightarrow s_a \leqslant s_b \quad \text{and} \quad s^*_a \leqslant s^*_b.$

LEMMA 32. *Let $s$ be an endpoint of $S$, and let $a \in K$ then $s \in a \Leftrightarrow a \leqslant s$.*

*Proof.* Let $a, b \in K$. Since $a \leqslant b$ implies $s_a \leqslant s_b$ we conclude that $a \leqslant s$ iff $s_a \leqslant s$. Since $s$ is an endpoint we conclude that $s_a \leqslant s \Leftrightarrow s \notin V - A \Leftrightarrow s \in A$.

LEMMA 33. $(T, \leqslant)$ *fulfills* (13).

*Proof*. Clearly, # must hold for points $t \in S$ or $t \in S^*$.

Let $a = A \cup B \in K$. We shall show that $\{a\}^+$ has property #. Let $E^+ \subseteq \{a\}^+$ and let $C \cup D$ be the endpoints of $E^+$. ($C \subseteq S$ and $D \subseteq S^*$.) Consider the element $b = (A - C) \cup (B - D)$. We distinguish several cases.

*Case* 1. $A - C \neq \emptyset$ and $B - D \neq \emptyset$.

Since for any endpoint $t$, $t \in b$ iff $b \leqslant t$ we get that $b \in \{a\}^+ - E^{+-}$. We shall now show that $b$ is the first element of this set. Let $t$ be an element such that $\{t\}^+ \cap E^{+-} = \emptyset$. We shall show that $b \leqslant t$. If $t \in K$ this follows from (32).

Assume that $t \in S$. Then the endpoints of $\{t\}^+$ are all in $A - C$ and so $\{t\}^+ \cap [V - (A - C)]^- = \emptyset$ which means that $s_b \leqslant t$.

*Case* 2. $A - C = \emptyset$.

In this case $E^{+-} \supseteq S$ and so we take as our $b$ the first element in $S^*$ of $S^* - E^{+-}$.

LEMMA 34. *Let* $t \in S \cup S^*$ *and let* $a \in K$ *then it is not the case that* $t \leqslant a$.

*Proof*. By definition.

THEOREM 35. *The Kreisel-Putnam logic has the disjunction property.*

*Proof*. Let $\not\vdash B$ and $\not\vdash C$; we shall show that $\not\vdash B \vee C$.

Let $B$ be refuted in a Kripke structure with a partially ordered system $S$ and let $C$ be refuted in a Kripke structure with a system $S^*$. Consider the following Kripke structure with the partially ordered system $T$: For any propositional variable let

$$\begin{aligned}[p]_t &= 0 \quad \text{for } t \in K,\\ &= \text{the value given in } S(S^*) \text{ for } t \in S (t \in S^*) \\ &\quad\ \text{respectively.}\end{aligned}$$

By (34) we get a Kripke structure. In this structure clearly $B \vee C$ cannot hold.

In this section we apply the Diego theorem to study a wide class of extensions of the intuitionistic propositional calculus.

DEFINITION 36. A formula $B$ is said to be *absolute* iff it is built up from a finite number of *negated* propositional variables by means of conjunction, implication and negation.

THEOREM 37. *Any extension of the intuitionistic propositional calculus with a (not necessarily finite) set of absolute formulae has the finite model property.*

*Proof.* Before we begin with the proof we need some definitions.

Let us define a translation from the intuitionistic propositional calculus into the monadic second order theory of partially ordered systems

$$(38) \qquad (p)_x = x \in C_p$$

where $C_p$ is a set variable and $x$ is an element variable of the theory of partially ordered sets.

$$(A \wedge B)_x = (A)_x \wedge (B)_x$$

$$(A \rightarrow B)_x = (\forall y \geqslant x)((A)_y \rightarrow (B)_y)$$

where the free variable of $(A)_y$ and of $(B)_y$ is $y$.

$$(A \vee B)_x = (A)_x \vee (B)_x.$$

We now turn to the proof of the theorem. Let $S \supseteq h$ be an extension of $h$ with absolute formulae only as axiom schemes. Let $A$ be a nonprovable formula of $S$. Let $A$ be refuted in a structure of $S$ with partially ordered set $(T, \leqslant)$ and let $\Theta_0$ be the set of all subformulae of $A$. Let $\Theta$ be the closure of $\Theta_0$ under conjunctions, negations and implications. We know that $\Theta$ is finite. Define $xRy$ iff $\forall \alpha \in \Theta([\alpha]_x = 1 \rightarrow [\alpha]_y = 1)$; $x \equiv y$ iff $xRy \wedge yRx$. $T$ is divided into a finite number of equivalence classes. Call this set $U$. We turn $(U, R)$ into a structure in the same manner as we did in the Kreisel-Putnam case (we need only consider propositional variables that appear in $\Theta$). We want to show that $(U, R)$ validates $S$. It is sufficient to show that the following is true in $(U, R)$

$$(39) \qquad \forall C_1 \ldots \forall C_n (\textstyle\bigwedge_i \forall x \forall y (x \in C_i \wedge xRy \rightarrow y \in C_i) \rightarrow \forall y (\alpha)_y)$$

where $\alpha$ is any absolute axiom scheme with $n$ propositional variables.

To show that this is indeed the case, recall that for any $\beta \in \Theta$ we have $[\beta]_{x/\equiv} = [\beta]_x$ for all $x \in T$. This fact together with the following observation (40) would yield the result:

(40) Let $C \subseteq U$, be a subset such that $\forall u \forall v(u \in C \wedge uRv \to v \in C)$. Then by what we did in the Kreisel-Putnam case there exists a formula $\beta = \bigvee_C \gamma_i$, $\gamma_i \in \Theta$ such that $C = \{u \in U \mid [\beta]_u = 1\}$.

Now since $\alpha$ is an absolute formula and $\neg\beta \in \Theta$ we get that $\alpha(\beta, \beta', \ldots) \in \Theta$, and therefore since $(T, \leqslant)$ validates $S$ we get that (39) holds in $(U, R)$.

COROLLARY 41. *S is complete for the class of all finite partially ordered sets that fulfill* (39) *for every axiom in S.*

COROLLARY 42. *Theorem* 37 *and Corollary* 41 *hold for disjunctions of absolute formulae as well.*

Medvedev considered the following extension of the Kreisel-Putnam logic. It is the system defined by all the partially ordered sets of the form $\{s \subseteq A \mid s \neq \emptyset$, $A$ a finite set with $s \leqslant s'$ iff $s \supseteq s'$.

This logic will appear later in connection with validity in post systems. It is the same logic as mentioned in conjecture 2 section 7 chapter 13. See also 4.11.

## 3. An Intermediate Logic without the F.M.P.

DEFINITION 1. Let $p$, $q$, $r$, $s$ be atoms. Define a sequence of wffs:

(a) Let $A_0 = p$, $B_0 = q$, $C_0 = r$

(b) Let $A_{i+1} = \neg s \wedge A_i \to B_i \vee C_i$

$B_{i+1} = \neg s \wedge B_i \to A_i \vee C_i$

$C_{i+1} = \neg s \wedge C_i \to A_i \vee B_i$

DEFINITION 2. Let $kf$ be the extension of $h$ with the schema

$(kf)$ $(A_2 \vee B_2 \vee C_2) \to (A_1 \vee B_1 \vee C_1 \vee \neg s)$

where $A_i$, $B_i$, $C_i$ are defined in (1) above.

DEFINITION 3. Define the following structure $(S, R, O, D)$ for a language with $p$, $q$, $r$, $s$.

$S$ is as in the diagram.

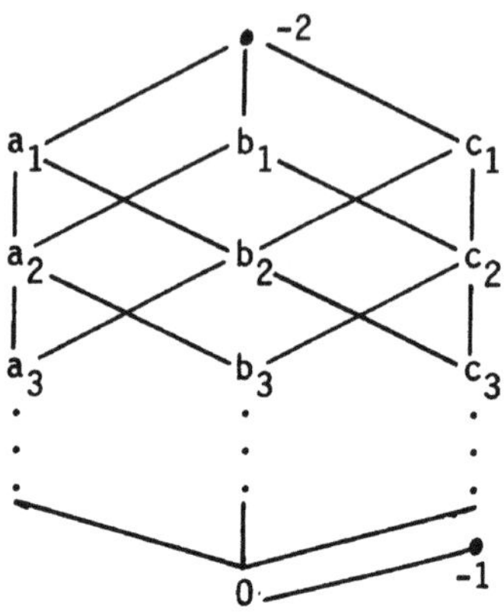


i.e., $S = \{0, -1, -2\} \cup \{a_n, b_n, c_n \mid n \in \omega\}$.

$R$ is as shown in the diagram, i.e., $R$ is reflexive and transitive and $a_1 R - 2$, $b_1 R - 2$, $c_1 R - 2$, $a_{n+1} R a_n$, $a_{n+1} R b_n$, $b_{n+1} R a_n$, $b_{n+1} R c_n$, $C_{n+1} R b_n$, $C_{n+1} R c_n$, and of course, 0 is the base point, and $0 R - 1$.

To define $D$, let $[p]_t = 1$ iff $t = a_i$ or $t = -2$. $[q]_t = 1$ iff $t = b_1$ or $t = -2$, $[r]_t = 1$ iff $t = c_1$ or $t = -2$ and $[s]_t = 1$ iff $t = -1$.

LEMMA 4. (a) *In* $(S, R, O, D)$, $[A_1 \vee B_1 \vee C_1 \vee \neg s]_0 = 0$.

(b) $(S, R, 0)$ *validates* $(kf)$ *for any assignment* $D'$.

*Proof.* (a) Easy to see, as e.g., $[A_1]_{a_1} = 0$ since $[\neg s \wedge p]_{a_1} = 1$ and $[q \vee r]_{a_1} = 0$. Similarly for $[B_1]_{b_1} = [C_1]_{c_1} = 0$. Also $[s]_{-1} = 1$.

(b) We show that any substitution instance of $(kf)$ is valid in $(S, R, 0, D')$. Since the variables of the schema $(kf)$ are atomic, it is sufficient to check for the wff $(A_2 \vee B_2 \vee C_2) \rightarrow (A_1 \vee B_1 \vee C_1 \vee \neg s)$. Let $t \in S$ be such that the antecedent of the above holds and the consequent is false. Thus

$$[A_1]_t = [B_1]_t = [C_1]_t = [\neg s]_t = 0.$$

So there exist $t_1, t_2, t_3, t_4$ with $tRt_j$, $j = 1 \ldots 4$ and $[\neg s \wedge p]_{t_1} = 1$,

$$[q \vee r]_{t_1} = 0, \qquad [\neg s \wedge q]_{t_2} = 1, \qquad [p \vee r]_{t_2} = 0,$$
$$[\neg s \wedge r]_{t_3} = 1, \qquad [p \vee q]_{t_3} = 0, \qquad [s]_{t_4} = 1.$$

The above shows that the points $t_i$ $i = 1 \ldots 4$ are all $R$-incomparable, and hence $t_4 = -1$, and for some $n$, $t_1 = a_n$, $t_2 = b_n$, $t_3 = c_n$ and $t = 0$.

But then if this is the situation,

$$[A_2]_{a_{n+1}} = 0, \qquad [B_2]_{b_{n+1}} = 0, \quad \text{and} \quad [C_2]_{c_{n+1}} = 0,$$

a contradiction.

LEMMA 5. *Let* $(S_0, R_0, *, D_0)$ *be any structure which validates* $(kf)$ *and in which* $[A_1 \vee B_1 \vee C_1 \vee \neg s]_* = 0$, *then* $S_0$ *is infinite.*

*Proof*. First observe by induction that for any $i > 0$, $[A_i \vee B_i \vee C_i]_* = 0$. This holds for $i = 1$. Assume it holds for $i$, show for $i+1$. But $(A_{i+1} \vee B_{i+1} \vee C_{i+1}) \rightarrow (A_i \vee B_i \vee C_i \vee \neg s)$ is a substitution instance of schema $(kf)$ and must hold at $*$ and hence the case for $i+1$ follows.

Since $[A_i]_* = 0$, for all $i$, there exist points $t_i$ such that $*R_0 t_i$ and $[\neg s \wedge A_{i-1}]_{t_i} = 1$ and $[B_{i-1} \vee C_{i-1}]_{t_i} = 0$.

We now show that for $i > j$, $t_i \neq t_j$. Since $[A_{j-1}]_{t_j} = 1$, $[B_j \wedge C_j]_{t_j} = 1$, and hence $[B_{j+1} \wedge C_{j+1}]_{t_j} = 1$ continuing in this manner we get $[B_{i-1} \wedge C_{i-1}]_{t_j} = 1$ and hence $[A_i]_{t_j} = 1$ and hence $t_j \neq t_i$.

Thus $S_0$ is infinite.

## 4. Exercises

**Exercise 1.** Let $s_n$ be the extension of $lc$ with the schema

$$\bigvee_{0 \leq i < j \leq n} A_i \leftrightarrow A_j.$$

$$\text{Let} \quad s_\omega = \bigcap_{n \geq 1} s_n$$

(a) Show that $s_\omega = lc$.

(b) Show that $s_n$ is characterised by a finite linear chain of $n$ elements.

(c) If $x$ is any intermediate logic between $c$ and $lc$ then $x = s_n$ for some $n$.

**Exercise 2.** Define a sequence of wffs.

$$A_1 = p_1 \vee \neg p_1 \qquad (\text{or } p_1 \vee (p_1 \rightarrow p_0)$$

$$A_{n+1} = p_{n+1} \vee (p_{n+1} \rightarrow A_n).$$

Let $b_n$ be the extension of $h$ with $A_n$. $b_\omega = \bigcap b_n$.

(a) Show that $b_\omega = h$.

(b) Show that $b_n$ is characterised by the class of all partially ordered sets in which every increasing chain has at most $n$-elements. In such a case we say the structure is of height $\leqslant n$.

(c) Show $b_n$ has f.m.p.

**Exercise 3.** Nagata sequence.
Let $P_1 = ((p_1 \to p_0) \to p_1) \to p_1$

$$P_{n+1} = ((p_{n+1} \to P_n) \to p_{n+1}) \to p_{n+1}$$

Show that $b_n$ of the previous exercise can be obtained by adding $P_n$ to $h$.

**Exercise 4.** We say an intermediate logic is in the *n-th slice* iff it is complete for a class of structures with height $\leqslant n$ but not complete for any class of structures of height $\leqslant n-1$.

(a) Show that $x$ is in the $n$-th slice iff $b_n \subseteq x \subseteq s_n$.

(b) Show that the $n$-th slice, $n \geqslant 2$ is infinite.

(c) Show that any intermediate logic belongs to some slice.

(d) Let $x$ be in the $n$-th slice, $n$ finite, then $x$ has f.m.p. iff it is complete for a class of partially ordered sets.

**Exercise 5.** Let $B_n = \bigvee_{i=1}^{n} (P_i \to \bigvee_{j \neq 1} P_j)$.

Let $W_n$ be the extension of $h$ with $B_n$. $W_\omega = \bigcap_n W_n$.

(a) Show that $W_n$ is characterised by the class of all partially ordered sets in which every anti-chain has at most $n$-elements.

(b) $W_n$ has f.m.p.

(c) Is $W_2$ equal to the extension of $h$ with $\bigvee_{i \neq j} P_i \to P_j$, $i, j = 1, 2, 3$?

**Exercise 6.** (a) Show that $kc$ is obtained from $h$ by adding the schema

$$(\neg A \to \neg B) \vee (\neg B \to \neg A)$$

(b) Let $D_n$ be the schema $\bigvee_{i=0}^{n} (\neg P_i \to \bigvee_{j \neq i} \neg P_j)$.
Let $d_n$ be the extension of $h$ with $\bigwedge_{0 \leqslant i < j \leqslant n} \neg\neg(p_i \vee p_j) \to D_n$.
Let $d_\omega = \bigcap d_n$.

(c) Show that $d_n$ is complete for the class of all partially ordered sets with $\exists x_1, \ldots, x_n \forall y (\bigvee_i yRx_i)$.

(d) Show $d_n$ has f.m.p.

(e) Show that $d_\omega = h$.

**Exercise 7.**

(a) Say that $x$ is in the $n$-th counterslice iff $W_n \subseteq x$ and not $W_{n-1} \subseteq x$.

(b) Say that $x$ is in the $(m, n)$ box iff it is in the $m$-th slice and $n$-th counterslice.

(c) Show that the $(m, n)$ box has a first element.

(d) For $m > 2$, the $(m, n)$ box is infinite.

(e) The $(2, n)$ box has only one element. What is it?

(f) Classify the $(3, 1)$ box.

(g) For $n > 2$, show that the $n$-th counterslice has no greatest element.

(h) Show that if $x$ has the disjunction property, then $x$ is not a member of any $(m, n)$ box.

**Exercise 8.** Let $x$ be the extension of $h$ with the schema

$$((\neg\neg A \to A) \to (\neg A \vee A)) \to \neg A \vee \neg\neg A.$$

Show that $x$ has the disjunction property. Does it have f.m.p.?

**Exercise 9.** Let $x$ be the extension of $h$ with the schema

$$((\neg\neg A \to A) \to (\neg A \vee \neg\neg A)) \to \neg A \vee \neg\neg A.$$

Does $x$ have the disjunction property? f.m.p.? Can you find a condition $F$ on partially ordered sets such that $x$ is characterised by $F$?

**Exercise 10.** Show that $KP^+$ (the fragment of $KP$ without $\vee$) is $h^+$.

**Exercise 11.** Let $T$ be a finite nonempty set. Let $S_T = 2^T - \{\emptyset\}$. Consider the logic $y$ determined by the class of all $S_T, \supseteq, T)$, $T$ finite.

(a) Show that $KP \subseteq y$.

(b) Show $x \subseteq y$, where $x$ is the logic of Exercise 8.

(c) Show (M. K. Andrews) that the following schema is strictly stronger than $KP + X$ of Exercise 8 and is a theorem of $y : ((\neg p \to s) \to q \vee r) \to ((\neg p \to s) \to q) \vee (\neg\neg p \to r)$.

(d) Is $h^+$ the disjunctionless fragment of $y$?

**Exercise 12.** Answer questions (1)–(6) of Section 1. Use results in this and the following chapters.

## NOTES

Based on two of the author's papers, one of them jointly with de Jongh and one paper of K. Fine (Section 3). Algebraic methods yielding many intermediate logics without f.m.p. were used by Cerciu-Kuznezov. M. K. Andrews has studied the Medvedev system.

## CHAPTER 6

# FORMULAS IN ONE VARIABLE

In Chapter 4 we saw that in $h^+$, there is only a finite number of inequivalent wffs built up from $n$ atomic wffs. If we add disjunction, we can show that there are infinitely many wffs built up from one atom. We shall also see that the fragment of wffs of $h^+$ with one atom is complete for finite binary trees.

## 1. Nishimura lattice

DEFINITION 1. Define the following sequence of wffs in one variable $p$:

$$P_0(p) = p \wedge \neg p$$

$$P_\infty(p) = p \to p$$

$$P_1(p) = p$$

$$P_2(p) = \neg p$$

$$\vdots$$

$$P_{2n+3}(p) = P_{2n+1}(p) \vee P_{2n+2}(p)$$

$$P_{2n+4}(p) = P_{2n+3}(p) \to P_{2n+1}(p).$$

$$\cdots\cdots\cdots\cdots\cdots\cdots\cdots\cdots\cdots$$

*Remark* 2. Let $(T, R, 0, D)$ be a structure for one atomic variable $p$. This structure can be described by writing down $(S, R, 0)$ and beside each node $x \in S$, writing down T or F, namely, whether $p$ is supposed to be true or false at $x$.

DEFINITION 3. We define the following sequence of structures for a language with one atom:

$$\chi_1 = \quad \bullet \ \mathrm{T}$$

$$\chi_2 = \quad \bullet \ \mathrm{F}$$

$\chi_3 =$ 

$\chi_n =$ 

*Example*

$\chi_4 =$ 

$\chi_5 =$ 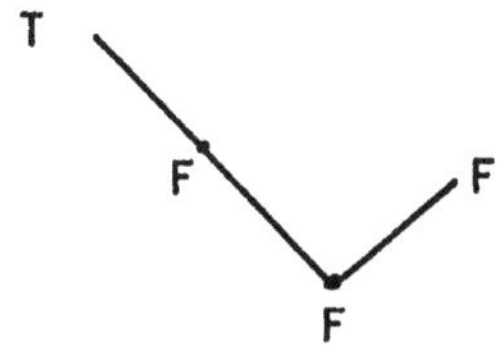


$\chi_7 =$ 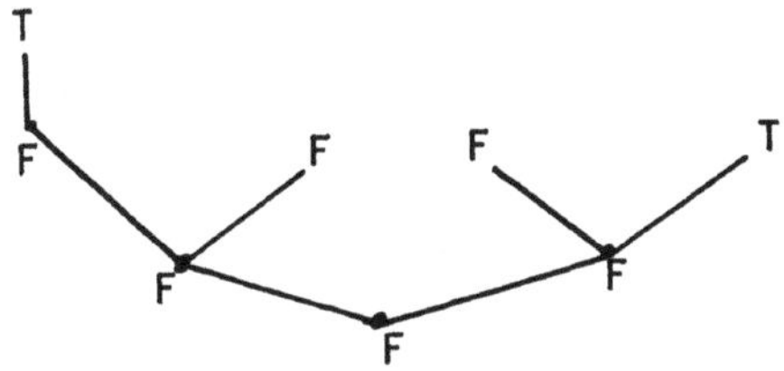


LEMMA 4. *If $m > i + 1$ then*

(a) *$\chi_i$ is a branch (substructure) of $\chi_m$.*

(b) *If $A(p)$ is false at $\chi_n$ is false at any $\chi_m$, $m \geqslant n + 1$.*

*Proof.* Easy.

DEFINITION 5. Let $\chi^*$ be the following structure

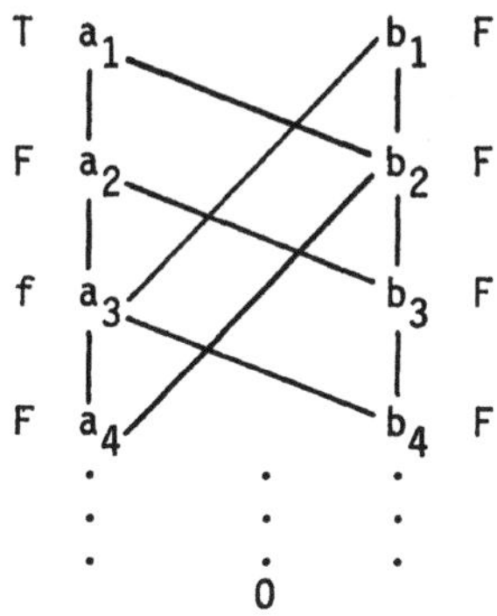


Let $\chi^*_{2n-1}$, $n = 1, 2, \ldots$ be the structure obtained from $\chi^*$ by taking all the points above $a_n$ and (including $a_n$) and let $\chi^*_{2n}$ be the structure obtained by taking all points above $b_n$ (including $b_n$).

LEMMA 6. *There exists a homomorphism $\alpha$ from $\chi_n$ onto $\chi^*_n$ (in the sense of Definition* 4.2.6).
*Proof.* By induction on $n$.

THEOREM 7. *Let A be any wff with one propositional variable p, then A is equivalent in h to a wff of the form $P_n$ of Definition* 1.
*Proof.* To show that every $A(p)$ is equivalent to a $P_n(p)$, we need only show that the set of formulae $P_n(p)$ is provably closed under the operations $\wedge$, $\vee$, $\rightarrow$, $\neg$.

First, notice that $P_k \vdash P_{2n+1}$ for $0 \leq k \leq 2n+1$. For, if $k = 2n$, $P_{2n+1} = P_{2n-1} \vee P_{2n}$ (unless $n = 0$ and trivially $P_0 \vdash P_1$), and $P_{2n} \vdash P_{2n+1}$. If $k < 2n$, $P_k \rightarrow P_{2n-1}$ and so $P_k \rightarrow P_{2n+1}$.

We now show $P_i \vdash P_j$ if $0 \leq i, j < \infty$ unless $i = 2n+1$ and $j = 2n+2$ or $i = 2n+2$ and $j = 2n+4$. Let $j = 2n+2$, $i = 2n-1$. Then

$$P_j = P_{2n+1} \rightarrow P_i, \quad \text{so} \quad P_i \vdash P_j.$$

Thus, for $k \leq i = 2n-1$, $P_k \rightarrow P_j$.

These comments now allow us to complete the proof of closure under $\wedge$, $\vee$, $\rightarrow$, $\neg$ by inspecting the cases.

First, consider $\neg$. Obviously, we have $\neg P_k = P_k \rightarrow P_0$. Thus, negation is handled by considering implication.

There are 16 cases for implication:

I1. $$\vdash (P_{2n+1} \rightarrow P_{2n+2}) \leftrightarrow P_{2n+2}.$$

For $n = 0$,

$$\begin{aligned} \vdash (P_1 \rightarrow P_2) &\leftrightarrow (p \rightarrow \neg p) \\ &\leftrightarrow \neg p \\ &\leftrightarrow P_2. \end{aligned}$$

For $n > 0$,

$$P_{2n+1} \rightarrow P_{2n+2} = P_{2n-1} \vee P_{2n} \rightarrow (P_{2n+1} \rightarrow P_{2n-1}).$$

But,

$$\vdash P_{2n+1} \rightarrow P_{2n+2}$$

(i.e., $P_{2n-1} \rightarrow (P_{2n+1} \rightarrow P_{2n-1})$), so

$$\vdash (P_{2n+1} \to P_{2n+2}) \leftrightarrow (P_{2n} \to (P_{2n+1} \to P_{2n-1}))$$
$$\leftrightarrow (P_{2n} \wedge P_{2n+1} \to P_{2n-1}).$$

But $P_{2n} \vdash P_{2n+1}$ and thus $\vdash P_{2n+1} \leftrightarrow P_{2n} \wedge P_{2n+1}$, whence

$$\vdash (P_{2n+1} \to P_{2n+2}) \leftrightarrow (P_{2n+1} \to P_{2n-1})$$
$$\leftrightarrow P_{2n+2}.$$

I2. $\vdash (P_{2n+2} \to P_{2n+1}) \leftrightarrow P_{2n+4}$.

We leave the proof to the reader.

I3. $\vdash (P_i \to P_j) \leftrightarrow P_\infty$, for $i \leqslant j$ except as in I1 and I2.

This was proved above.

I4. $\vdash (P_1 \to P_0) \leftrightarrow P_2$.

I5. $\vdash (P_2 \to P_0) \leftrightarrow P_4$.

I6. $\vdash (P_3 \to P_0) \leftrightarrow P_0$.

I7. $\vdash (P_4 \to P_2) \leftrightarrow P_2$.

I8. $\vdash (P_4 \to P_0) \leftrightarrow P_2$.

I4–I8 are obvious.

I9. $\vdash (P_i \to P_0) \leftrightarrow P_0$, for $i \geqslant 5$.

This is because $P_i$ is a classical tautology for $i \geqslant 5$.

I10. $\vdash (P_{2n+3} \to P_{2n+1}) \leftrightarrow P_{2n+4}$.

I11. $\vdash (P_{2n+2} \to P_{2n+1}) \leftrightarrow P_{2n+4}$.

By I3, $P_{2n+2} \vdash P_{2n+3}$, so $\vdash P_{2n+4} \to (P_{2n+2} \to P_{2n+1})$. For the converse, $P_{2n+3} = P_{2n+2} \vee P_{2n+1}$ and assume $P_{2n+3}$ and $P_{2n+2} \to P_{2n+1}$.

*Case* 1. $P_{2n+2}$. Then clearly $P_{2n+1}$.

*Case* 2. $P_{2n+1}$. Then even more clearly $P_{2n+1}$.

Thus, from $(A \to C) \wedge (B \to C) \vdash (A \vee B \to C)$, we have $P_{2n+3} \vdash P_{2n+1}$, i.e., $P_{2n+4}$.

I12. $\vdash (P_{2n+4} \to P_{2n+1}) \leftrightarrow P_{2n+6}$.

I13. $\vdash (P_{2n+5} \to P_{2n+1}) \leftrightarrow P_{2n+1}$.

I14. $\vdash (P_{2n+6} \to P_{2n+1}) \leftrightarrow P_{2n+4}$.

I15. $\vdash (P_i \to P_{2n+1}) \leftrightarrow P_{2n+1}$, for $i \geqslant 2n+7$.

I16. $\vdash (P_i \to P_{2n+2}) \leftrightarrow P_{2n+2}$, for $i \geqslant 2n+3$.

I12–I16 are left as exercises.

Conjunction requires only 5 verifications:

C1. $\vdash P_1 \wedge P_2 \leftrightarrow P_0$.

This is obvious.

C2. $\vdash P_{2n+3} \wedge P_{2n+4} \leftrightarrow P_{2n+1}$.

By I10, $(P_{2n+3} \to P_{2n+1}) \leftrightarrow P_{2n+4}$.

C3. $\vdash P_2 \wedge P_4 \leftrightarrow P_0$.

Again obvious.

C4. $\vdash P_{2n+4} \wedge P_{2n+6} \leftrightarrow P_{2n+1}$.

Use I12.

C5. $\vdash P_i \wedge P_j \leftrightarrow P_i$, for $i \leqslant j$ except as in C1–C4.

Use I3.

Disjunction only has 3 cases:

D1. $\vdash P_{2n+1} \vee P_{2n+2} \leftrightarrow P_{2n+3}$.

D2. $\vdash P_{2n+2} \vee P_{2n+4} \leftrightarrow P_{2n+5}$.

By definition, $\vdash P_{2n+5} \leftrightarrow P_{2n+3} \vee P_{2n+4} \leftrightarrow P_{2n+1} \vee P_{2n+2} \vee P_{2n+4}$. But $P_{2n+1} \vdash P_{2n+4}$, so $P_{2n+5} \leftrightarrow P_{2n+2} \vee P_{2n+4}$.

D3. $\vdash P_i \vee P_j \leftrightarrow P_j$, for $i \leqslant j$ except as in D1, D2.

C5 yields this and the Theorem.

THEOREM 8.

(a) $[P_0]_{X_i} = 0$, $[P_\omega]_{X_i} = 1$, *for all* $i \geqslant 1$.

(b) *For* $n \geq 1$ *we have*

$$[P_{2n}]_{\chi_i} = \begin{cases} 1 & i \leq n \\ 0 & i > n \end{cases}$$

$$[P_{2n-1}]_{\chi_i} = \begin{cases} 1 & i < n \\ 0 & i = n \\ 1 & i = n+1 \\ 0 & i > n+1 \end{cases}$$

*Proof.* That $[P_0]_{\chi_i} = 0$, $[P_\omega]_{\chi_i} = 1$ for all $i \geq 1$ is obvious. The rest of the proof is by induction on $n$.

By definition $P_{2(n+1)-1} = P_{2n+1} \rightarrow P_{2n}$ and assuming the theorem for indices $\leq n$, we have

$$[P_{2n-1}]_{\chi_{n+1}} = 1 \qquad [P_{2n}]_{\chi_{n+1}} = 0$$

and accordingly $[P_{2n+1}]_{\chi_{n+1}} = 0$, so that from Lemma 4 $[P_{2n+1}]_{\chi_i} = 0$, $i > n+2$. Next we show that $[P_{2n+1}]_{\chi_i} = 1$ for $i < n+1$, $i = n+2$. Note that $\chi_1, \chi_2, \ldots, \chi_n, \chi_{n+2}$ and all substructures of $\chi_{n+2}$ and from the induction hypothesis for no $i \in \{1, \ldots, n, n+2\}$ is it true that $[P_{2n-1}]_{\chi_i} = 1$ and $[P_{2n}]_{\chi_i} = 0$. It follows therefore that $[P_{2n+1}]_{\chi_{n+2}} = 1$, and, a fortiori, $P_{2n+1}$ must hold in every substructure of $\chi_{n+2}$, that is $[P_{2n+1}]_{\chi_i} = 1$, for $i \in \{1, \ldots, n, n+2\}$.

To complete the proof we have by definition that $P_{2n+2} = P_{2n-1} \vee P_{2n}$ and from the induction hypothesis that both $[P_{2n-1}]_{\chi_i} = [P_{2n}]_{\chi_i} = 0$ iff $i > n+1$. So we have $[P_{2(n+1)}]_{\chi_i} = 1$ for $i = n+1$ and $[P_{2(n+1)}]_{\chi_i} = 0$ for $i > n+1$.

LEMMA 10. *Every sentence A with one variable which is not a theorem of h has one or two minimal structures* $\chi_n$ *in which it is false. If A is* $P_{2n}$ *the two minimal structures are* $\chi_{n+1}$, $\chi_{n+2}$, $n \geq 0$. *If A is* $P_{2n-1}$, *the single minimal structure is* $\chi_n$.

*Proof.* Follows easily from Lemma 4 and Theorem 8.

LEMMA 11. *Let* $(S, R, 0, D)$ *be any finite tree structure for one variable p. Then it is homomorphic on some* $\chi_n$, *for some n.*

*Proof.* By induction on the number of points of $S$. If $\bar{\bar{S}} = 1$, then the only possible structures are $\chi_1$ and $\chi_2$. The structure $|^{\mathrm{T}}_{\mathrm{F}}$ is $\chi_3$ and $|^{\mathrm{F}}_{\mathrm{F}}$ is homomorphic on $\chi_2$.

Suppose $S$ is of the form

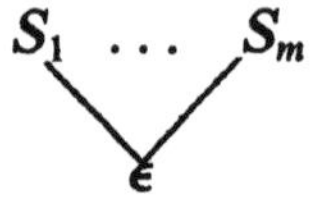


then if $\epsilon = T$, then $S$ is homomorphic on $\chi_1$. If $\epsilon = F$, let $S_i$ be homomorphic on $\chi_{n_i}$, $i = 1, \ldots, m$ and let $\alpha_i$ be the homomorphism. Assume $n_1 \geqslant n_2 \geqslant n_3 \ldots$.

Let $n_j$ be the first with $j$ such that $n_j > n_{j+1}$. If (a) $n_j - n_{j+1} = 1$, then clearly by the definition of $\chi_n$, $n \in \omega$, any $\chi_{n_k}$, $k \geqslant j+2$, is a substructure of either $\chi_{n_j}$ or $\chi_{n_{j+1}}$. If (b) $n_j - n_{j+1} \geqslant 2$, then any $\chi_{n_k}$, $k \geqslant j+1$, is a substructure of $\chi_{n_j}$.

In case (a) let $\alpha$ map $S$ on $\chi_{n_j+2} = \chi_{n_j} \quad \chi_{n_j-1}$.

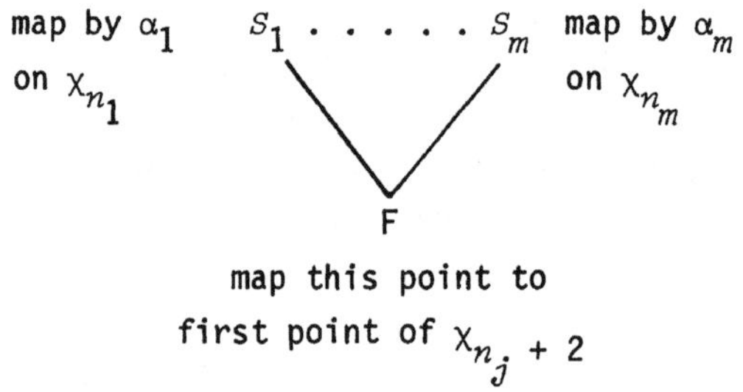


Not that all $\chi_{n_k}$, $k = 1, \ldots, m$ are substructures of $\chi_{n_j+2}$.

In case (b) let $\alpha$ map $S$ onto $\chi_{n_j}$ as in the diagram

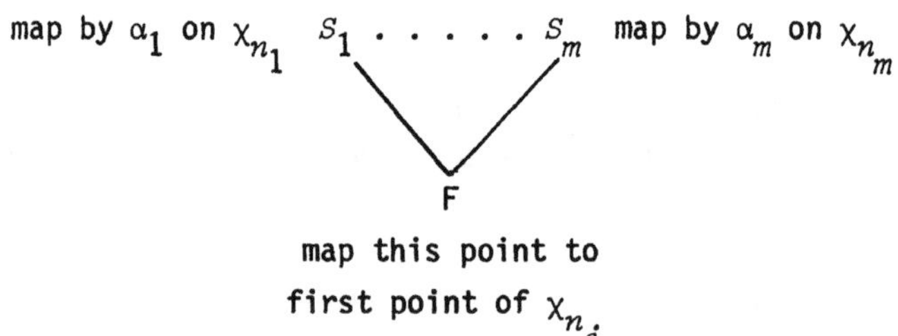


In this case as well, any $\chi_{n_k}$ is a substructure of $\chi_{n_j}$.

The reader can verify that we get a proper homomorphism.

COROLLARY 11. *The fragment of h of all wffs with one variable is complete for the class of finite binary trees.*

*Proof.* By Lemma 10.

DEFINITION 12. Let $\mathbf{M} = (S, R, 0, D)$ be a tree structure, $S$ finite and let $A$ be any wff with possibly several atomic variables. Let $p$ be any atom and consider the structure $(S, R, 0, D')$ for one variable obtained by $D'(t, p) = [A]_t$. Denote this structure by $(A:\mathbf{M})$. This structure is homomorphic to a smallest $\chi_n$, $n$ minimal. We indicate this by writing $(A:\mathbf{M}) \sim \chi_n$.

DEFINITION 13. Let $y_n$ be the extension of $h$ with the schema $P_n$.

THEOREM 14. $\bigcap y_n = h$.

*Proof.* If $h \nVdash A$, then there exists a finite tree structure $\mathbf{M}$ in which $A$ is false. Let $\chi_n$ be such that $(A:\mathbf{M}) \sim \chi_n$. Clearly $A$ can be made false in $\chi_n$. But by Lemma 10, $P_{2n}$ is always valid in $\chi_n$ and hence $y_{2n} \nVdash A$.

**Exercise 15.** Show that

(a) $y_0 = y_1 = y_2 = y_3$

(b) $y_4 = y_5$

(c) $y_6 = y_7$

(d) $y_i = y_j$ iff $i = j$ for either $i > 7$ or $j > 7$.

THEOREM 16. *There is an infinite number of non-equivalent (over h) wffs with the single atomic variable p.*

*Proof.* Follows from Theorem 8.

## 2. THE DISJUNCTION PROPERTY FOR $y_n$

It is clear that $y_{2n}$ does not have the disjunction property, because the axiom of $y_{2n}$ is $P_{2n-3} \vee P_{2n-2}$ and none of the disjuncts is provable, as follows from the lemma below.

LEMMA 1.

(a) $y_i \vdash P_j$ *iff* $\vdash_h P_i \to P_j$ *for* $i, j \geq 8$

(b) $y_{2i} \nvdash P_j$ *for* $j < 2i, i > 1$.

*Proof.* (a) If $h \vdash P_i \to P_j$, then obviously $y_i \vdash P_j$. Now assume $h \nvdash P_i \to P_j$, then for some minimal $\chi_{j^*}$, $P_i$ holds in $\chi_{j^*}$ and $P_j$ does not hold (see 1.1a). Since $j \geqslant 8$, note that $j^* \geqslant 5$.

We want to use the above to show that $y_i \nvdash P_j$. Clearly if $y_i \vdash P_j$ then for some finite number of substitutions instances $P_i(A_k)$, $k = 1, 2, \ldots, m$, $\bigwedge_k P_i(A_k) \vdash_h P_j$. We can assume that $A_k$ are built up from the $p$, the same $p$ that appears in $P_j$ (for we can substitute anything for any other atom). So to show that $y_i \nvdash P_j$, it is sufficient if we show that in $\chi_{j^*}$ every substitution instance $P_i(A)$ is valid, where $A = A(p)$ is built up from $p$.

We distinguish 4 cases, for $A(p)$ in classical logic:

(1) $A(p)$ is a classical tautology.

(2) $A(p)$ is a classical contradiction.

(3) $A(p)$ is classically true for $p$ true and classically false for $p$ false.

(4) $A(p)$ is classically false for $p$ true and classically true for $p$ false.

We take each alternative in turn and show that either $(A(p):\chi_{j^*}) \sim \chi_l$, $l \leqslant 4$ in which case $P_i(A(p))$ holds in $\chi_{j^*}$ (since $i \geqslant 8$) or $(A(p):\chi_{j^*}) \sim \chi_{j^*}$ in which case $P_i(A(p))$ holds in $\chi_{j^*}$ by our assumption.

(1) In this case $(A(p):\chi_{j^*})$ is a structure with $T$ at each endpoint. It is homomorphic to *a* $\chi_n$, for some $n$, the only structures $\chi_n$ with $T$ at all endpoints are $\chi_1, \chi_3$.

(2) In this case $(A(p):\chi_{j^*})$ takes $F$ on all nodes and so $\sim \chi_2$.

(3) The minimal $\chi_n$ rejecting $A(p)$ is $\chi_2$ or possibly $\chi_3$ and hence $\vdash_h A \leftrightarrow P_3 = p$ or $\vdash_h A \leftrightarrow P_4 = (\neg p \to p)$.

In the first case $(p:\chi_{j^*}) \sim \chi_{j^*}$ and in the second case $(P_4:\chi_3) \sim \chi_1$, $(P_4:\chi_2) \sim \chi_2$ and hence $(P_4:\chi_{j^*}) \sim \chi_n$ for $\chi_n$ that does not contain $\chi_2$ as substructure, that is $\chi_n$ can be only $\chi_1$, $\chi_2$ or $\chi_4$.

(4) The only minimal $\chi_n$ rejecting $A(p)$ is $\chi_1$ in this case and so $\vdash_h A(p) \leftrightarrow \neg p$. Since $(\neg p:\chi_3) \sim \chi_2$, $(\neg p:\chi_2) \sim \chi_1$. So $(\neg p:\chi_{j^*}) \sim \chi_n$, where $\chi_n$ has no substructure of the form $\chi_3$ and hence $\chi_n$ can be either $\chi_1$, $\chi_2$ or $\chi_4$.

(b) Since $j < 2i$, the minimal structure rejecting $P_j$ is some $\chi_k$, $k < i$ which has fewer points than either $\chi_{i+1}$ or $\chi_{i+2}$, the minimal rejecting structures of $P_{2i}$. Thus $P_{2i}(A(p))$ is always valid in $\chi_k$.

THEOREM 2. *$y_{2n+1}$ has the disjunction property for $n \geqslant 4$, $n \neq 6$.*

*Proof.* Let $T$ be the set of all theorems of $y_{2n+1}$. Well-order $T$ in such a way that wffs with fewer connectives appear earlier in the order than wffs with more connectives. Define a function $f(A)$ on elements of $T$, $f$ giving values $+$ or $-$, by induction on the well ordering and on the structure of the wff $A$.

(1) $\quad f(A \to B) = -\quad$ iff $\quad f(A) = +\quad$ and $\quad f(B) = -$.

(2) $\quad f(A \wedge B) = -\quad$ iff $\quad f(A) = -\quad$ or $\quad f(B) = -$.

(3) $\quad f(A \vee B) = -\quad$ iff $\quad f(A) = f(B) = -$.

Let $T^* = \{A \in T \mid f(T) = +\}$.

LEMMA 3. *If $h \vdash A$ then $A \in T^*$.*

*Proof.* By verifying that every substitution instance of an axiom of $h$ gets $+$ and that this property is preserved by modus ponens. See (2.3.8).

LEMMA 4. *Any substitution instance of $P_{2n+1}$ is in $T^*$.*

The theorem follows from Lemma 4, since Lemma 4 implies that $T^* = T$ and clearly $A \vee B \in T^*$ iff $A \in T^*$ or $B \in T^*$.

We shall therefore continue to prove Lemma 4.

*Proof of Lemma* 4. Assume that for some substitution instance of $P_{2n+1}$, namely $P_{2n+1}(A)$, the value of $f$ is $-$. Thus $P_{2n+1}(A) = P_{2n-1}(A) \to P_{2n}(A)$, and thus $f(P_{2n-1}(A)) = +$ and $f(P_{2n}(A)) = -$. In order that $f(P_{2n}(A))$ be defined, $P_{2n}(A)$ must be in $T$, that is $y_{2n+1} \vdash P_{2n}(A)$. A may be built up from $m$-atoms $x_1, \ldots, x_m$. Notice also that neither $A$ nor $\neg A$ can be classical tautologies. For if $A$ is a tautology then for any structure $\mathbf{M}$, $(A:\mathbf{M}) \sim \chi_1$ or $\sim \chi_3$. But then $[P_{2n+1}(A)]_{\mathbf{M}} = [P_{2n+1}(p)]_{(A:\mathbf{M})} = 1$. Thus by the completeness theorem $h \vdash P_{2n+1}(A)$ and hence $P_{2n+1}(A) \in T^*$. Similarly if $\neg A$ is a tautology then for any $\mathbf{M}$, $(A:\mathbf{M}) \sim x_2$ and hence $[P_{2n+1}(A)]_{\mathbf{M}} = [P_{2n+1}(p)]_{(A:\mathbf{M})} = 1$ and again $\vdash_h P_{2n+1}(A)$ and again $P_{2n+1}(A) \in T^*$.

Thus if neither $A$ nor $\neg A$ are tautologies let $\mu$ be a one point structure with $(A:\mu) \sim \chi_1$ and $\nu$ be one point structure with $(A:\nu) \sim \chi_2$ ($\mu$, $\nu$ are structures for the language of $A$, with atoms $x_1, \ldots, x_m$).

We now continue with the proof of Lemma 4 by looking at cases $n = 4$, $n \geqslant 7$ and $n = 5$.

*Case $n = 4$.* We assumed $P_9(A) \notin T^*$ and hence $y_9 \vdash P_7(A)$. Consider the structure for $x_1, \ldots, x_n$

$\chi_4^0 =$ 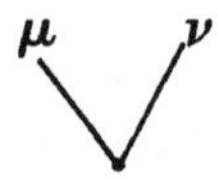


$x_i = \mathsf{F}$ at this point, $i = 1 \ldots n$.

Clearly $(A : \chi_4^0) \sim \chi_4$. Thus $[P_7(A)]_{\chi_4^0} = [P_7(p)]_{(A : \chi_4^0)} = [P_7(p)]_{\chi_4} = 0$.

But the minimal rejecting structure for $P_9(p)$ is $\chi_5$ and hence every substitution instance of $P_9(p)$, e.g., $P_9(B)$ satisfies $[P_9(B)]_{\chi_4^0} = [P_q(p)]_{(B : \chi_4^0)} = 1$ since $(B : \chi_4^0)$ has only 3 points and $\chi_5$ has 4 points. Thus $y_9 \not\vdash P_7(A)$, contradiction.

*Case* $n \geqslant 7$. In this case we obtain a contradiction by showing $y_{2n+1} \vdash P_{2n-3}(A)$. By definition $P_{2n}(A) = P_{2n-3}(A) \vee P_{2n-2}(A) = P_{2n-3}(A) \vee P_{2n-5}(A) \vee P_{2n-4}(A)$. We have $P_{2n}(A) \not\in T^*$ and hence $P_{2n-3}(A)$, $P_{2n-5}(A)$, $P_{2n-4}(A) \not\in T^*$. Assuming for the moment that $P_{2n-3}(A) \in T$, we show how we get a contradiction. Thus $f$ gives $-$ as value to $P_{2n-3}(A) = P_{2n-5}(A) \to P_{2n-4}(A)$ and hence $f(P_{2n-5}(A)) = +$, i.e., it is in $T^*$, a contradiction.

*Proof that* $y_{2n+1} \vdash P_{2n-3}(A)$:

Since $y_{2n+1} \vdash P_{2n-1}(A)$, let $\bigwedge_i P_{2n+1}(B_i) \vdash_h P_{2n-1}(A)$. Let $\mathbf{M}$ be a minimal structure in which $\bigwedge_i P_{2n+1}(B_i)$ holds and $P_{2n-3}(A)$ is false. We want to obtain a contradiction and hence deduce that it is a theorem of $h$.

$$[P_{2n-3}(A)]_{\mathbf{M}} = [P_{2n-3}(p)]_{(A : \mathbf{M})} = [P_{2n-3}(p)]_{\chi_{n-1}} = 0$$

since $(A : \mathbf{M}) \sim \chi_{n-1}$, as $\chi_{n-1}$ is the minimal structure in which $P_{2n-3}(p)$ is false. Since $n > 4$, $\chi_{n-1}$ contains $\chi_3$ as a substructure. Since $(A : \mathbf{M}) \sim \chi_{n-1}$, there must be a substructure $\xi$ of $\mathbf{M}$ such that $(A : \xi) \sim \chi_3$. We show that for some $B_i$ $[P_{2n+1}(B_i)]_\xi = 0$, which is a contradiction. For this end, define a sequence

$$\chi_1^0 = \mu, \qquad \chi_2^0 = \nu, \qquad \chi_3^0 = \xi$$

$\chi_l^0 =$ 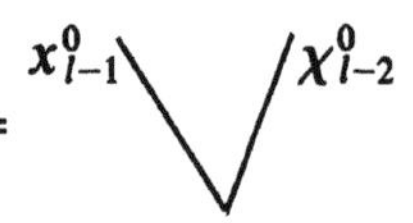


$x_1 = x_2 = \cdots = x_n = \mathsf{F}$ at this point.

$\chi_l^0$ is a structure for the language of $A$. It is easy to see that $(\mathbf{A} : \chi_l^0) = \chi_l$ and hence $\chi_n^0$ is a minimal structure in which $P_{2n-1}(A)$ is false. Since $\bigwedge_i P_{2n+1}(B_i) \vdash_h P_{2n-1}(A)$, for some $B_i$, $[P_{2n+1}(B_i)]\chi_n^0 = 0$. Therefore $(B_i : \chi_n^0) \sim \chi_j$ for $j \geqslant n+1$.

Now from Lemma * below, $(B_i : \chi_3^0)$ cannot be homomorphic to any $\chi_l$ for $l \leqslant 5$ and hence $(B_i : \chi_3^0) \sim \chi_l$, $l > 5$ and hence again by Lemma * below, $(B_i : \chi_3^0) = (B_i : \xi) = (B_i : \chi_n^0)$ and hence $[P_{2n+1}(B_i)]_\xi = 0$ and hence $[P_{2n+1}(B_i)]_{\mathbf{M}} = 0$.

We therefore conclude that $\vdash_h \bigwedge_i P_{2n+1}(B_i) \rightarrow P_{2n-3}(A)$ and hence $y_{2n+1} \vdash P_{2n-3}(A)$.

This completes the proof of case $n \geqslant 7$ provided we have proved Lemma *.

LEMMA *. *Let* $\mu$, $\nu$, $\xi$ *be tree structures for the language of* $A(x_1, \ldots, x_n)$ *such that* $\mu$, $\nu$ *are one-point trees. Define* $\chi_l^0$ *by* $\chi_1^0 = \mu$, $\chi_2^0 = \nu$, $\chi_3^0 = \xi$,

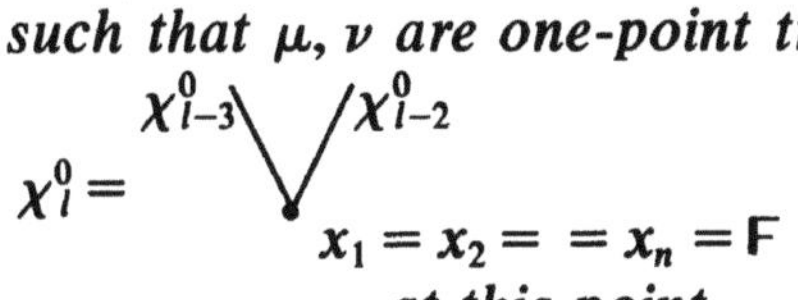


*then*

(a) *If* $(A : \xi) \sim \chi_k$, $k \geqslant 6$ *then* $(A : \chi_i^0) \sim (A : \xi) \sim \chi_k$ *for all* $i \geqslant 7$.
(b) *If* $(A : \xi) \sim \chi_5$ *then for all* $i \geqslant 7$, $(A : \chi_i^0) \sim \chi_7$.
(c) *If* $(A : \xi) \sim \chi_4$ *then for all* $i \geqslant 7$, $(A : \chi_i^0) \sim \chi_6$.
(d) *If* $(A : \xi) \sim \chi_1, \chi_2$ *or* $\chi_3$ *then for all* $i \geqslant 1$, $(A : \chi_i^0) \sim \chi_j$, $j \leqslant i$.

*Proof.* Exercise.

*Case* $n = 5$. In this case we show that $y_{11} \vdash P_5(A)$ (i.e., $y_{2n+1} \vdash P_{2n-5}(A)$).

By definition $P_{10}(A) = P_7(A) \vee P_8(A) = P_7(A) \vee P_5(A) \vee P_3(A) \vee P_4(A)$ and since $P_{10}(A) \notin T^*$, $P_7(A)$, $P_5(A)$, $P_3(A)$, $P_4(A) \notin T^*$. But if $P_5(A) = P_3(A) \rightarrow P_4(A)$ is in $T$ then $P_5(A) \in T^*$ implies $P_3(A) \in T^*$ a contradiction. In essence the proof is the same as for $n \geqslant 7$. We have $P_9(A) \in T$, i.e., $y_{11} \vdash P_9(A)$ and hence for some $B_i$, $\vdash_h \bigwedge_i P_{11}(B_i) \rightarrow P_9(A)$. We claim $\vdash_h \bigwedge_i P_{11}(B_i) \rightarrow P_5(A)$. Otherwise let $\mathbf{M}$ be a minimal structure validating $\wedge P_{11}(B_i)$ in which $P_5(A)$ is false. Clearly $(A : \mathbf{M}) \sim \chi_3$. Let $\xi = \mathbf{M}$ and define the sequence $\chi_l^0$ as in the previous case. Thus $\chi_5^0$ is a minimal structure rejecting $P_9(A)$ and so for some $B_i$, $[\bigwedge_i P_{11}(B_i)]_{\chi_5^0} = 0$ and hence $(B_i : x_5^0) \sim x_j$ for some $j \geqslant 6$. But

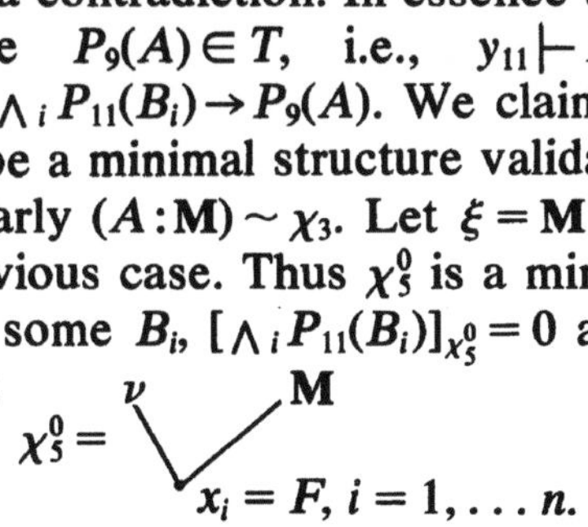

As $\nu$ is a one point structure, it is immediate that $(A:\mathbf{M}) \sim x_j$ and thus $[P_{11}(B_i)]_{\mathbf{M}} = 0$. Thus we get a contradiction and case $n = 5$ is finished.

## NOTE

Based on the thesis of J.G. Anderson and some refinements in the thesis of C. Smorynski.

## CHAPTER 7

# PROPOSITIONAL CONNECTIVES

## 1. Negation

Let $h^-$ be the Hilbert system in a propositional language with $\wedge, \vee, \rightarrow$, obtained from $h$ by Definition 4.1.1 by dropping axiom (g). So $h^-$ is the negationless fragment of $h$. The deduction theorem holds for $h^-$, the proof being as in 2.1.13 and $h^-$ is complete for the class of propositional Kripke structures of Definition 4.1.8 with $[A]_t$ defined as in 4.1.8 for any wff $A$. The completeness proof is the same as the one given in Section 1 of Chapter 4.

We want to examine the possibilities of introducing negation into $h^-$. For this purpose we examine the general question of introducing negation into an arbitrary Tarski system $\vdash$.

DEFINITION 1. (a) A refutability system is a pair $(\vdash, \Delta)$, where $\vdash$ is a Tarski consequence system and $\Delta$ a set of wffs of the language of $\vdash$, closed under substitution.

(b) A refutability system is not consistent if for some $A \in \Delta$, $\vdash A$ holds.

(c) For any wff $B$, $B$ is said to be negated in the refutability system if $B \vdash A$, for some $A \in \Delta$.

(d) A refutability system is said to be an absurdity system iff for any $B$.
For all $A$, $B \vdash A$ iff for some $A \in \Delta$, $B \vdash A$.

(e) A refutability system is said to be complete iff for all $B$:

$$\vdash B \text{ or for some } A \in \Delta,\ B \vdash A.$$

*Remark* 2. Refutability systems are obtained from Tarski systems by specifying a set $\Delta$ of non-theorems. The weakest form of negation of a wff $B$ is to say that $B \vdash A$ for some non-theorem $A$. If we have a refutability system that is an absurdity, then if $B$ is negated then it proves everything.

Completeness means that every wff is either a theorem or non-theorem.

Given a refutability system $(\vdash, \Delta)$, we can extend the language with an additional new propositional constant $\perp$ and seek a Tarski system

$\vdash^*$ in the new language with the property that

(3a) $\varphi \vdash^* B$ iff $\varphi \vdash B$

(3b) $\varphi \vdash^* \bot$ iff for some $A \in \Delta$ $\varphi \vdash A$.

Where $\varphi$, $B$ are in the language of $\vdash$.

LEMMA 4. *For any* $(\vdash, \Delta)$ *a* $\vdash^*$ *exists with properties* (3a–3b).

*Proof.* Let $\Vdash$ be the minimal Scott consequence system agreeing with $\vdash$. Consider a language with the additional new propositional variable $\bot$ and let $\Vdash_0$ be a conservative minimal extension of $\Vdash$ to the new language, i.e., $\Vdash_0$ is the minimal Scott system satisfying

$$\varphi \Vdash_0 \psi \quad \text{iff} \quad \varphi \Vdash \psi$$

for all $\varphi$, $\psi$ in the language without $\bot$. $\bot$ has no special properties in $\Vdash_0$ since it is a variable. Let $\mathbf{T}_{\Vdash_0}$ be the canonical semantics of $\Vdash_0$ and let $\Delta^*$ denote all substitution instances of elements of $\Delta$ in the new language. For $\mathbf{t} \in \mathbf{T}_{\Vdash_0}$, we say $\mathbf{t}$ is admissible iff

$$t(\bot) = 0 \Rightarrow t(A) = 0 \quad \text{for all} \quad A \in \Delta^*$$

$$t(\bot) = 1 \Rightarrow t(A) = 1 \quad \text{for some} \quad A \in \Delta^*.$$

Let $\mathbf{T}_*$ be the set of all admissible models and let $\Vdash^*$ be $\Vdash_{\mathbf{T}_*}$.

Let $\varphi \vdash^* B$ iff $\varphi \Vdash^* B$. We show that $\vdash^*$ satisfies Condition 3. Let $\varphi$, $B$ be in the language without $\bot$.

*Case* 3*a*. (1) If $\varphi \nVdash^* B$ then for some $\mathbf{t} \in \mathbf{T}_*$, $\mathbf{t}(A) = 1$ for all $A \in \varphi$, $t(B) = 0$. Consider the restriction $\mathbf{t}'$ of $\mathbf{t}$ to the wff without $\bot$. Since $\mathbf{t} \in \mathbf{T}_{\Vdash_0}$, the restriction is in $\mathbf{T}_{\Vdash}$ and thus $\varphi \nVdash B$.

(2) Assume $\varphi \nVdash B$. Then for some $\mathbf{t}' \in T_{\Vdash}$, $\mathbf{t}'$ gives all members of $\varphi$ the value 1 and gives $B$ the value 0.

(a) If $\varphi \Vdash A$ for some $A \in \Delta$, define $\mathbf{t} \in T_{\Vdash^*}$ on any wff $C(\bot)$ $\mathbf{t}(C(\bot)) = \mathbf{t}'(C(A))$. We want to show that $\mathbf{t} \in T_{\Vdash_0}$. If not, then for some $\varphi'$, $\psi'$ in the language with $\bot$, $\varphi' \Vdash_0 \psi'$ but for all $C(\bot) \in \varphi'$, $\mathbf{t}(C(\bot)) = 1$ and all $C(\bot) \in \psi'$, $\mathbf{t}(C(\bot)) = 0$. But since $\mathbf{t}(C(\bot)) = \mathbf{t}'(C(A))$ and $\mathbf{t}' \in \mathbf{T}_{\Vdash}$ this implies that $\varphi'_A \nVdash \psi'_A$, where $\varphi'_A$ and $\psi'_A$ are obtained from $\varphi'$, $\psi'$ by substituting $A$ for $\bot$. This is clearly impossible since $\varphi' \Vdash_0 \psi'$ implies $\varphi'_A \Vdash_0 \psi'_A$ and hence $\varphi'_A \Vdash \psi'_A$. Thus $\mathbf{t} \in \mathbf{T}_{\Vdash_0}$ and since $\mathbf{t}'(A) = 1$ also $t(\bot) = 1$ and therefore $\mathbf{t} \in T_*$ and hence $\varphi \nVdash^* B$.

(b) If $\varphi \nVdash A$, for all $A \in \Delta$, then the theory $(\varphi, \Delta \cup \{B\})$ is $\Vdash$ consistent (recall that $\Vdash$ is the *minimal* consequence relation agreeing with $\vdash$ and hence inconsistency means that some $A \in \Delta \cup \{B\}$ is $\vdash$ provable from $\varphi$) and hence we can assume that $\mathbf{t}'(A) = 0$, for all $A \in \Delta \cup \{B\}$. Fix $A \in \Delta$ and proceed to define $\mathbf{t}$ as above. Show again that $\mathbf{t} \in \mathbf{T}_*$, since $t(\bot) = 0$ and $\mathbf{t}(A) = 0$ for all $A \in \Delta$.

*Case* 3*b*. Follows from the condition on the members of $\mathbf{T}_*$.

Thus Lemma 4 is proved.

We thus see that formally the act of introducing negation $\neg$ into a Tarski system $\vdash$ boils down to choosing a propositional constant $\bot$ and letting $\vdash \neg A$ mean $A \vdash \bot$. If negation is absurdity, we want the rule $\bot \vdash B$, $B$ arbitrary, to be valid.

DEFINITION 5. (a) Let $\bot$ be a fixed propositional variable of the language of $h^-$. Regard $\bot$ as a propositional constant. Define $\neg A$ as $A \to \bot$. Call the logical system thus obtained by *hm*.

(b) Let *h* be the extension of *hm* with the axiom $\bot \to A$.

(c) Let *hd* be the extension of *hm* with the axiom $A \vee (A \to \bot)$.

(d) Let *he* be the extension of *hm* with $((A \to B) \to A) \to A$.

(e) Let *c* be the extension of *h* with the axiom $A \vee (A \to \bot)$.

*Remark* 6. *hm* is the system obtained from $h^-$ by adding the notion of negation corresponding to a refutability system. The schema $(A \to \neg B) \to (B \to \neg A)$ is a theorem of *hm*. *h* itself corresponds to the absurdity system with the axiom $\bot \to A$ or equivalently $\neg A \to (A \to B)$. The notion of completeness of a refutability system is reflected in our case in the axiom $A \vee \neg A$. *hd* and *c* are obtained using this axiom.

**Exercise 7.** (a) Consider propositional Kripke structures of the form $(N, S, R, 0, D)$, where $(S, R, 0, D)$ is a propositional Kripke structure and $N \subseteq S$ a set with the property $x \in N \wedge xRy \to y \in N$, for all $x$, $y$. Define $[A]_t$ in these structures as in Chapter 4, Section 1, with the difference that the clause for $\neg$ being $[\neg A]_t = 1$ iff for all $s$, $tRs$ and $[A]_s = 1$ implies $s \in N$.

Show that *hm* is complete for this interpretation.

Show that the axiom $A \vee \neg A$ corresponds to the condition $N \supseteq S - \{0\}$.

**Exercise 8.** Characterise the conditions on the structures of (7a) corresponding to each of the schemas below, when added to *hm*.

(1) $((A \to B) \to A) \to A.$

(2) $A \vee \neg A.$

(3) $\neg A \vee \neg\neg A.$

(4) $\bot \to (A \to B) \vee (B \to A).$

(5) $A \vee (A \to B).$

(6) $\bot \to A \vee (A \to B).$

(7) $(A \to B \vee \bot) \vee (B \to A \vee \bot).$

(8) $A \vee (\bot \to A).$

## 2. Strong negation

Some authors introduced another type of negation into *HPC*, called strong negation. It has the properties ($-$ denotes strong negation):

(0) $--A \leftrightarrow A.$

(1) $-(A \to B) \leftrightarrow A \wedge -B.$

(2) $-(A \wedge B) \leftrightarrow -A \vee -B.$

(3) $-(A \vee B) \leftrightarrow -A \wedge -B.$

(4) $A \wedge -A \to B.$

DEFINITION 5. Let *nh* be the propositional Hilbert system in a language with the connectives $\wedge$, $\vee$, $\to$, $-$ and the axioms and rules of Definition 4.1.1. with the exception of axiom (g) and the additional axioms (0), (1), (2), (3), (4) above.

LEMMA 6. *The deduction theorem holds for nh.*

*Proof.* As in 2.1.13.

**Exercise 7.** (a) Note that $\vdash A \leftrightarrow B$ does not imply $\vdash -A \leftrightarrow -B$.

(b) Let $A \equiv B$ mean $\vdash (A \to B) \wedge (-B \to -A)$, then is $\vdash (A \equiv B) \to C(A) \equiv C(B)$ true? With $C(p)$ any wff?

DEFINITION 8. (a) A strong propositional Kripke structure has the form $(S, R, 0, \alpha)$ where $(S, R, 0)$ is a partially ordered set with first element $0 \in S$ and $\alpha$ is a function with the property that for each $t \in S$ and atomic $q$, $\alpha(t, q) \in \{-1, 0, 1\}$. We require that $tRs \wedge \alpha(t, q) \neq 0 \rightarrow \alpha(t, q) = \alpha(s, q)$.

(b) The truth value $[A]_t$ of a wff $A$ at a point $t$ of a strong propositional Kripke structure is defined by induction.

(1) $[A]_t = \alpha(t, A)$ for $A$ atomic.

(2) $[A \wedge B]_t = \min([A]_t, [B]_t)$.

(3) $[A \vee B]_t = \max([A]_t, [B]_t)$.

(4) $[A \rightarrow B]_t = 1$ iff for all $s$, $tRs \wedge [A]_s = 1$ imply $[B]_s = 1$.
$[A \rightarrow B]_t = -1$ iff $[A]_t = 1$ and $[B]_t = -1$.

(5) $[-A]_t = 1$ iff $[A]_t = -1$.

We say $A$ is valid in the structure if $[A]_0 = 1$. A theory $(\Delta, \Theta)$ is valid in the structure iff all wffs of $\Delta$ are valid and none of the wffs of $\Theta$ are valid.

LEMMA 9. *For any wff* $A$, $[A]_t \neq 0 \wedge tRs \Rightarrow [A]_t = [A]_s$.

*Proof.* By induction on $A$.

DEFINITION 10. Let $\varphi \Vdash \psi$ be defined as $\wedge\varphi \vdash_{nh} \vee\psi$ $(\wedge\emptyset = p \rightarrow p$, $\vee\emptyset = p \wedge -p)$.

PROPOSITION 11. (a) $\Vdash$ *is a Scott consequence system that agrees with nh.*

(b) $\wedge$ *and* $\vee$ *are classical in* $\Vdash$.

*Proof.* As in (2.2.4)–(2.2.6).

THEOREM 12. $\Vdash$ *of Definition* 10 *is identical with the consequence system arising from the interpretation in the strong propositional Kripke structures. In particular* $nh \vdash A$ *iff* $A$ *is valid in every such structure.*

*Proof.* Since $\wedge$ and $\vee$ are classical in $\Vdash$, we take $\mathbf{T}_{\Vdash}$, the canonical semantics for $\Vdash$. Define $\mathbf{t}R\mathbf{s}$ iff for all $A$, $\mathbf{t}(A) = 1 \Rightarrow \mathbf{s}(A) = 1$.

LEMMA 13. *If* $\mathbf{t}(A \rightarrow B) = 0$ *then for some* $\mathbf{s}$, $\mathbf{t}R\mathbf{s}$ *and* $\mathbf{s}(A) = 1$ *and* $\mathbf{s}(B) = 0$.

*Proof.* Since $\mathbf{t}(A \to B) = 0$, then by the deduction theorem $(\{A' \mid \mathbf{t}(A') = 1\} \cup \{A\}, \{B\})$ is $\Vdash$ consistent and hence can be extended to a complete $\Vdash$ saturated theory $(\Delta, \Theta)$. Let $\mathbf{s}$ be defined by $\mathbf{s}(A') = 1$ iff $A' \in \Delta$. Thus $\mathbf{t}R\mathbf{s}$ holds and $\mathbf{s}(A) = 1$ and $\mathbf{s}(B) = 0$.
To turn $(S, R)$ into a structure let for atomic $A$, $\alpha(\mathbf{t}, A) = 1$ if $\mathbf{t}(A) = 1$ and $\alpha(\mathbf{t}, A) = -1$ if $\mathbf{t}(-A) = 1$.

LEMMA 14. *In $(S, R, \alpha)$ we have that for any $A, \mathbf{t}$,*

$$[A]_{\mathbf{t}} = 1 \quad \textit{iff} \quad \mathbf{t}(A) = 1.$$

$$[A]_{\mathbf{t}} = -1 \quad \textit{iff} \quad \mathbf{t}(-A) = 1.$$

*Proof.* By induction on $A$. Since $\wedge$, $\vee$ are classical in $\Vdash$ and axioms (0)–(4) are available we get

$$\mathbf{t}(-(A \wedge B)) = 1 \quad \text{iff} \quad \mathbf{t}(-A) = 1 \quad \text{or} \quad \mathbf{t}(-B) = 1.$$

$$t(-(A \vee B)) = 1 \quad \text{iff} \quad t(-A) = 1 \quad \text{and} \quad \mathbf{t}(-B) = 1.$$

$$\mathbf{t}(--A) = \mathbf{t}(A).$$

$$\mathbf{t}(-(A \to B)) = 1 \quad \text{iff} \quad \mathbf{t}(A \wedge -B) = 1.$$

$$t(A) \cdot t(-A) = 0.$$

Using these properties together with Lemma 13 one can prove Lemma 14.

The completeness theorem follows from Lemma 14.

**Exercise 15.** (a) Show that *nh* is complete for the class of finite strong propositional Kripke structure.

(b) Let $f$ be a fixed proposition in *nh* and let $\neg A$ be $A \to (f \wedge -f)$. Show that for any wff $B$ of $h$, $nh \vdash B$ iff $h \vdash B$.

(c) Let *nc* be the extension of *nh* with the schema $((A \to B) \to A) \to A$. Is $A \vee -A$ a theorem of *nc*?

(d) What interpretation theorems can you find from *nc* to *nh*? from *c* to *nh*?

**Exercise 16.** Let us look at the connectives $\wedge$, $\vee$, $\to$ only and at the system $h^-$. Add a new unary connective $\sim$ and let $\sim h^-$ be the extension of $h^-$ with the axiom

(a) $\qquad \sim p \to (p \to q).$

Clearly if we regard $\sim$ as a negation $\neg$, then $\sim h^-$ is a weaker system than $h$ itself.

(1) Show that if we add to $\sim h^-$

(b) $\quad (p \to \sim p) \to \sim p$

then we obtain the system $h$ with $\sim$ acting as $\neg$.

(2) Show that (b) is not a theorem of $\sim h^-$.

Suppose we introduce in $\sim h^-$ the definition

(c) $\quad \neg p = \mathrm{def}(p \to \sim p).$

(3) Show that under this definition $h \vdash A$ implies $\sim h^- \vdash A$.

(4) Show that

$$\sim h^- \vdash \neg p \to \sim p$$

but

$$\sim h^- \nvdash \sim p \to \neg p.$$

(5) Show that the following are not theorems of $\sim h^-$

(d) $\quad (p \to q) \to (\sim q \to \sim p).$

(e) $\quad \sim \sim p \to p.$

(6) Show that (b) is not provable even in $\sim h^- + (\mathrm{d}) + (\mathrm{e})$.

(7) Define the following 'modal' operators:

(f) $\quad \Box p = \mathrm{def} \sim (p \to \sim p).$

$\quad \Diamond p = \mathrm{def} \sim p \to p.$

What can you say about their properties in $\sim h^-$?

## 3. Functional completeness

We know that in the classical propositional calculus $c$, the connectives $\neg$, $\to$ are functionally complete, that is, for any function $\tau \in 2^{2^n}$ ($\tau$ is an $n$-ary truth table) there exists a wff $A$ built up from $p_1, \ldots, p_n$, $\neg$, $\to$ such that $\tau$ is the truth table of $A$. We are interested in similar results for the case of the intuitionistic propositional calculus $h$.

Consider a language, in the monadic classical second order predicate calculus, suitable for the theory of partially ordered sets. That is, we have equality $=$, order $\leqslant$, variables ranging over elements, $x$, $y$, $z$, $t$, $s$, ... and variables ranging over subsets $P_1, P_2, \ldots$. We also quantify over element and subset variables. We can assume that 0 is a constant. The theory of partial order with a first element 0 can be easily formalised in this language. We want to define the notion of an $n$-place 1st order truth table wff $\gamma(t, P_1, \ldots, P_n)$. This we take as any wff of this language with element quantifiers and the free variables $t$, $P_1, \ldots, P_n$ that satisfies

$$(1) \qquad \bigwedge_{i=1}^{n} \forall xy(P_i(x) \wedge x \leqslant y \rightarrow P_i(y)) \rightarrow (\gamma(t, P_i) \leftrightarrow (\forall s \geqslant t)\gamma(s, P_i)).$$

Given a wff $A(p_1, \ldots, p_n)$ of $h$ with the variables $p_1, \ldots, p_n$, we define a wff $\gamma_A(t)$ of second order monadic partially ordered sets as follows:

$$(2a) \qquad \gamma_{p_i}(t) = P_i(t), \quad p_i \text{ atomic.}$$

$$(2b) \qquad \gamma_{A \wedge B}(t) = \gamma_A(t) \wedge \gamma_B(t).$$

$$\gamma_{A \vee B}(t) = \gamma_A(t) \vee \gamma_B(t).$$

$$(2c) \qquad \gamma_{A \rightarrow B}(t) = (\forall s \geqslant t)(\gamma_A(s) \rightarrow \gamma_B(s)).$$

$$(2d) \qquad \gamma_f(t) = t \neq t.$$

LEMMA 3. *For each wff $A$, $\gamma_A$ satisfies* (1).
*Proof.* By induction on $A$.

*Remark* 4. It is easy to see that in a given propositional Kripke structure $(S, R, 0, D)$, if $\leqslant$ is taken as $R$ and for atomic $P_i$, $P_i(t) = D(t, P_i)$, then $\gamma_A(t)$ is $[A]_t$.

DEFINITION 5. *A* wff $\gamma(t, P_1, \ldots, P_n)$ is said to be definable by a wff or *regular* iff for some $A(p_1, \ldots, p_n)$

$$\bigwedge_{i=1}^{n} \forall xy(P_i(x) \wedge x \leqslant y \rightarrow P_i(y)) \rightarrow (\gamma \leftrightarrow \gamma_A).$$

PROPOSITION 6. *Let $\gamma(t)$ be a first order truth wff. Then $\gamma(t)$ is regular iff it is built up from atomic wff of the form $P(x)$ or $x \neq x$ using the connectives and bounded quantification of the form $\forall x \geqslant y$ or $\exists x \geqslant y$.*

*Proof.* Clearly any regular wff $\gamma_A$, satisfies the requirement of the theorem.

Assume $\gamma(t)$ is built up from atoms $P(x)$ or $x \neq x$ using the connectives and bounded quantification. We can assume that only $\forall$ $\vee$, $\neg$, $\wedge$, $\rightarrow$ are used.

LEMMA 7. *Each subformula* $\alpha(t_1, \dots, t_n)$ *of* $\gamma(t)$, *with the free variables* $t_1, \dots, t_n$ *is obtained using* $\neg$, $\wedge$, $\vee$, $\rightarrow$ from wffs of the form $\gamma_A(t_i)$.

*Proof.* By induction on the number of quantifiers in $\alpha$. If there are no quantifiers, then $\alpha$ is built up from $P_j(t_i)$. Certainly $P_j(t_i)$ is $\gamma_{p_j}(t_i)$. $x \neq x$ is $\gamma_f(x)$.

Assume that $\alpha(s, t_1, \dots, t_n)$ is built up from wffs of this form. Let $\alpha$ be written as $\alpha = \bigwedge_i (\bigvee_j \gamma_{A_{ij}}(t_{ij}) \vee \bigvee_k \neg \gamma_{B_{ik}}(t'_{i}))$, where $t_{ij}$, $t'_{ik} \in \{t_1, \dots, t_n, s\}$.

Consider $(\forall s \geq t)\alpha$. Let $\alpha_i$ be the $i$-th conjunct. Then $\alpha \leftrightarrow \bigwedge_i (\forall s \geq t)\alpha_i$. Consider any conjunct $\alpha_i$. It can be written as $\alpha_i = \alpha'_i \vee \alpha''_i \vee \alpha'''_i$. Where $\alpha'_i$ does not contain $s$ free. $\alpha''_i$ is a (possibly empty) disjunction of elements of the form $\neg \gamma_{A_{ij}}(s)$ and $\alpha'''_i$ is a possibly empty disjunction of element of the form $\gamma_{B_{ik}}(s)$.

Since $\neg \gamma_A(s) \vee \neg \gamma_B(s) \leftrightarrow \neg \gamma_{A \wedge B}(s)$ and $\gamma_A(s) \vee \gamma_B(s) \leftrightarrow \gamma_{A \vee B}(s)$, we can write $\alpha_i$ as $\alpha'_i \vee \neg \gamma_{C'_i}(s) \vee \gamma_{C''_i}(s)$. If $\alpha''_i$ is the empty disjunction we take $C'_i = f \rightarrow f$. If $\alpha'''_i$ is the empty disjunction we take $C_i$ as $f$. Clearly $(\forall s \geq t)\alpha_i$ is equivalent to $\alpha'_i \vee (\forall s \geq t)(\neg \gamma_{C'_i}(s) \vee \gamma_{C''_i}(s))$. So we get that $\alpha_i$ is equivalent to $\alpha'_i \vee \gamma_{C'_i \rightarrow C''_i}(t)$. This completes the induction step.

We therefore conclude that $\gamma(t)$ itself has the form $\bigwedge_i (\bigwedge_j \gamma_{A_{ij}}(t) \rightarrow \bigvee_k \gamma_{B_{ik}}(t))$ or equivalently if $A_i = \bigwedge_j A_{ij}$, $B_i = \bigvee_k B_{ik}$, then $\gamma(t)$ is equivalent to $\bigwedge_i (\gamma_{A_i}(t) \rightarrow \gamma_{B_i}(t))$. Now since $\gamma(t)$ is a first order truth wff so are $\gamma_{A_i}(t) \rightarrow \gamma_{B_i}(t)$ and hence $\gamma(t)$ is equivalent to $\bigwedge_i \gamma_{A_i \rightarrow B_i}(t)$ which is equivalent to $\gamma_{\bigwedge_i (A_i \rightarrow B_i)}(t)$.

This completes the proof of Proposition 6.

**Exercise 8.** Let $\gamma(t)$ be $(\forall s \geq t)(\exists s' \geq s)(P_1(s) \vee \neg P_2(s') \rightarrow (\exists s'' \geq s')P_3(s''))$.

Show that $\gamma(t)$ is obtained from $A = \neg(p_2 \wedge \neg p_3) \rightarrow p_1$.

**Example 9.** The wffs $\forall x P(x)$ and $\exists x P(x)$ are not regular.

## 4. New intuitionistic connectives

We want to investigate the general question of adding to $h$ new connectives. The case of strong negation is one particular instance. Of course, we may have some intended interpretation in mind for the new connective, but here we study these connectives from a formal point of view.

DEFINITION 1. Let $H$ be a Hilbert system in a certain language. Let $\#(x_1, \ldots, x_n)$ be a new $n$-ary connective and consider the Hilbert system $H(\Delta)$, in the new language obtained by adding to the axioms of $H$ an additional set of axioms $\Delta(\#)$ (involving $\#$).

(1) We say that $H(\Delta)$ is a conservative extension of $H$ iff for any $\varphi$, $B$, in a language without $\#$ we have

$$\varphi \vdash_H B \quad \text{iff} \quad \varphi \vdash_{H(\Delta)} B.$$

(2) We say that $\Delta$ determines $\#$ uniquely iff the following holds. Let $\#_1$, $\#_2$ be two new $n$-ary connectives and let $H^*$ be the extension of $H$ in the language with $\#_1$ and $\#_2$ and with the set of axioms $\Delta(\#_1)$ and $\Delta(\#_2)$ (i.e., we add the axioms of $\Delta(\#)$ written for $\#_1$ and for $\#_2$) then $\#_1 \vdash_{H^*} \#_2$ and $\#_2 \vdash_{H^*} \#_2$.

**Example 2.** (a) Consider $h^+$, this is a Hilbert system in a language with $\wedge$, $\rightarrow$, $f$ and without $\vee$. Let $\Delta(\vee)$ be all instances of

$$A \rightarrow A \vee B, \quad B \rightarrow A \vee B,$$
$$(A \rightarrow C) \rightarrow ((B \rightarrow C) \rightarrow (A \vee B \rightarrow C)).$$

Then clearly $h^+(\Delta) = h$.

To check whether $\Delta$ determines $\vee$ uniquely, we consider the extension $h^*$ of $h^+$ with two disjunction symbols $\vee_1$, $\vee_2$ and the additional 6 axiom schemas obtained from the three above by replacing $\vee$ by $\vee_1$ and by $\vee_2$. Since the deduction theorem is true is $h^*$ we can show that $h^* \vdash A \vee_1 B \leftrightarrow A \vee_2 B$.

(b) Consider $h^-$ and let $\sim$ be strong negation. Let $\Delta(\sim)$ be all axioms for strong negation then $h^-(\Delta) = nh$. Is $\sim$ uniquely determined by its axioms?

In the search for new intuitionistic connectives, one may define connectives in many ways, via different interpretations. We shall see in a later chapter that new connectives may be defined in second order propositional $h$ or second order $H$ (predicate calculus) or if we allow infinite disjunction, etc. In this section, we study one such type of a new connective. Let $p$ be a variable of $h$ and consider the following wff $\gamma(t)$, which can be taken as a one place truth table wff in the sense of the preceding section.

$$(3) \qquad \gamma(t, P) = (N(t) \wedge (\forall s \underset{\neq}{\geq} t)P(s)) \vee (\sim N(t) \wedge P(t))$$

where $N(t) = \exists s(s \neq t \wedge s \geqslant t)$.

This condition says that if $t$ is not an endpoint, $p$ starts being true immediately after $t$, otherwise it is true at $t$. This condition is weaker than the condition corresponding to $p$ is true at $t$ and stronger than the condition corresponding to $\neg\neg p$ holds at $t$, i.e.,

$$\gamma_p(t) \rightarrow \gamma(t, P) \rightarrow \gamma_{\neg\neg p}(t).$$

DEFINITION 4. Let $J$ be an extension of $h$ in a language with a new connective $\#(x_1, \ldots, x_n)$. $J$ is said to be an intuitionistic extension iff it satisfies the following conditions:

(1) $J$ is a conservative extension of $h$.

(2) $J$ has the disjunction property.

(3) If we add to $J$ the schema $((A \rightarrow B) \rightarrow A) \rightarrow A$ then we obtain an extension $J^*$ equivalent to $c$ and $\#(x_1, \ldots, x_n)$ collapses to some wff of $c$.

(4) The additional axioms of $J$ determine $\#$ uniquely.

(5) For no $A(x_1, \ldots, x_n)$ do we have $J \vdash \# \leftrightarrow$ A.

*Remark* 5. Notice that for any extension $J$ of $h$, the following can be shown, as was done in Chapter 4, Section 1.

(a) The deduction theorem holds for $J$.

(b) Define $\varphi \Vdash_J \psi$ by $\varphi \vdash_J \vee \psi$, then $\Vdash_J$ is a Scott system agreeing with $\vdash_J$ in which $\wedge$, $\vee$, $f$ are classical.

THEOREM 6. *Let $B(x_1, \ldots, x_n)$ be a formula of $h$ with propositional atoms $x_1, \ldots, x_n$, such that $h \nvdash \neg\neg B \rightarrow B$. Let $JB$ be the extension of $h$ in a language with $\#(x_1, \ldots, x_n)$ with the axiom schemas.*

(a) $\#(x_1, \ldots, x_n) \to C \vee (C \to B(x_1, \ldots, x_n))$.

(b) $(B(x_1, \ldots, x_n) \to B(y_1, \ldots, y_n)) \to$
$\to (\#(x_1, \ldots, x_n) \to \#(y_1, \ldots, y_n))$.

(c) $B(x_1, \ldots, x_n) \to \#(x_1, \ldots, x_n)$.

(d) $\#(x_1, \ldots, x_n) \to \neg\neg B(x_1, \ldots, x_n)$.

(e) $(\#(x_1, \ldots, x_n) \to B(x_1, \ldots, x_n)) \wedge \neg\neg B(x_1, \ldots, x_n) \to$
$\to B(x_1, \ldots, x_n)$.

*Then JB is an extension of h satisfying the conditions of* (4).

COROLLARY 7. *For the special case of $B(x) = x$, we get the following system $J_0$:*

(a) $\#A \to C \vee (C \to A)$.

(b) $(A \to B) \to (\#A \to \#B)$.

(c) $A \to \#A$.

(d) $\#A \to \neg\neg A$.

(e) $(\#A \to A) \wedge \neg\neg A \to A$.

We want to prove Theorem 5. From now on we stop indicating the atoms of # and $B$ whenever possible.

LEMMA 8. *Let $J^*B$ be the extension of h obtained by adding to h two new connectives $\#_1, \#_2$ with axioms* (a)–(e) (*of Theorem* 5) *for each of them. Then $J^*B \vdash \#_1 = \#_2$.*

*Proof.* Assume $\#_1(x_1, \ldots, x_n)$ and show $\#_2(x_1, \ldots, x_n)$.
By axiom (a) we get $\#_1 \vee \#_2 \to B$.
(1) If $\#_2$, then certainly $\#_2$.
(2) If $\#_2 \to B$, then we now show that $\#_2$ by arguing as follows:

(2a) $\neg\neg B$ (by axiom (d) and the main assumption).

(2b) $B$ (by (2a) and axiom (e)).

(2c) $\#_2$ (by (2c) and axiom (c).

From (1) and (2) we get that $\#_1 \to \#_2$. By symmetry we get $\#_2 \to \#_1$.
Let $JB^-$ be the system obtained by dropping axiom (e) of Theorem 5.

We are interested in providing $JB^-$ with a sound and complete Kripke type interpretation.

The structures for $JB^-$ have the form $(N, S, \leq, 0, D)$ where $(S, \leq, 0, D)$ is an ordinary Kripke structure for propositional intuitionistic logic and $N \subseteq S$. That is, $S$ is the set of possible worlds, $\leq$ is a reflexive and transitive (accessibility) relation on $S$, $0 \in S$ is the first element of $(S, \leq)$ and for each atomic $x$ and $t \in S$, $D(t, x)$ is a truth value. It is also required that the following holds:

$$(9) \qquad x \in S \quad \text{and} \quad \forall y(x \leq y \rightarrow x = y) \quad \text{imply} \quad x \notin N.$$

The truth table for the connectives $\wedge$, $\vee$, $\rightarrow$, $\neg$ is defined as usual (for a Kripke structure) and the truth table for # is defined using $N$ and the truth values for $B$ (as $B$ does not contain #, its truth value is obtained as usual).

DEFINITION 10. Let $A$ be a formula of $JB$, then the truth value of $A$ at a point $t \in S$ (is denoted by $[A]_t$ and) is defined by induction:

(a) $[A]_t = D(t, A)$, for atomic $A$.

(b) $[A \wedge C]_t = 1$ iff $[A]_t = [C]_t = 1$.

(c) $[A \vee C]_t = 1$ iff $[A]_t = 1$ or $[C]_t = 1$.

(d) $[A \rightarrow C]_t = 1$ if for all $s$, $t \leq s$ and $[A]_s = 1$ imply $[C]_s = 1$, $[f]_t = 0$, for all $t$.

(e) $[\#(x_1, \ldots, x_n)]_t = 1$ iff:
(1) $t \notin N$ and $[B(x_1, \ldots, x_n)]_t = 1$.
(2) $t \in N$ and for all $s$, $t \lneq s$ implies $[B(x_1, \ldots, x_n)]_s = 1$.

LEMMA 11. *For any $A$, $t$, $[A]_t = 1$ and $t \leq s$ imply $[A]_s = 1$.*
*Proof.* By induction.

LEMMA 12. *All axioms of $JB^-$ are valid in this interpretation.*
*Proof.* Simple to verify.

LEMMA 13. *$JB$ is valid in any structure $(N, S, \leq, 0, D)$ with $S$ finite, and with $N$ being its set of non-endpoints.*
*Proof.* Simple to verify.

DEFINITION 14. A consistent $JB^-$ theory $\Theta$ is said to be saturated iff whenever $A \vee A' \in \Theta$, then either $A \in \Theta$ or $A' \in \Theta$.

LEMMA 15. *Let $\Theta$ be a consistent saturated $JB^-$ theory such that $\not\Vdash_{JB^-} A \vee (A \to C)$ $A, C$ arbitrary. There exists a saturated $JB^-$ theory $\Theta \subsetneq \Theta'$ with the properties:*

(a) $\Theta' \supseteq \Theta \cup \{A\} \cup \{B(x_1, \ldots, x_n) \mid \#(x_1, \ldots, x_n) \in \Theta\}$,

(b) $\Theta' \not\Vdash_{JB^-} C$.

*Proof.* We claim that for the following theory $\Theta_0$ we have $\Theta_0 \not\Vdash_{JB^-} C$:

$$\Theta_0 = \Theta \cup \{A\} \cup \{B(x_1, \ldots, x_n) \mid \#(x_1, \ldots, x_n) \in \Theta\}.$$

For otherwise, for some $E \in \Theta$ we have:

$$JB^- \vdash (E \to \wedge B(x_1^i, \ldots, x_n^i) \to (A \to C)),$$

hence $\wedge B(x_1^i, \ldots, x_n^i) \to (A \to C) \in \Theta$. We can assume that $\bigvee_i B(x_1^i, \ldots, x_n^i) \notin \Theta$. From axiom (a) we get that $B(x_1^1, \ldots, x_n^1) \to (\bigwedge_{i>1} B(x_1^i, \ldots, x_n^i) \to (A \to C))$ implies in $JB^-$ that

$$\#(x_1^1, \ldots, x_n^1) \to A \vee \left(A \to \left(\bigwedge_{i>1} B(x_1^i, \ldots, x_n^i) \to (A \to C)\right)\right)$$

and since $\#(x_1^1, \ldots, x_n^1) \in \Theta$, $A \notin \Theta$, we get that $\bigwedge_{i>1} B(x_1^i, \ldots, x_n^i) \to (A \to C) \in \Theta$.

Continuing in this manner we get that $A \vee (A \to (A \to C)) \in \Theta$, a contradiction, since $A \vee (A \to C)$ was assumed not in $\Theta$.

Thus $\Theta_0 \not\Vdash_{JB^-} C$ and can therefore be extended to a saturated theory $\Theta'$ fulfilling the properties of Lemma 15, by Lemma 1.4.3 for $\Vdash_{JB^-}$.

(16). Let $\Delta$ be a consistent and saturated $JB^-$ theory, we show that it holds in some structure. Let $S$ be the set of all consistent and saturated theories containing $\Delta$. Let $0 = \Delta$. To define $D$, let $D(\Theta, q) = 1$ iff $q \in \Theta$, for $q$ atomic. To define $\leqslant$ let $\Theta \leqslant \Theta'$ iff (a) or (b) hold, where:

(a) $\Theta = \Theta'$.

(b) $\Theta \subsetneq \Theta'$ and whenever $\#(x_1, \ldots, x_n)$ is in $\Theta$ then $B(x_1, \ldots, x_n)$ is in $\Theta'$.

Define $N$ by $\Theta \in N$ iff for some $\#(x_1, \ldots, x_n) \in \Theta$,

$B(x_1, \ldots, x_n) \notin \Theta$. Notice that condition (8) of the interpretation holds. For let $\Theta$ be such that for all $\Theta'$, $\Theta \leq \Theta'$ implies $\Theta = \Theta'$. We must show that $\Theta \notin N$. Assume otherwise, that for some $\#(x_1, \ldots, x_n) \in \Theta$, $B(x_1, \ldots, x_n) \notin \Theta$. We first claim that for all $A$, $C$, $A \vee (A \to C) \in \Theta$, for if not, by Lemma 16, there exists a $\Theta'$, $\Theta \leq \Theta'$, $A \in \Theta'$, contradicting the assumption. Thus letting $C = E \wedge \neg E$, $A = B$, we get that $B \vee \neg B \in \Theta'$ and since $B(x_1, \ldots, x_n) \notin \Theta$, we have $\neg B(x_1, \ldots, x_n) \in \Theta$ and therefore $\neg \#(x_1, \ldots, x_n) \in \Theta$, (by axiom (d)), a contradiction.

LEMMA 17. *In the structure $(N, S, \leq, 0, D)$ just defined, we have for all $A$, $t : [A]_t = 1$ iff $A \in t$.*

*Proof.* For atomic $A$ this holds by definition. The cases of $\wedge$, $\vee$ present no difficulties. The case of $\neg$ follows from the case of $\to$.

*Case of* $\to$. Assume $(A \to A') \in \Theta$, then clearly, for any $\Theta'$, $\Theta \leq \Theta'$ and $A \in \Theta'$ imply $A' \in \Theta'$, thus $[A \to A']_\Theta = 1$.

Assume $(A \to A') \notin \Theta$. If $A \in \Theta$, then clearly $[A \to A']_\Theta = 0$. If $A \vee (A \to A') \notin \Theta$, then by Lemma 15, there exists a $\Theta'$ with $\Theta \leq \Theta'$ and $A \in \Theta'$ and $A' \notin \Theta'$. Thus again, $[A \to A']_\Theta = 0$.

*Case of* $\#$. Assume that $\#(x_1, \ldots, x_n) \in \Theta$. Then, if $\Theta \notin N$, by definition of $N$, $B(x_1, \ldots, x_n) \in \Theta$. If $\Theta \in N$, then by definition of $\leq$ (Case (b)) if $\Theta \leq \Theta'$, $\Theta \neq \Theta'$, then $B(x_1, \ldots, x_n) \in \Theta'$.

Assume that $\#(x_1, \ldots, x_n) \notin \Theta$, then by the axioms, $B(x_1, \ldots, x_n) \notin \Theta$. So if $\Theta \notin N$, then $[\#]_\Theta = 0$, because $B \notin \Theta$. If $\Theta \in N$, then for some $Y_1, \ldots, Y_n$, $\#(Y_1, \ldots, Y_n) \in \Theta$ and $B(Y_1, \ldots, Y_n) \notin \Theta$. We claim that $B(Y_1, \ldots, Y_n) \to B(x_1, \ldots, x_n) \notin \Theta$, for otherwise by axiom (b) and the fact that $\#(Y_1, \ldots, Y_n) \in \Theta$, we get $\#(x_1, \ldots, x_n) \in \Theta$, which is impossible. Thus we conclude that $B(Y_1, \ldots, Y_n) \vee (B(Y_1, \ldots, Y_n) \to B(x_1, \ldots, x_n)) \notin \Theta$. Therefore by Lemma 15, there exists a $\Theta'$ such that $B(Y_1, \ldots, Y_n) \in \Theta'$ and $B(x_1, \ldots, x_n) \notin \Theta'$ and $\Theta \leq \Theta'$ hold. Therefore by the induction hypothesis, $[\#(x_1, \ldots, x_n)]_\Theta = 0$.

Thus Lemma 17 is proved.

LEMMA 18. *$JB^-$ is complete for the proposed interpretation.*

*Proof.* Follows from Lemmas 12 and 17.

LEMMA 19. *$JB^-$ is a conservative extension of h.*

*Proof.* Follows from Lemma 18.

LEMMA 20. $JB \vdash A_1 \vee A_2$ implies $JB \vdash A_1$ or $JB \vdash A_2$.

*Proof*. Suppose $JB \not\vdash A_1$ and $JB \not\vdash A_2$. We show that $JB \not\vdash A_1 \vee A_2$.

Since $JB \not\vdash A_i$, and $\{C \mid JB \vdash C\}$ is a consistent $JB^-$ theory, there exist structures $\mathbf{M}_i = (N_i, S_i, \leqslant_i, 0_i, D_i)$ such that all theorems of $JB$ hold and $[A_i]_{0_i} = 0$. We can assume that $S_1 \cap S_2 = 0$. We define a structure $\mathbf{M} = (N, S, \leqslant, 0, D)$ as follows:

$$S = S_1 \cup S_2 \cup \{0\}.$$

$$\leqslant\, = \,\leqslant_1 \cup \leqslant_2 \cup \{0\} \times S.$$

$$D = D_1 \cup D_2 \cup \{((0, q), 0) \mid q \text{ atomic}\}.$$

$$N = N_1 \cup N_2 \cup \{0\}.$$

Clearly $\mathbf{M}$ is a $JB^-$ structure. Also $[A_1 \vee A_2]_0 = 0$, since $[A_i]_{0_i} = 0$ (Lemma 1.1). All we have to show now is that $\mathbf{M}$ validates $JB$. The only axiom we have to check is axiom (e). Assume that for some $x_1, \ldots, x_n$ we have, $[\#(x_1, \ldots, x_n) \to B(x_1, \ldots, x_n)]_0 = 1$, and $[\neg\neg B(x_1, \ldots, x_n)]_0 = 1$. Now by Lemma 11 these two sentences are true in $0_1$ and $0_2$ and since $JB$ is valid in $0_i$ we get that $[B(x_1, \ldots, x_n)]_{0_i} = 1$. But now by definition, since $0 \in N$, we get $[B(x_1, \ldots, x_n)]_0 = 1$.

Thus Lemma 20 is proved. To prove Theorem 6 we need now a completeness theorem for $JB$. We start with:

*Remark* 21. A structure is called regular iff: $t \notin N$ iff $\forall y(t \leqslant y \to t = y)$. We show that if $\Delta$ is a consistent and saturated $JB$ theory, then it is valid in some regular structure. For let $(S, \leqslant, \Delta, D)$ be constructed as in (16) above. Let $N$ be defined by $t \in N$ iff $\exists s(t \lneqq s)$. We thus get a regular structure for $\Delta$, which is identical with the structure obtained in (16) except for $N$.

We now want to show that for all $A$,

$$[A]_\Theta = 1 \quad \text{iff} \quad A \in \Theta.$$

The proof is by induction. All cases except $\#$ follow as before. The only case we have to check is that of $\#$, since $N$ is a different set now.

Assume $\#(x_1, \ldots, x_n) \in \Theta$. Then if $\Theta \in N$, then certainly by definition of $\leqslant$, if $\Theta \lneqq \Theta'$ then $B(x_1, \ldots, x_n) \in \Theta'$. If $t \notin N$, then $\Theta$ is a classical theory and hence $\#(x_1, \ldots, x_n) \leftrightarrow B(x_1, \ldots, x_n) \in \Theta$ and so $B(x_1, \ldots, x_n) \in \Theta$. In both cases $[\#(x_1, \ldots, x_n)]_\Theta = 1$.

Assume $\#(x_1, \ldots, x_n) \notin \Theta$. If $\#(x_1, \ldots, x_n) \to B(x_1, \ldots, x_n) \notin \Theta$, then by construction 16, there exists a $\Theta'$, $\Theta \lneq \Theta'$, $B(x_1, \ldots, x_n) \notin \Theta'$. If $\#(x_1, \ldots, x_n) \to B(x_1, \ldots, x_n) \in \Theta$, then since $\#(x_1, \ldots, x_n) \notin \Theta$, and $\vdash_{JB} B(x_1, \ldots, x_n) \to \#(x_1, \ldots, x_n)$, then $B(x_1, \ldots, x_n) \notin \Theta$ and since $\Theta$ is a *JB* theory, $(\#(x_1, \ldots, x_n) \to B(x_1, \ldots, x_n)) \wedge \neg\neg B(x_1, \ldots, x_n) \to B(x_1, \ldots, x_n) \in \Theta$ and hence $\neg\neg B(x_1, \ldots, x_n) \notin \Theta$.

Consider $C = \neg B(x_1, \ldots, x_n) \vee \neg\neg B(x_1, \ldots, x_n)$. If $\Theta \nVdash C$, then again by construction, there exists a $\Theta'$, $\Theta \lneq \Theta'$ and $\neg B(x_1, \ldots, x_n) \in \Theta'$ and hence $B(x_1, \ldots, x_n) \notin \Theta'$. If $C \in \Theta$, then $\neg B(x_1, \ldots, x_n) \in \Theta$. So if $\Theta \notin N$, we are finished since $\neg B(x_1, \ldots, x_n) \in \Theta$ and hence $[\#(x_1, \ldots, x_n)]_\Theta = 0$ by the induction hypothesis. If $\Theta \in N$, then for some $\Theta'$, $\Theta \lneq \Theta'$ and $B(x_1, \ldots, x_n) \notin \Theta'$.

Thus we found that at all cases, either $\Theta \notin N$ and $B(x_1, \ldots, x_n) \notin \Theta$ and hence $[\#(x_1, \ldots, x_n)]_\Theta = 0$ or for some $\Theta'$, $\Theta \lneq \Theta'$ and $B(x_1, \ldots, x_n) \notin \Theta'$ and so again by the induction hypothesis $[\#(x_1, \ldots, x_n)]_\Theta = 0$.

THEOREM 21. *JB is complete for the class of all regular finite $JB^-$ structures.*

*Proof.* By Lemma 13 all theorems of *JB* are valid in all regular finite $JB^-$ structures.

Suppose $JB \nVdash A$, we want to show that for some regular finite structure $(N^*, S^*, \leqslant^*, 0^*, D^*)$, $[A]_{0^*} = 0$. By Lemma 20, *JB* has the disjunction property and hence $\{C \mid JB \vdash C\}$ is a saturated $JB^-$ theory and therefore by the completeness theorem, given above, for some $JB^-$ regular structure $(N, S, \leqslant, 0, D)$ we have for all $C$, $[C]_0 = 1$ iff $JB \vdash C$. In particular $[A]_0 = 0$.

We now proceed to define a finite regular $(N^*, S^*, \leqslant^*, 0^*, D^*)$ with $[A]_{0^*} = 0$. Let $\Theta_0$ be the set of all subformulas of $A$ and let $\Theta_1 = \{B(C_1, \ldots, C_n) \mid \#(C_1, \ldots, C_n) \in \Theta_0\}$. Let $\Theta_2$ be the set of all subformulas of elements of $\Theta_0 \cup \Theta_1$ and let $\Theta$ be the closure of under the application of $\neg$. Clearly $\Theta$ is finite, and satisfies the condition that if $\#(C_1, \ldots, C_n) \in \Theta$ then $B(C_1, \ldots, C_n) \in \Theta$.

Define a partial function $f$ on $S$ as follows.

(1) Let $C \in \Theta$ be a wff of the form $C_1 \to C_2$ or of the form $\neg C_1$ (regarded as $C_1 \to$ *falsity*) such that $[C_1]_t = 0$, and $[C]_t = 0$. Let $f(t, C) \in S$ be a point with the property $t \lneq f(t, C)$ and such that:

$$[C_1]_{f(t,C)} = 1, \qquad [C_2]_{f(t,C)} = 0, \qquad ([C_1]_{f(t,C)} = 1 \text{ for the case of negation}).$$

(2) For each $t \in S$ and $C = \#(C_1, \ldots, C_n) \in \Theta$ such that $[\neg B(C_1, \ldots, C_n)]_t = 0$ and $[C]_t = 0$ let $f(t, C)$ be a point with $t \lneq f(t, C)$ and either $[\neg B(C_1, \ldots, C_n)]_{f(t,C)} = 1$ or $[C]_{f(t,C)} = 1$ and $[B(C_1, \ldots, C_n)]_{f(t,C)} = 0$.

$f(t, C)$ can be defined in case 2 above because if $[C]_t = 0$, then by axiom (e) either $[\neg\neg\neg B]_t = 0$ or $[C \to B]_t = 0$ and hence $f(t, C)$ can be chosen. Notice that indeed $t \lneq f(t, C)$ in this case.

Let $S^* \subseteq S$ be defined as the smallest set such that $0 \in S^*$ and if $t \in S^*$ and $f(t, C)$ is defined for some $C$ then $f(t, C) \in S^*$. $S^*$ is finite since in any $s = f(t, C)$, $f(s, C)$ is no longer defined.

Let $0^* = 0$ and $D^*, \leqslant^*$ be the restrictions of $D$ (respectively $\leqslant$) to $S^*$. Let $N^*$ be the set of endpoints of $S^*$.

LEMMA $*$. *If $t \in S^*$ and $t$ is an endpoint of $S^*$ and $\#(C_1, \ldots, C_n) \in \Theta$ and $[\#(C_1, \ldots, C_n)]_t = 1$ then $[B(C_1, \ldots, C_n)]_t = 1$.*

*Proof.* Otherwise, we know that $B(C_1, \ldots, C_n) \in \Theta$ and hence so is $\neg\neg B(C_1, \ldots, C_n)$. Now $[\neg\neg B(C_1, \ldots, C_n)]_t = 1$ and hence $[\neg B(C_1, \ldots, C_n)]_t = 0$, and hence for $C = \neg B(C_1, \ldots, C_n)$, $f(t, C)$ is defined and so $t$ cannot be an endpoint.

Notice that by definition if $t \in N^*$ then $t \in N$.

LEMMA 22. *For any $A \in \Theta$, $t \in S^*$*

$$[A]_t^* = [A]_t$$

*where $[A]_t^*$ is the value in $(N^*, S^*, \leqslant^*, 0^*, D^*)$.*

*Proof.* By double induction on the number of *nested* occurrences of $\#$ in $A$ and on the connectives of $A$. For $A$ atomic this holds by definition. $\wedge$, $\vee$, $f$, $\neg$ present no difficulties.

*Case of $C = C_1 \to C_2$.* If $[C_1 \to C_2]_t = 1$, clearly $[C_1 \to C_2]_t^* = 1$. If $[C_1 \to C_2]_t = 0$, then if $[C_1]_t = 1$, then since $[C_2]_t = 0$, we get by the induction hypothesis $[C_1]_t^* = 1$, $[C_2]_t^* = 0$ and hence $[C]_t^* = 0$. If $[C_1]_t = 0$, then $f(t, C)$ is defined and in $S^*$ and by construction and the induction hypothesis $[C_1]_{f(t,C)}^* = 1$, $[C_2]_{f(t,C)}^* = 0$ and hence $[C]_t^* = 0$.

*Case of $C = \#(C_1, \ldots, C_n)$.* (1) Assume $[C]_t = 1$.

(1a) If $t \notin N$, then by Lemma $*$, $[B(C_1, \ldots, C_n)]_t = 1$ and by the induction hypothesis, since $B(C_1, \ldots, C_n)$ has less nested occurrences of $\#$, $[B(C_1, \ldots, C_n)]_t^* = 1$ and hence $[C]_t^* = 1$.

(1b) If $t \in N$, then if $t \in N^*$, then clearly $[B(C_1, \ldots, C_n)]_s = 1$ in every $s \gneq t$ and hence by the induction hypothesis $[B(C_1, \ldots, C_n)]_s^* =$

1 for every $s \in S^*$, $s \gneq^* t$ and so $[C]_t^* = 1$. If $t \notin N^*$, then by definition of $N^*$, since $C = \#(C_1, \ldots, C_n) \in \Theta$, $[B(C_1, \ldots, C_n)]_t = 1$ and hence $[B(C_1, \ldots, C_n)]_t^* = 1$ and so $[C]_t^* = 1$.

(2) Now assume $[C]_t = 0$.

(2a) If $t \notin N$, then $t \notin N^*$ and $[B(C_1, \ldots, C_n)]_t = 0$ and hence by the induction hypothesis $[C]_t^* = 0$.

(2b) If $t \in N$ and $t \notin N^*$, then certainly $[B(C_1, \ldots, C_n)]_t = 0$ and hence by the induction hypothesis $[C]_t^* = 0$. If $t \in N$ and $t \in N^*$, then if $[\neg B(C_1, \ldots, C_n)]_t = 1$, then since $t \in \mathcal{N}^*$ is not an endpoint, there exists an $s \in S^*$, $t \lneq s$ and so $[B(C_1, \ldots, C_n)]_s = 0$ and by the induction hypothesis $[B(C_1, \ldots, C_n)]_s^* = 0$ and so $[C]_t^* = 0$. If $t \in N$, $t \in N^*$ and $[\neg B(C_1, \ldots, C_n)]_t = 0$ then since $[B(C_1, \ldots, C_n)]_t = 0$ as well, we are in a case where $f(t, C)$ is defined, and so $[B(C_1, \ldots, C_n)]_{f(t,C)} = 0$ and $t \lneq f(t, C)$ and $f(t, C) \in S^*$ and hence by the induction hypothesis $[C]_t^* = 0$.

This completes the proof of Lemma 22.

Since $A \in \Theta$, $[A]_{0^*}^* = 0$ and hence we found a finite regular structure in which $A$ is false. Thus Theorem 21 is proved.

*Remark* 23. We can now show that properties (1)–(5) mentioned in (4) hold for # of *JB*. We check them one by one.

(1) *JB* is a conservative extension of $h_1$ since it is complete for the same kind of interpretation as $h_1$, namely finite trees.

(2) To show that for no $C$ of $h_1$ can be we have $JB \vdash \# \leftrightarrow C$ we notice the following fact. Let $(N, S, \leqslant, 0, D)$ be a regular structure with $(S, \leqslant, 0)$ a finite tree. Let $*$ be a new point and let $S^* = S \cup \{*\}$ and let $\leqslant^* = \leqslant \cup (\{*\} \times S^*)$.

Thus what we are doing is adding a new point $*$ at the bottom of $S$, i.e., $* \leqslant^* 0$. Let $D^*$ be like $D$ on $S$ and let $D^*(*, q) = D(0, q)$, and let $N^* = N \cup \{*\}$.

Prove the following:

($\Box$) For any $C$ of $h_1$, $[C]_* = [C]_0$, in the new structure. This does not generally hold for #, since condition (e2) of Definition 10 can be destroyed. For let $q_1, \ldots, q_n$ be atoms and suppose we have a structure with $[\#(q_1, \ldots, q_n)]_0 = 1$ but $[B(q_1, \ldots, q_n)]_0 = 0$. Then by Definition 10 clause (e2), $[\#(q_1, \ldots, q_n)]_* = 0$, thus # does not satisfy ($\Box$).

(3) Disjunction property was proved in Lemma 20.

(4) Clear from the axioms of *JB*.
(5) Shown in Lemma 1.

## 5. Modality

There is no agreement as to the notion of intuitionistic modality. Following Curry, we consider here the system *s4h* of intuitionistic modality. The language contains $\wedge$, $\vee$, $\to$, $f$ and $\Box$, $\Box$ being a unary modal operator.

DEFINITION 1. Let *s4h* be the extension of *h* in a language with an additional unary operator $\Box$, with the additional axiom schemas and rules below:
*axiom schemas*:

(a) $\Box A \to A.$

(b) $\Box A \to \Box\Box A.$

(c) $\Box(A \to B) \to (\Box A \to \Box B).$

*provability rule*:

$$\frac{\vdash A}{\vdash \Box A}.$$

LEMMA 2. (a) *The deduction theorem holds for s4h.*

(b) *Let* $\varphi \Vdash_{s4h} \psi$ *be defined as* $\varphi \vdash_{s4h} \vee \psi$, *then* $\Vdash_{s4h}$ *is a consequence relation that agrees with s4h in which* $\wedge$, $\vee$, $f$ *are classical.*

*Proof.* Similar to the proofs in Chapter 4, Section 1.

DEFINITION 3. (1) By a modal propositional *s4h* structure we mean a structure of the form $(S, R, \leqslant, 0, D)$ where $R \subseteq S^2$, $\leqslant \subseteq S^2$ are two transitive and reflexive relations $0 \in S$ and $D$ a function assigning a value in $\{0, 1\}$ to each pair $(t, q)$, $t \in S$, $q$ atomic. The following holds:

(a) $t \leqslant s \wedge D(t, q) = 1$ imply $D(s, q) = 1.$

(b) $t \leqslant s \wedge sRs'$ imply $tRs'.$

(2) Define the value $[A]_t$, for a wff $A$ and $t \in S$ by induction.

(a) $[A]_t = D(t, A)$ for $A$ atomic.

(b) $[A \wedge B]_t = 1$ iff $[A]_t = [B]_t = 1$

$[A \vee B]_t = 0$ iff $[A]_t = [B]_t = 0$ $[f]_t = 0$.

(c) $[A \to B]_t = 1$ iff for all $s$, $t \leqslant s$ and $[A]_s = 1$ imply $[B]_s = 1$.

(d) $[\Box A]_t = 1$ iff for all $s$, $tRs$ implies $[A]_s = 1$.

$A$ is said to be valid in the structure iff $[A]_0 = 1$.

THEOREM 4. *$s4h \vdash A$ iff $A$ is valid in every structure.*

*Proof.* One direction is easy to verify, namely, that if $s4h \vdash A$ then $A$ is valid in every structure. For the other direction, we need some lemmas. The notion of a consistent and saturated theory is the same as that of Chapter 4, Section 1. Here we use the Scott consequence $\Vdash_{s4h}$. Thus by Chapter 1, Section 4 every $\Vdash_{s4h}$ consistent theory can be extended to a saturated theory.

LEMMA 5. *Let $(\Delta, \Theta)$ be a theory such that $\Delta \nVdash_{s4h} A \to B$, then there exists a saturated theory $(\Delta', \Theta')$, with $\Delta' \supseteq \Delta \cup \{A\}$ such that $B \notin \Delta'$.*

*Proof.* By the deduction theorem $(\Delta \cup \{A\}, \{B\})$ is $\Vdash_{s4h}$ consistent and can therefore be extended to a saturated theory.

LEMMA 6. *Let $(\Delta, \Theta)$ be a consistent theory such that $\Delta \vdash_{s4h} A$, then there exists a saturated theory $(\Delta', \Theta')$ such that $\Delta' \nvdash_{s4h} A$ and for all $B$ if $\Box B \in \Delta$ then $B \in \Delta'$.*

*Proof.* We show that the following theory is consistent

$$(\{B \mid \Delta \vdash_{s4h} \Box B\}, \{A\}).$$

For otherwise for some $B_1, \ldots, B_n$

$$\vdash_{s4h} B_1 \to (\cdots \to (B_n \to A) \ldots)$$

hence

$$\vdash_{s4h} \Box (B_1 \to (\cdots \to (B_n \to A) \ldots))$$

and hence

$$\vdash_{s4h} \Box B_1 \to (\cdots \to (\Box B_n \to \Box A) \ldots)$$

and hence, since $\Delta \vdash_{s4h} \Box B_i$, we get $\Delta \vdash_{s4h} \Box A$, a contradiction. Now since this theory is consistent it can be extended to a saturated theory $(\Delta', \Theta')$.

LEMMA 7. *Define* $\Delta R \Delta'$ *iff for all* $B$, $\Delta \vdash \Box B$ *implies* $\Delta' \vdash B$ *then* $R$ *is reflexive and transitive.*

*Proof.* Since $\vdash \Box A \to A$, we get reflexivity. Transitivity follows from $\vdash \Box A \to \Box\Box A$.

LEMMA 8. $\Delta \subseteq \Delta' \wedge \Delta' R \Delta''$ *imply* $\Delta R \Delta''$.

*Proof.* Easy.

To prove the completeness theorem, let $(\Delta_0, \Theta_0)$ be a consistent saturated theory. We shall construct a structure that validates exactly all members of $\Delta_0$. Let $S = \{\Theta \mid (\Theta, \Theta')$ is a consistent saturated theory of $\Vdash_{s4h}\}$. For $t \in S$, let $D(t, q) = 1$ iff $q \in t$. Consider $(S, R, \subseteq, \Delta_0, D)$. This is a modal structure.

LEMMA 9. *For any* $A, t \in S$

$$[A]_t = 1 \quad \textit{iff} \quad A \in t.$$

*Proof.* By induction on $A$.

For atomic $A$ this follows by definition. The cases of $\wedge$, $\vee$, $f$ present no difficulty. The case of $\to$ follows from Lemma 5. The case of $\Box$ follows from Lemma 6.

This proves the completeness theorem.

**Exercises 10.** (a) Let $s4c$ be the extension of $s4h$ with the schema

$$((A \to B) \to A) \to A.$$

Show that the $s4c$ is complete for the class of all structures satisfying $x \leqslant y \to x = y$.

(b) Find a faithful translation of $s4h$ into $s4c$.

**Exercise 11.** Write a finitely axiomatizable Hilbert system (an extension of $h$) that is complete for the following interpretation.

The structures have the form $(S \times T, \leqslant, (0,0), D)$, where $\leqslant S^2$ is reflexive and transitive. $(0,0) \in S \times T$, $D((x, y), q) = 1$ and $x \leqslant x'$ implies $D((x', y), q) = 1$. The truth values $[A]_{(t,s)}$ is defined by induction by

(a) $\quad [A]_{(t,s)} = D((t, s), A)$ for $A$ atomic.

(b) $\quad [A \wedge B]_{(t,s)} = 1$ iff $[A]_{(t,s)} = [B]_{(t,s)} = 1$

$\quad [A \vee B]_{(t,s)} = 0$ iff $[A]_{(t,s)} = [B]_{(t,s)} = 0$

$\quad [f]_{(t,s)} = 0.$

(c) $[A \to B]_{(t,s)} = 1$ iff for all $t' \geq t$,

$[A]_{(t's)} = 1$ implies $[B]_{(t',s)} = 1$.

(d) $[\Box A]_{(t,s)} = 1$ iff for all $t' \geq t$ and all $s'$, $[A]_{(t',s')} = 1$.

**Exercise 12.** Consider the class of all $s4h$ structures with the property that $tRs \Leftrightarrow t \lneq s$.

What system is complete for this class?

**Exercise 13.** *The tense logical nature of the intuitionistic connectives.*

Imagine a flow of time $(S, R)$, where $S$ is a set and $R \subseteq S^2$ is reflexive and transitive. Then we can regard $S$ as the set of moments of time and $R$ as the earlier–later relation. We assume that with each moment of time $t \in S$ a function $D_t$ is associated that tells us which atomic $q$ are known to be true at $t$. We understand that as time goes on, more and more atoms become known to be true, i.e., $tRs \wedge D_t(q) = 1 \to D_s(q) = 1$.

This interpretation is really the propositional Kripke interpretation if we give the Kripke inductive truth table for $\wedge$, $\vee$, $\to$, $\neg$, $f$. In particular we recall that

(1) $[A \to B]_t = 1$ iff $\forall s(tRs \wedge [A]_s = 1 \to [B]_s = 1)$.

So the table for $\to$ goes into the future. We are interested in adding another connective that goes into the past. Let $T$ be such connectives with truth table.

(2) $[TA]_t = 1$ iff for some $s \neq t$ such that $sRt$,
$[A]_s = 1$.

DEFINITION. Let $th$ be the Hilbert system in a language with $\wedge$, $\vee$, $\to$, $\neg$, $T$ and the following axioms, in addition to the axioms and rules of $h$, taken for this language.

(a) $TA \to A$.

(b) $T(A \vee B) \to TA \vee TB$.

(c) $T(A \wedge B) \to TA \wedge TB$.

(d) $A \to (B \vee (B \to TA))$.

The provability rules:

$$\frac{\vdash A \to B}{\vdash TA \to TB}, \frac{\vdash A}{\vdash TA}.$$

THEOREM. *th is complete for the above interpretation.*

*Proof.* Give a Henkin completeness proof and use the following lemmas:

(1) *Let* $(\Delta, \Theta)$ *be a th consistent saturated theory S.t.* $\Delta \vdash A \vee (A \rightarrow B)$, *then the following theory is th consistent*:

$$(\Delta \cup \{A\} \cup \{TC \mid C \in \Delta\}, \{B\}).$$

(2) *Let* $(\Delta, \Theta)$ *be a consistent and saturated theory such that* $TA \in \Delta$, *then* $(\{C \mid TC \in \Delta\}, \Theta)$ *is consistent and saturated.*

**Exercise 14.** Is $T$ of the previous exercise the dual of $\rightarrow$?

Is it the dual of $\#$ of Corollary 4.7?

See (1.3.15) for the notion of dual connectives.

## NOTES

The material on negation is a refinement of the presentation in Curry's book. Strong negation, introduced by Nelson has been studied by many authors. We used here Y. Gurevitz and R. Thomason's papers. See H. Rasiowa's book for algebraic treatment. Section 3 is based on a paper of McCullough. Section 4 is based on a paper by the author.

CHAPTER 8

# THE INTERPOLATION THEOREM

The interpolation theorem for Heyting's predicate calculus states that if $\vdash_H A \to B$ then there is a wff $C$ containing only the constants and predicates appearing in both $A$ and $B$ such that $\vdash_H A \to C$ and $\vdash_H C \to B$.

In classical logic, the interpolation theorem is equivalent to the consistency theorem, which reads:

## 1. General consistency theorem

*Let $L_1$, $L_2$ be two languages differing only in the atomic predicates and constants they contain, but otherwise containing the connectives and quantifiers of $C(\wedge, \vee, \to, f, \exists, \forall)$. Let $\Delta_1$ and $\Delta_2$ be two consistent theories in the languages $L_1$ and $L_2$ resp. such that $\Delta_1 \cap \Delta_2$ is complete in the common language. Then $\Delta_1 \cup \Delta_2$ is consistent.*

Before we describe the results for the case of $H$ recall the following (Chapters 1, 2, 3):

DEFINITION 1. By an intuitionistic (or $H$) theory we mean a pair $(\Delta, \Gamma)$ of sets of wffs.

$(\Delta, \Gamma)$ is said to be complete iff for all $A$ either $A \in \Delta$ or $A \in \Gamma$. $(\Delta, \Gamma)$ is said to be consistent iff for no $\varphi \subseteq \Delta$ $\psi \subseteq \Gamma$ do we have $\varphi \vdash_H \vee \psi$. A Kripke structure validates $(\Delta, \Gamma)$ iff all wffs of $\Delta$ hold in the structure and all wff of $\Gamma$ fail to hold. In Chapter 3 we proved:

THEOREM 2. $(\Delta, \Gamma)$ *is consistent iff it is valid in a structure.*

We now give the general form of Robinson's consistency statement.

(3) Let $(\Delta_1, \Gamma_1)$, $(\Delta_2, \Gamma_2)$ be two consistent theories in the languages $L_1$ and $L_2$ respectively, with the property that the common theory (in the common language) $(\Delta_1 \cap \Delta_2, \Gamma_1 \cap \Gamma_2)$ is complete, then $(\Delta_1 \cup \Delta_2, \Gamma_1 \cup \Gamma_2)$ is consistent.

The following is true:

THEOREM 4. (a) *The general form of Robinson's consistency statement* (3) *fails to hold, see* (34) *below.*

(b) *The version of* (3) *for the case of* $\Gamma_1 \subseteq \Gamma_2$ *(and hence* $\Gamma_1$ *is in the common language) is equivalent to the Craig interpolation theorem.*

The proofs will be given below.

## 2. Robinson's Theorem

In this section we give the proof of Robinson's theorem for the case (4b) (i.e. $\Gamma_1 \subseteq \Gamma_2$).

Let $(\Delta_0, \Gamma_0)$, $(\Theta_0, \Gamma_0^*)$ be two consistent theories in the languages $L_0$ and $M_0$ respectively such that

(5) $\Gamma_0$ is a set of formulae in the common language $L_0 \cap M_0$ and $\Gamma_0 \subseteq \Gamma_0^*$.

(6) $(\Delta_0 \cap \Theta_0, \Gamma_0)$ is a complete theory in the language $L_0 \cap M_0$.

We intend to construct a Kripke structure of $(\Delta_0 \cup \Theta_0, \Gamma_0^*)$. We assume our languages to be countable and to *contain no individual constants which are not in the common language.*

Let $C_1, C_2 \ldots$ be a sequence of pairwise disjoint countable sets of individual constants. Define:

$$L_1 = L_0 + C_1$$
$$M_1 = M_0 + C_1$$
$$\ldots\ldots\ldots$$
$$L_{n+1} = L_n + C_n$$
$$M_{n+1} = M_n + C_n$$
$$\ldots\ldots\ldots$$
$$L_\omega = \bigcup_n L_n \quad M_\omega = \bigcup_n M_n,$$

where $L + C$ is the language obtained from $L$ by adding all the constants of $C$.

(7) Recall that: A theory $(\Delta, \Gamma)$ is said to be saturated iff

(a) $A, B \in \Delta \Rightarrow A \wedge B \in \Delta$.

(b) $A \vee B \in \Delta \Rightarrow A \in \Delta$ or $B \in \Delta$.

(c) $\exists x A(x) \in \Delta \Rightarrow$ for some $u$ of the language $A(u) \in \Delta$.

(d) $\Delta \vdash A \Rightarrow A \in \Delta$.

In the definition of saturation $\Gamma$ plays no role.

It is known (Chapter 1) that any theory $(\Delta, \Gamma)$ can be extended to a saturated and complete theory $(\Delta, \Gamma')$ in a language with $\aleph_0$ more individual variables such that: $\Delta \subseteq \Delta'$ and $\Gamma \subseteq \Gamma'$.

We now turn to the construction of the structure.

First note that $(\Delta_0, \Gamma_0)$ is not necessarily saturated. We may extend $(\Delta_0, \Gamma_0)$ to a complete and saturated theory $(\Delta_1, \Gamma_1')$ in the language $L_1$. We are not interested, however, in $\Gamma_1'$ and so we will only regard $(\Delta_1, \Gamma_0)$. It is a saturated theory in the language $L_1$.

Note that $\Delta_0 \subseteq \Delta_1$, $\Gamma_0 \subseteq \Gamma_1'$. We now turn to prove some basic lemmas:

LEMMA 8. *The following theory is consistent*: $(\Theta_0 \cup \{E \mid E \in \Delta_1 \cap M_1\}, \Gamma_0^*)$.

*Proof*. Otherwise for some $\theta_0 \in \Theta_0$, some $E_1 \in \Delta_1 \cap M_1$ and some $\gamma_0^* \in \Gamma_0^*$ we have: $\vdash \theta_0 \wedge E_1 \rightarrow \gamma_0^*$.

Note that $\Delta_1$ and $\Theta_0$ are closed under conjunctions and $\Gamma_0^*$ under disjunctions.

So

$$\vdash E_1 \rightarrow (\theta_0 \rightarrow \gamma_0^*)$$

$$\vdash (\exists c_1 \ldots) E_1 \rightarrow (\theta_0 \rightarrow \gamma_0^*)$$

where $c_1 \ldots$ are the new variables that appear in $E_1$ and not in $\theta_0 \rightarrow \gamma_0^*$.

Now $(\exists c_1 \ldots) E_1$ is a sentence of language $L_0 \cap M_0$. Now since the $L_0 \cap M_0$-language-part of $(\Theta_0, \Gamma_0^*)$ is complete we get that $(\exists c_1 \ldots) E_1$ is either in $\Theta_0$ or in $\Gamma_0^*$. If it is in $\Gamma_0^*$ then it is in $\Gamma_0$ by (6). Now since $\Gamma_0 \subseteq \Gamma_1'$ we get that $(\exists c_1 \ldots) E_1 \notin \Delta_1$ which contradicts $E_1 \in \Delta_1$.

So we conclude that $(\exists c_1 \ldots) E_1 \in \Theta_0$, and therefore $(\theta_0 \rightarrow \gamma_0^*) \in \Theta_0$, but this contradicts the consistency of $(\Theta_0, \Gamma_0^*)$. Therefore we conclude that the above theory is consistent.

(9) Now we extend the theory of (8) to a saturated and complete theory in the language $M_2$. Let us denote this theory by $(\Theta_2, \Gamma_2^*)$ and let $\Gamma_2$ stand for $\Gamma_2^* \cap L_2$. We have that $\Gamma_0 \subseteq \Gamma_2$.

LEMMA 10. *The following is consistent*: $(\Delta_1 \cup \{E_2 | E_2 \in \Theta_2 \cap L_2\}, \Gamma_2)$.

*Proof.* Otherwise for some $\delta_1 \in \Delta_1$, $E_2 \in \Theta_2 \cap L_2$ and $\gamma_2 \in \Gamma_2$ we have:

$$\vdash \delta_1 \wedge E_2 \rightarrow \gamma_2$$

$$\vdash \delta_1 \rightarrow (E_2 \rightarrow \gamma_2)$$

$$\vdash \delta_1 \rightarrow (\forall c_2 \ldots)(E_2 \rightarrow \gamma_2)$$

where $c_2 \ldots$ are the new constants. Now since this is a sentence $L_1$ and $\delta_1 \in \Delta_1$ we get that $(\forall c_2 \ldots)(E_2 \rightarrow \gamma_2) \in \Delta_1 \cap M_1$ and so by our construction in (8) we get $(\forall c_2 \ldots)(E_2 \rightarrow \gamma_2) \in \Theta_2$. But this is a contradiction since $E_2 \in \Theta_2$ and $\gamma_2 \in \Gamma_2^*$; and $(\Theta_2, \Gamma_2^*)$ is consistent.

(11) Now extend the theory of (10) to a saturated and complete theory $(\Delta_3, \Gamma_3')$ in the language $L_3$, with the properties that $\Gamma_2 \subseteq \Gamma_3'$ and $\Delta_2 \cup \{E_2\text{'s}\} \subseteq \Delta_3$. Again we are interested only in $(\Delta_3, \Gamma_2)$.

LEMMA 12. *The following is consistent*: $(\Theta_2 \cup \{E_3 | E_3 \in \Delta_3 \cap M_3\}, \Gamma_2^*)$.

*Proof.* Otherwise for some $\theta_2 \in \Theta_2$, $E_3 \in \Delta_3 \cap M_3$ and $\gamma_2^* \in \Gamma_2^*$ we get $\vdash \theta_2 \wedge E_3 \rightarrow \gamma_2^*$, $\vdash \exists c_3 E_3 \rightarrow (\theta_2 \rightarrow \gamma_2^*)$.

Now if $\exists c_3 E_3 \in \Gamma_2^*$ then it is in $\Gamma_2$ and since $\Gamma_2 \subseteq \Gamma_3'$ it cannot be in $\Delta_3$ which contradicts $E_3 \in \Delta_3((\Delta_3, \Gamma_3')$ is complete!).

So $\exists c_3 E_3 \in \Theta_2$ and so $(\theta_2 \rightarrow \gamma_2^*) \in \Theta_2$ which is a contradiction.

Now using the set of (12) we construct $(\Theta_4, \Gamma_4^*)$ and continue to construct $(\Delta_5, \Gamma_5')$, $(\Delta_5, \Gamma_4)$ and so on. We obtain the following sequences.

(13) $\quad (\Delta_0, \Gamma_0), \quad (\Delta_1, \Gamma_0), \quad (\Delta_3, \Gamma_2) \ldots$

$\quad (\Theta_0, \Gamma_0^*), \quad (\Theta_2, \Gamma_2^*), \quad (\Theta_4, \Gamma_4^*) \ldots$

The following holds:

(a) $(\Theta_{2n}, \Gamma_{2n}^*)$ is a complete and saturated theory.

(b) $(\Delta_{2n+1}, \Gamma_{2n})$ is a saturated theory.

(c) $\Theta_{2n} \subseteq \Theta_{2n+2}$, $\Gamma_{2n}^* \subseteq \Gamma_{2n+2}^*$ and $\Gamma_{2n} = \Gamma_{2n}^* \cap L_{2n}$.

(d) $\Delta_0 \subseteq \Delta_1 \subseteq \Delta_3 \subseteq \ldots$.

Now let $\Delta_\omega = \bigcup_n \Delta_n$, $\Gamma_\omega = \bigcup_n \Gamma_n$, $\Gamma_\omega^* = \bigcup_n \Gamma_n^*$, $\Theta_\omega = \bigcup_n \Theta_n$. We now show that

(e) $(\Delta_\omega, \Gamma_\omega)$, $(\Theta_\omega, \Gamma^*_\omega)$ fulfill properties (5) and (6) and are saturated in the language $L_\omega$, $M_\omega$ respectively.

*Proof.* Clearly these theories are saturated and $\Gamma_\omega \subseteq \Gamma^*_\omega$. Now to show that $(\Delta_\omega \cap \Theta_\omega, \Gamma_\omega)$ is complete $L_\omega \cap M_\omega$ theory let $\alpha$ be a sentence in this language. So $\alpha \in L_{2n} \cap M_{2n}$ for some $n$. Now $(\Theta_{2n}, \Gamma^*_{2n})$ is complete and so either $\alpha \in \Theta_{2n}$ or $\alpha \in \Gamma^*_{2n}$. If $\alpha \in \Theta_{2n}$ it will be also in $\Delta_{2n+1}$ and if $\alpha \in \Gamma^*_{2n}$ it is certainly in $\Gamma_{2n}$ and so either $\alpha \in \Delta_\omega \cap \Theta_\omega$ or $\alpha \in \Gamma_\omega$.

(14) Now we conclude that we may assume that our original pair $(\Delta_0, \Gamma_0)$, $(\Theta_0, \Gamma^*_0)$ is a pair of *saturated* theories and we shall prove the theorem for this case.

We now turn to another series of lemmas and constructions.

(15) Let $F = (F_1 \to F_2) \notin \Delta_0$. So clearly $(\Delta_0 \cup \{F_1\}, \{F_2\})$ is consistent. We now extend this theory to a complete and saturated theory $(\Delta_1(F), \Gamma_1(F)^*)$ in the language $L_1$. Let $\Gamma_1(F)$ be $\Gamma_1(F)^* \cap M_1$.

LEMMA 16. *The following is consistent*: $(\Theta_0 \cup \{E \mid E \in \Delta_1(F) \cap M_1\}, \Gamma_1(F))$.

*Proof.* Otherwise for some $\theta_0 \in \Theta_0$, $E_1 \in \Delta_1(F) \cap M_1$ and $\gamma_1 \in \Gamma_1(F)$ we have

$$\vdash \theta_0 \wedge E_1 \to \gamma_1$$

$$\vdash \theta_0 \to (E_1 \to \gamma_1)$$

$$\vdash \theta_0 \to (\forall c_1 \ldots)(E_1 \to \gamma_1).$$

Now since $\theta_0 \in \Theta_0$ we get that $(\forall c_1 \ldots)(E_1 \to \gamma_1) \in \Theta_0$ and so $(\forall c_1)(E_1 \to \gamma_1) \in \Delta_0 \subseteq \Delta_1(F)$ which is a contradiction.

Notice that we did not use the fact that we started with $(\Delta_0, \Gamma_0)$ and then transferred the $E$'s to $\Theta_0$. We could have started with $F_1 \to F_2 \notin \Theta_0$ and transferred to $\Delta_0$. It does not matter whether we have $\Gamma_0$ or $\Gamma^*_0$ as second coordinate!

(17) Now extend the above pair of (16) to a saturated theory in the language $M_2$. Let us denote this theory by $(\Theta^F_2, \Gamma_2(F))$. We are only interested in $((\Theta^F_2, \Gamma_1(F))$.

LEMMA 18. *The following is consistent*: $(\Delta_1(F) \cup \{E_2 | E_2 \in \Theta_2^F \cap L_2\}, \Gamma_1(F)^*)$.

*Proof*. Otherwise for some $\delta_1 \in \Delta_2(F)$, $E_2 \in \Theta_2^F \cap L_2$ and $\gamma_1^* \in \Gamma_1(F)^*$ we have

$$\vdash \delta_1 \wedge E_2 \rightarrow \gamma_1^*$$

$$\vdash \exists c_2 E_2 \rightarrow (\delta_1 \rightarrow \gamma_1^*).$$

Now if $\exists c_2 E_2$ (being in $L_1$) is not in $\Delta_1(F)$ then it is in $\Gamma_1(F)^*$, and since it is in $M_1$ as well, it must be in $\Gamma_1(F)$ but this contradicts $E_2 \in \Theta_2^F$.

So $\exists c_2 E_2 \in \Delta_1(F)$ and so $\delta_1 \rightarrow \gamma_1^* \in \Delta_1(F)$ but $\gamma_1^* \in \Gamma_1(F)^*$ and $\delta_1 \in \Delta_1(F)$, a contradiction.

(19) Now extend this pair to a saturated and complete theory $(\Delta_3(F), \Gamma_3(F)^*)$.

We can proceed now and show that the following is consistent: $(\Theta_2^F \cup \{E_3 | E_3 \in \Delta_3 \cap M_3\}, \Gamma_3(F))$ for otherwise for some $\theta_2$, $E_3$, $\gamma_3$

$$\vdash \theta_2 \wedge E_3 \rightarrow \gamma_3$$

$$\vdash \theta_2 \rightarrow (\forall c_3)(E_3 \rightarrow \gamma_3).$$

So since $\theta_2 \in \Theta_2^F$ we get that $\forall c_3(E_3 \rightarrow \gamma_3) \in \Theta_2^F$ and so $\forall c_3(E_3 \rightarrow \gamma_3) \in \Delta_3(F)$ which is a contradiction.

We continue in this manner to obtain the following sequences

$$(20) \qquad (\Delta_1(F), \Gamma_1(F)^*), \qquad (\Delta_3(F), \Gamma_3(F)^*), \ldots$$
$$(\Theta_2^F, \Gamma_1(F)), \qquad (\Theta_4^F, \Gamma_3(F)), \ldots.$$

These sequences have properties (13) (a)–(e) stated for our case and in addition to this we have that

$$(21) \qquad F_2 \in \Gamma_1(F)^*.$$

We therefore have that in the limiting theories $(\Delta_\omega(F), \Gamma_\omega(F)^*)$, $(\Theta_\omega^F, \Gamma_\omega(F))$ we have $F_1 \in \Delta_\omega(F)$, $F_2 \in \Gamma_\omega(F)^*$.

(22) We may therefore conclude the following: Given $\Delta_0 \not\vdash F_1 \rightarrow F_2$ we can construct a saturated and complete theory $(\Delta_1(F), \Gamma_1(F)^*)$, which we may assume to be in the language $L_1$, such that $F_1 \in \Delta_1(F)$

and $F_2 \in \Gamma_1(F)^*$, and $\Delta_0 \subseteq \Delta_1(F)$. At the same time we can construct a saturated theory $(\Theta_1^F, \Gamma_1(F))$ which is in the language $M_1$ with the properties

(a) $\Theta_0 \subseteq \Theta_1^F$, $\Gamma_1(F) = \Gamma_1(F)^* \cap M_1$

(b) $(\Delta_1(F) \cap \Theta_1^F, \Gamma_1(F))$ is a complete and saturated $M_1 \cap L_1$-theory.

A similar construction may be done for a $F = F_1 \to F_2 \notin \Theta_0$. We then denote the theories by $(\Theta_1(F), \Gamma_1(F)^*)$, $(\Delta_1^F, \Gamma_1(F))$.

(23) Now assume that for some $\delta = \forall x \beta(x)$ we have that $\Delta_0 \not\Vdash \forall x \beta(x)$ then for some $u$ in $L_1$ we have $\Delta_0 \not\Vdash \beta(u)$ and so $(\Delta_0, \{\beta(u)\})$ is a consistent theory. We can therefore repeat the process of (15)–(21) to obtain two pairs $(\Delta_1(\delta), \Gamma_1(\delta)^*)$, $(\Theta_1^\delta, \Gamma_1(\delta))$ fulfilling (20) and also $\beta(u) \in \Gamma_1(\delta)^*$.

(24) Moreover, in the construction of $(\Delta_1(F), \Gamma_1(F)^*)$ we used no more than the fact that $(\Delta_0, \Gamma_0)$ and $(\Theta_0, \Gamma_0^*)$ fulfill (6) (used in (16)).

Now $(\Delta_1(F), \Gamma_1(F)^*)$ and $(\Theta_1^F, \Gamma_1(F))$ also fulfill (6) and therefore we can construct for example $(\Delta_2(F)(F'), \Gamma_2(F)^*(F')^*)$ etc.

(25) For any pair $(\Delta, \Gamma)$, $(\Theta, \Gamma^*)$ in the languages $L_n$, $M_n$, respectively, such that (6) holds and for any $F = F_1 \to F_2$ (or $\delta = \forall x \beta(x)$, such that $\Delta \not\Vdash F$ (or $\Delta \not\Vdash \delta$) we can construct theories $(\Delta_{n+1}(F), \Gamma_{n+1}(F)^*)$ and $\Theta_{n+1}(F)$, $\Gamma_{n+1}(F))$ etc. as in (22) and (23).

Now before constructing our structures let us recall that all our theories are saturated.

(26) We now construct two structures. One is associated with the $\Delta$-theories and one is associated with the $\Theta$-theories, we shall prove that these two structures are isomorphic in the common language $L_0 \cap M_0$.

We now define two sets $S_\Delta$ and $S_\Theta$ composed of ordered pairs whose first member is a theory and whose second member is a finite string of formulas. We shall also give a $1-1$ map: $S_\Delta \to S_\Theta$.

(a) Let $((\Delta_0, \Gamma_0), \langle 0 \rangle) \in S_\Delta$ and $((\Theta_0, \Gamma_0^*), \langle 1 \rangle) \in S_\Theta$ and say that these two are *related*.

(b) For any $F = F_1 \to F_2(\forall x \beta(x))$ such that $\Delta_0 \not\Vdash F_1 \to F_2(\Delta_0 \not\Vdash \forall x \beta(x))$,

put $((\Delta_1(F), \Gamma_1(F)^*), \langle 0 \rangle \hat{*} \langle F \rangle) \in S_\Delta$ and $((\Theta_1^F, \Gamma_1(F)), \langle 1 \rangle \hat{*} \langle F \rangle) \in S_\Theta$ and say that these two elements are related. Similarly for the case of $\forall x\beta(x)$, $\langle x\rangle$ denotes the sequences whose element is $x$ and $\hat{*}$ denotes concatenation.

(c) For any $F = F_1 \to F_2 \not\in \Theta_0$ put $((\Theta_1(F), \Gamma_1(F)^*), \langle 1, F\rangle) \in S_\Theta$ and $((\Delta_1^F, \Gamma_1(F)), \langle 0, F\rangle) \in S_\Delta$ and say that they are related. Similarly for $\forall x\beta(x) \not\in \Theta_0$, we say that these elements were constructed in stage 1.

(d) Induction clause: Let $((\Delta, \Gamma_1, \langle 0\rangle \hat{*} t)$ and $((\Theta, \Gamma_2, \langle 1\rangle \hat{*} t)$ be any pair of related elements constructed in stage $n$. We assume the respective pair of theories to fulfill (6). Repeat the construction of (b) and (c) above for this pair of theories. The right-hand members of the new elements of $S_\Delta$ and $S_\Theta$ will be of the form $\langle 0\rangle \hat{*} t \hat{*} \langle F\rangle$ or $\langle 1\rangle \hat{*} t \hat{*} \langle F\rangle$ for $F$ as in (b) and (c).

Now $S_\Delta$ and $S_\Theta$ are our sets of possible worlds which we shall use to build our two structures as done in Chapter 3.

Note that from our construction it follows that if $(\Delta, \Gamma_1)$, $(\Theta, \Gamma_2)$ are left-hand members of related elements then $(\Delta \cap \Theta, \Gamma_1 \cap \Gamma_2)$ is a complete and saturated theory of the common language.

Define on $S_\Delta$: $((\Delta, \Gamma), t) \leq ((\Delta', \Gamma'), t')$ iff $\Delta \subseteq \Delta'$ and $t$ is initial segment of $t'$.

Now let $[P(x_1, \ldots, x_n)]_{((\Delta,\Gamma),t)} = 1$ iff $P(x_1, \ldots, x_n) \in \Delta$ for atomic $P$. A similar definition would turn $S_\Theta$ into a structure.

CLAIM. *For any F in L*

(27) $\quad [F]_{((\Delta,\Gamma),t)} = 1 \quad \text{iff} \quad F \in \Delta.$

*Proof.* For atomic $F$ this holds by definition. $\wedge$, $\vee$, $\exists$ present no difficulties since the theories are saturated. For $F_1 \to F_2 \not\in \Delta$ take $\Delta(F)$. We have that $F_1 \in \Delta(F)$ and $F_2 \notin \Delta(F)$ since $F_2 \in \Gamma(F)^*$. This yields the case of $\to$, and similarly for $\forall$. We regard $\neg F$ as $F \to f$.

(28) A similar lemma holds for the $\Theta$-structure.

Now to show that these two structures are isomorphic in the language $L_0 \cap M_0$ take the isomorphism as '*being related to*' and identity as the 1–1 map which preserves the common language part. In fact we have constructed the sister-theories so that we shall have isomorphism. Recall that all constants of $M_0$ or $L_0$ are in the common language.

By (27)–(28) we conclude that e.g. $S_\Delta$ validates $(\Delta_0 \cup \Theta_0, \Gamma_0^*)$.

## 3. Craig's theorem

THEOREM 29. *Robinson's statement* (3) *for theories such that* $\Gamma_1 \subseteq \Gamma_2$ *holds for the minimal and intuitionistic logics.*

*Proof.* Our proof above works for both cases. In fact it works for positive logic as well.

THEOREM 30. *Robinson's statement in* (29) *is equivalent to Craig's interpolation theorem.*

*Proof.* I: Assume Craig's theorem. Now if $(\Delta_1 \cup \Delta_2, \Gamma_2)$ is not consistent then for $F_1 \in \Delta_1$, $F_2 \in \Delta_2$, $\gamma_2 \in \Gamma_2$ we have

$$\vdash F_1 \wedge F_2 \to \gamma_2$$
$$\vdash F_1 \to (F_2 \to \gamma_2).$$

By Craig's theorem we have that for some $F \in L_1 \cap L_2$

$$\vdash F_1 \to F, \qquad \vdash F \to (F_2 \to \gamma_2).$$

We conclude that $F \in \Delta_1$ and therefore $F \in \Delta_2$ and so $(F_2 \to \gamma_2) \in \Delta_2$ which is a contradiction.

II: Assume Robinson's theorem. Let $\vdash F_1 \to F_2$. Define

$$\Delta_0 = \{F \in L_1 \cap L_2 \,|\, \vdash F_1 \to F\}.$$

If for some $F \in \Delta$ we have $\vdash F \to F_2$ we are finished. Otherwise $(\Delta_0, \{F_2\})$ is consistent. Let us extend this pair to a complete $L_2$-theory $(\Delta_2, \Gamma_2)$. Let $\Delta, \Gamma$ be the common $L_1 \cap L_2$ part. We claim that $(\Delta \cup \{F_1\}, \Gamma)$ is a consistent theory. Otherwise for some $F \in \Delta$, $\gamma \in \Gamma$ we have

$$\vdash F \wedge F_1 \to \gamma$$
$$\vdash F_1 \to (F \to \gamma)$$

and so $(F \to \gamma) \in \Delta_0 \subseteq \Delta_2$ which is impossible.

Now $(\Delta \cup \{F_1\}, \Gamma)$, $(\Delta_2, \Gamma_2)$ are consistent with $\Gamma \subseteq \Gamma_2$ and $\Delta, \Gamma$ complete in the common language. So $(\Delta_2 \cup \{F_1\}, \Gamma_2)$ is consistent. But this contradicts $F_2 \in \Gamma_2$.

THEOREM 31. *Craig's theorem holds for HPC.*

To complete the readers' information, let us mention the following theorem:

THEOREM 32. *The following version # of Robinson's theorem is equivalent to Craig's theorem for $A \to \neg B$:*

#: *Let $(\Delta_i, \emptyset)$ $i = 1, 2$ be two consistent theories in the languages $L_1$, $L_2$ respectively. Then if $((\Delta_1 \cap L_2) \cup (\Delta_2 \cap L_1), \emptyset)$ is consistent so is $(\Delta_1 \cup \Delta_2, \emptyset)$.*

*Proof.* Exercise.

**Exercise 33.** Prove the interpolation theorem for the system *MH* (the extension of *H* with the schema $\forall x \neg\neg A(x) \to \neg\neg \forall x A(x)$).

**Exercise 34.** We want to refute the strong consistency theorem for *H*, i.e. prove Theorem 4a.

Consider a language with one unary predicate $P(x)$. Let $S$ be the set of natural numbers and $R$ be the usual ordering on $S$.

Let $A_n$, $B_n$ be a sequence of pairwise disjoint sets of elements. For $n \in S$ define the domains $U_n$ as follows:

$$U_0 = A_0 \cup B_0$$

$$U_1 = U_0 \cup A_1$$

$$U_2 = U_1 \cup A_2 \cup B_2$$

$$U_{2n+1} = U_{2n} \cup A_{2n+1}$$

$$U_{2n+2} = U_{2n+2} \cup A_{2n+2} \cup B_{2n+2}$$

. . . . . . . . . . . . . . . . . . . . .

$$\text{Let } D(n, P) = \begin{cases} A_0 & n = 0 \\ U_1 & n = 1 \\ U_1 \cup A_2 & n = 2 \\ U_n & n = 2m + 1 \\ U_{n-1} \cup A_n & n = 2m + 2 \end{cases}$$

We thus get a Kripke structure $(S, \leqslant, 0, D, U)$. It is easy to see that $(S^{2n+1}, \leqslant, 2n + 1, D, U)$ are all isomorphic and $(S^{2n}, \leqslant, 2n, D, U)$ are all isomorphic, where $S^m = \{x \in S \mid x \geqslant m\}$ and $\leqslant$, $D$, $U$ are restricted to $S^m$, respectively.

It therefore follows, that for any $A$, $[A]_{2n+1} = [A]_{2n+2}$ for any $n$. We now give two consistent theories $(\Delta_1, \Theta_1)$, $(\Delta_2, \Theta_2)$ such that $(\Delta_i, \Theta_i)$ is in a language with the two unary predicates $P$, $P_i$, $i = 1, 2$ and such that the common theory $(\Delta, \Theta)$ is complete but $(\Delta_1 \cup \Delta_2, \Theta_1 \cup \Theta_2)$ is not consistent.

Let $(\Delta, \Theta)$ be the complete theory of the structure $(S, \leqslant, 0, D, U)$. Let

$$\Delta_1 = \Delta \cup \{\forall x(P(x) \vee P_1(x));\ \exists x P_1(x)\}$$

$$\Delta_2 = \Delta \cup \{\forall x(P_2(x) \vee P(x))\}$$

$$\Theta_1 = \Theta \cup \{\exists x(P_1(x) \wedge P(x))\}$$

$$\Theta_2 = \Theta \cup \{\exists x(P_2(x) \wedge P(x));\ \exists x P_2(x)\}$$

$(\Delta_i, \Theta_i)$ both say that in the structure $[P(x)] = 1$ iff $[P_i(x)] = 0$.

$(\Delta_1, \Theta_1)$ is consistent since it is satisfied in the following structure. Take $(S, \leqslant, 0, D, U)$ defined above, and extend it to a structure for the language with $P_1$ by letting $[P_1(x)]_0 = 1$ iff $[P(x)]_0 = 0$ (i.e., $D(0, P_1) = B_0$) and $[P_1(x)]_m = 1$, $m \geqslant 1$ (i.e., $D(m, P_1) = U_m$, $m \geqslant 1$).

$(\Delta_2, \Theta_2)$ is consistent because it can be validated in a structure obtained from $(S^1, \leqslant, 1, D, U)$ by letting $[P_2(x)]_1 = 1$ iff $[P(x)]_1 = 0$ and $[P_2(x)]_m = 1$ for all $m \geqslant 2$.

However, it is clear that the union $(\Delta_1 \cup \Delta_2, \Theta_1 \cup \Theta_2)$ is not consistent.

THEOREM 35. *The strong Robinson consistency theorem is not true for CD.*

*Proof.* Consider the following Kripke structure with constant domains. The set $S$ of possible worlds is $\omega_0$, the set of positive integers. $R$ is the natural ordering $\leqslant$. Let $\omega_0 = \bigcup_{n \in \omega_0} B_n$, $B_n$ is a sequence of pairwise disjoint infinite sets. Let $L_0$ be a language with the unary predicates $P$, $P'$, and consider the following extensions for $P$, $P^1$ at the world $m$:

(a) $P$ is true on $\bigcup_{i \leqslant 2n+1} B_i$ and $P'$ is true on $\bigcup_{i \leqslant 2n+1} B_i$ for $m = 2n + 1$

(b) P is true on $\bigcup_{i \leqslant 2n} B_i$ and $P'$ for $\bigcup_{i \leqslant 2n+1} B_i$ for $m = 2n$.

Let $(\Delta, \Theta)$ be the complete theory of this structure. Consider another unary predicate $Q$ and let $L$ be the language with $P$, $Q$ and $M$ be the language with $P'$, $Q$.

Consider the theories

$$\Delta_1 = \Delta \cup \{\forall x(P(x) \vee Q(x)); \exists x(P'(x) \wedge Q(x))\}$$

$$\Theta_1 = \Theta \cup \{\exists x(P(x) \wedge Q(x))\}$$

$$\Delta_2 = \Delta \cup \{\forall x(P(x) \vee Q(x))\}$$

$$\Theta_2 = \Theta \cup \{\exists x(P(x) \wedge Q(x)); \exists x(P'(x) \wedge Q(x))\}$$

clearly $(\Delta_1, \Theta_1)$, $(\Delta_2, \Theta_2)$ are consistent but not their union.

As before we can show that the weak Robinson theorem is equivalent to the Craig theorem and we proceed to prove the weak Robinson theorem. For this purpose we use an extended completeness theorem for *CD*.

DEFINITION 36. Recall the following: (a) A theory $(\Delta, \Theta)$ is *CD* consistent iff for no finite $\Delta_0 \subseteq \Delta$, $\Theta_0 \subseteq \Theta$ do we have $CD \vdash \bigwedge \Delta_0 \to \bigwedge \Theta_0$.

(b) A theory $(\Delta, \Theta)$ is said to be saturated (in a certain language) iff

(1) $A \vee B \in \Delta$ implies $A \in \Delta$ or $B \in \Delta$

(2) $\Delta \vdash A$ implies $A \in \Delta$

(3) $\exists x A(x) \in \Delta$ implies for some variable $u$ of the language $A(u) \in \Delta$.

(c) A theory $(\Delta, \Theta)$ is said to be of constant domains (in a certain language) iff whenever $(\Delta, \Theta \cup \{\forall x A(x)\})$ is consistent, then for some variable $u$ of the language $(\Delta, \Theta \cup \{A(u), \forall x A(x)\})$ is consistent.

LEMMA 37. (a) *Let* $(\Delta, \Theta)$ *be a consistent CD theory, then* $(\Delta, \Theta)$ *can be extended to a complete saturated and consistent theory of constant domains in a language with possibly* $\aleph_0$ *more variables.*

(b) *Let* $(\Delta, \Theta)$ *be a complete saturated and consistent theory of constant domains in a certain language. Let* $\alpha = A \to B \notin \Delta$. *Then there exists a complete saturated and consistent theory of constant domains* $(\Delta_\alpha, \Theta_\alpha)$ *with* $\Delta \cup \{A\} \subseteq \Delta_\alpha$, $B \in \Theta_\alpha$.

*Proof.* See Chapter 3, Section 3.

We are now in a position to prove the restricted version of the consistency theorem (3) for the case $\Gamma_1 \subseteq \Gamma_2$.

THEOREM 38. *Let* $(\Delta_1, \Gamma_1)$, $(\Delta_2, \Gamma_2)$ *be two CD consistent theories in the languages* $L_1$ *and* $L_2$ *respectively with* $\Gamma_1 \subseteq \Gamma_2$ *and with the property that* $(\Delta_1 \cap \Delta_2, \Gamma_1 \cap \Gamma_2)$ *is complete, then* $(\Delta_1 \cup \Delta_2, \Gamma_1 \cup \Gamma_2)$ *is consistent.*

*Proof.* Repeat the construction of the two isomorphic structures defined in (26). In the course of the construction we used various lemmas and definitions in Section 2 of Chapter 1. A key fact was the remark to the effect that if $(\Delta, \Theta)$ is a consistent theory it can be extended to a saturated and complete theory in a language with $\aleph_0$ more constants.

In the course of the various constructions of Section 2, we used the above remark to extend various theories to saturated and complete theories. When we repeat the construction of (8)–(25) above for the case of the logic *CD* we use Lemma 37a instead. This ensures that we get two sets of Henkin theories $S_\Delta$, $S_\Theta$ as in (26) with the additional property that *all the theories involved are of constant domains*![1]

Note we have to change the definitions of $(D_\omega, \Gamma_\omega)$ and $(\Theta_\omega, \Gamma^*_\omega)$ of 13d. These theories were defined using unions. But unions do not preserve the *CD*-property of the theories. We thus have to take the ultraproducts (over a non-principal-ultrafilter $F$) of the sequences. See chapter 10 section 3 for the definition. Thus for our case:

$$(\Delta^{CD}_\omega, \Gamma^{CD}_\omega) = \Pi(\Delta_n, \Gamma_n)|_F$$

$$(\Theta^{CD}_\omega, \Gamma^{*CD}_\omega) = \Pi(\Theta_n, \Gamma^*_n)|_F$$

(13e) is still valid for these theories.

(1) Since (27) is true we can conclude that for e.g. the structure with $S_\Delta$ and all wff of $L_1$

$$[\forall x A(x)]_{((\Delta, \Gamma), t)} = 0$$

iff for some $u$ *in the language of* $(\Delta, \Gamma)$ *itself*,

$$[A(u)]_{((\Delta, \Gamma), t)} = 0.$$

We have thus proved the following lemma.

LEMMA 39. *Under the assumptions of Theorem* 38, *there exists a*

[1] Actually we don't need in the construction of $S_\Delta$, $S_\Theta$ the clauses about the universal quantifier but it does no harm to perform them.

*structure (not necessarily of constant domains) for the language $L_1 \cup L_2$ such that $(\Delta_1 \cup \Delta_2, \Gamma_1 \cup \Gamma_2)$ is valid in this structure and such that for any $t$ and any $A$ which is in the language $L_1$ or $L_2$ (but not mixed!)*

$$(*) \qquad [\forall x A(x)]_t = 0 \quad \textit{iff for some } u \text{ at the world } t \quad [A(u)]_t = 0.$$

*Remark.* We don't yet know if the logic *CD* is valid in this structure. The above just shows that $(\Delta_1 \cup \Delta_2, \Gamma_1 \cup \Gamma_2)$ is consistent as an *HPC* (intuitionistic) theory, but not necessarily as a *CD* theory.

*Proof of Theorem* 38. By (39) there exists a structure satisfying (*) of (39) for $L_1$ and for $L_2$. Take this structure and look at $\lceil\ \rceil$ truth valuation of (3.3.11). We get a structure of *CD* by (3.3.13). By (*) of (39) the values of $(\Delta_1 \cup \Delta_2, \Gamma_1 \cup \Gamma_2)$ do not change. Thus this theory is *CD* consistent.

*Remark* 40. The same method can be used to obtain Craig's theorem for the systems $CD + (34)$ and $CD + (35)$ and $CD + (34) + (35)$.

## NOTES

The interpolation theorem for *H* was obtained syntactically by K. Schutte 1962 and refined by Nagashima, 1966. The present semantic proof is based on the author's papers. The method yields interpolation for other systems, e.g., *CD, MH.*

Meanwhile the paper of Maximova, 1977, has appeared. According to her (using algebraic methods), interpolation for propositional intermediate logic is true for only 8 logics.

$H_1$ = intuitionistic propositional calculus $h$

$H_2 = h + \neg A \vee \neg\neg A$

$H_3 = h + A \vee (A \to (B \vee \neg B))$

$H_4 = H_3 + (A \to B) \vee (B \to A) \vee (A \leftrightarrow \neg B)$

$H_5 = H_3 + H_2$

$H_6 = h + (A \to B) \vee (B \to A)$

$H_7 = h + A \vee \neg A$

$H_8 = h + t \leftrightarrow f$

Failure of interpolation was obtained by others as well, such as Y. Komorie and G. R. Renardel. J. Zucker has some very interesting results about fragments.

CHAPTER 9

# SECOND ORDER PROPOSITIONAL CALCULUS

## 1. The system $2h$

We add to the language of the intuitionistic propositional calculus with connectives $\wedge$, $\vee$, $\rightarrow$, $f$ (falsity) and atomic propositions $P = \{p_1, p_2, \ldots\}$ the intuitionistic propositional quantifiers $\forall$ and $\exists$ and the set $V = [v_1, v_2, \ldots]$ of propositional variables.

We take as axiom schemas, besides the axioms and rules of the intuitionistic propositional calculus, the following axioms and rules: ($x$, $y$ are propositional variables)

(1) $(\forall x)A(x) \rightarrow A(y)$, (axiom).

(2) $A(y) \rightarrow \exists x A(x)$, (axiom).

(3) $\dfrac{\vdash A(x) \rightarrow B}{\vdash \exists x A(x) \rightarrow B}$, $x$ not free in $B$, (provability rule).

(4) $\dfrac{\vdash B \rightarrow A(x)}{\vdash B \rightarrow (\forall x)A(x)}$, $x$ not free in $B$, (provability rule).

(5) $\forall x(B \vee A(x)) \rightarrow (B \vee \forall x A(x))$, $x$ not free in $B$, (axiom).

(6) $(\exists x)(x \leftrightarrow A)$, $A$ any formula, $x$ not free in $A$, (axiom).

Let $2h$ be the system obtained by taking all the axioms except axiom 6. Let $C2h$ be $2h$ with axiom 6. We adopt axiom 5 for two reasons: (a) the proofs are the same for both systems, with and without (5), (b) in the presence of (6), anyway, (5) is a single axiom.

Turning to the appropriate interpretation we define Kripke structures of 2nd order as follows: The 2nd order structure have the form $(S, R, 0, D, g)$, where $(S, R, 0)$ is a partially ordered set $S$ with ordering $R$ and a first element $0 \in S$. D is a subset of $2^S$ such that the empty set is in D and $X \in D$ implies that $t \in X \wedge tRs \rightarrow s \in X$. $g$ is a function $g$: $V \cup P \rightarrow D$, with $g(f) =$ empty set.

We extend $g$, for a given $(S, R, 0, D, g)$, to a function $\| \|_g$: $wff \rightarrow D$, by induction as follows:

1. $\|A\|_g = g(A)$, for $A \in P \cup V$.
2. $\|A \wedge B\|_g = \|A\|_g \cap \|B\|_g$.

3. $\|A \vee B\|_g = \|A\|_g \cup \|B\|_g$.

4. $\|A \to B\|_g = \{t \in S \mid \text{for all } s, \text{ if } tRs \text{ and } s \in \|A\|_g \text{ then } s$ $s \in \|B\|_g\}$.

5. $\|\neg A\|_g = \{t \in s \mid \text{for all } s \text{ such that } tRs \text{ we have that}$ $s \notin \|A\|_g\}$ obtained from $\neg A = A \to f$).

6. $\|\exists x A(x)\|_g = \cup\, y \in \mathrm{V}\, \|A(y)\|_g$.

7. $\|\forall x A(x)\|_g = \cap\, y \in \mathrm{V}\, \|A(y)\|_g$.

We shall show that:

THEOREM A. *For any $A$, $2h \vdash A$ iff $\|A\|_g = S$, for all structures* $(S, R, 0, \mathrm{D}, g)$.

A structure is called a $C$-structure iff for all wff $A$, there exists $x \in V$ such that $\|A\|_g = g(x)$.

THEOREM B. *For any $A$, $C2h \vdash A$ iff $\|A\|_g = S$, for all $C$-structures,* $(S, R, 0, \mathrm{D}, g)$.

THEOREM C. *The deduction theorem holds for $C2h$.*

*Proof.* As in (2.1.13).

## 2. The completeness theorem

We begin with some definitions.

(a) If $L$ is either $2h$ or $C2h$, then by an $L$-theory we mean a pair $(\Delta, \Theta)$ of sets of formulas.

(b) $(\Delta, \Theta)$ is said to be $L$-consistent iff for no $A_i \in \Delta$, $B_j \in \Theta$, do we have $L \vdash \wedge A_i \to \vee B_j$.

(c) $(\Delta, \Theta)$ is said to be complete (in a certain language) iff for all *wff* $A$ of this language, either $A \in \Delta$ or $A \in \Theta$.

(d) A theory $(\Delta, \Theta)$ is said to be saturated, in a certain language iff

1. $A \vee B \in \Delta$ implies $A \in \Delta$ or $B \in \Delta$.

2. $\Delta \vdash A$ implies $A \in \Delta$.

3. $\exists x A(x) \in \Delta$ implies that, for some propositional variable $u$ of the language, $A(u) \in \Delta$.

(e) $(\Delta, \Theta)$ is said to be of constant domains in a certain language if whenever $(\Delta, \Theta \cup \{(\forall x)B(x)\})$ is consistent, then for some propositional variable $u$ of the language, $(\Delta, \Theta \cup \{B(u), (\forall x)B(x)\})$ is consistent.

(f) $(\Delta^*, \Theta^*)$ is said to extend $(\Delta, \Theta)$ iff $\Delta \subseteq \Delta^*$ and $\Theta \subseteq \Theta^*$.

We now continue with a series of lemmas that will enable us to prove Theorems A, B. Recall that Theorem C is available.

LEMMA 1. *Let* $(\Delta, \Theta)$ *be a consistent 2h theory, then* $(\Delta, \Theta)$ *can be extended to a complete, saturated and consistent theory* $(\Delta^*, \Theta^*)$ *of constant domains in a language with possibly* $\aleph_0$ *more propositional variables.*

*Proof.* Let V be a new set of $\aleph_0$ additional new propositional variables, and let $B_1, B_2, B_3, \ldots$ be an enumeration of all formulas of the new language. We define a sequence of theories $(\Delta_n, \Theta_n)$ in the new language such that $\Delta_n \subseteq \Delta_{n+1}$, $\Theta_n \subseteq \Theta_{n+1}$, and $(\Delta_n, \Theta_n)$ is consistent, as follows:

$$(\Delta_0, \Theta_0) = (\Delta, \Theta).$$

Suppose $(\Delta_n, \Theta_n)$ is given and is consistent; we define $(\Delta_{n+1}, \Theta_{n+1})$. We look at $B_n$, and distinguish several cases:

*Case* 1. $(\Delta_n, \Theta_n \cup \{B_n\})$ is inconsistent. In the case, $(\Delta_n \cup \{B_n\}, \Theta_n)$ must be consistent. We distinguish several subcases.

(1a) $B_n = A_1 \vee A_2$, then by the axioms, either $(\Delta_n \cup \{B_n, A_1\}, \Theta_n)$ or $(\Delta_n \cup \{B_n, A_2\}, \Theta_n)$ is consistent. We take $(\Delta_{n+1}, \Theta_{n+1})$ to be the first, if consistent, and otherwise the second.

(1b) $B_n = A_1 \wedge A_2$. Let $\Theta_{n+1} = \Theta_n$, $\Delta_{n+1} = \Delta_n \cup \{B_n, A_1, A_2\}$.

(1c) $B_n = A_1 \rightarrow A_2$. Let $\Theta_{n+1} = \Theta_n$, $\Delta_{n+1} \cup \{B_n\}$.

(1d) $B_n = (\exists x)A(x)$. Let $\Theta_{n+1} = \Theta_n$, $\Delta_{n+1} = \Delta_n \cup \{\exists x A(x), A(u)\}$, where $u \in$ V, is the first of the new variables that hasn't been used before [i.e. doesn't appear in $(\Delta_n, \Theta_n)$]. Again, the axioms and rules will yield consistency.

(1e) $B_n = \forall x A(x)$. Let $\Theta_{n+1} = \Theta_n$, $\Delta_{n+1} = \Delta_n \cup \{B_n\}$.

*Case* 2. $(\Delta_n, \Theta_n \cup \{B_n\})$ is consistent. We distinguish several subcases:

(2a) $B_n = A_1 \vee A_2$. Let $\Theta_{n+1} = \Theta \cup \{B_n, A_1, A_2\}$, $\Delta_{n+1} = \Delta_n$.

(2b) $B_n = A_1 \wedge A_2$. Then either $(\Delta_n, \Theta_n \cup \{A_1, B_n\})$ or $(\Delta_n, \Theta_n \cup \{A_2, B_n\})$ is consistent.

Let $(\Delta_{n+1}, \Theta_{n+1})$ be the first that is consistent.

(2c) $B_n = A_1 \rightarrow A_2$. Let $\Delta_{n+1} = \Delta_n$, $\Theta_{n+1} = \Theta_n \cup \{B_n\}$.

(2d) $B_n = \exists x A(x)$. Let $\Delta_{n+1} = \Delta_n$, $\Theta_{n+1} = \Theta_n \cup \{B_n\}$.

(2e) $B_n = \forall x A(x)$. Let $u$ be the first new variable not used before and let $\Theta_{n+1} = \Theta_n \cup \{B_n, A(u)\}$, $\Delta_{n+1} = \Delta_n$.

We claim that $(\Delta_{n+1}, \Theta_{n+1})$ is consistent. Otherwise for some $A'_i \in \Delta_n$, $A''_j \in \Theta_n$ we have

$$2h \vdash \wedge A'_i \rightarrow \vee A''_j \vee B_n \vee A(u).$$

$$2h \vdash \wedge A'_i \rightarrow \forall u(\vee A''_j \vee B_n \vee A(u))$$

$$2h \vdash \wedge A'_i \rightarrow \vee A''_j \vee B_n \vee (\forall u) A(u)$$

which implies that $(\Delta_n, \Theta_n \cup \{B_n\})$ was inconsistent, contrary to our assumption.

The case of $B_n = f$ is simple. Put $f \in \Theta_{n+1}$.

Having completed the inductive definition of the sequence $(\Delta_n, \Theta_n)$, we take $\Delta^* = \cup \Delta_n, \Theta^* = \cup \Theta_n$ and claim that $(\Delta^*, \Theta^*)$ is the theory required for Lemma 1. Clearly $(\Delta^*, \Theta^*)$ is consistent, saturated and complete. The reader can verify that it is of constant domains in the new language.

LEMMA 2. *Suppose that* $(\Delta, \Theta)$ *is a consistent theory of constant domains in a certain language and further assume that* $\Theta$ *is finite. Let* $A$, $B$ *be two formulas of this language such that* $(\Delta', \Theta') = (\Delta \cup \{A\}, \Theta \cup \{B\})$ *is consistent. Then* $(\Delta', \Theta')$ *is of constant domain in the language.*

*Proof.* Assume that the following theory is consistent $(\Delta \cup \{A\}, \Theta \cup \{B, \forall x F(x)\})$. We want to show that for some variable $u$ of the language.

$$(\Delta \cup \{A\}, \Theta \cup \{B, \forall x F(x), F(u)\})$$

is consistent. First notice that $(\Delta, \Theta \cup \{A \rightarrow \vee \Theta \vee B \vee \forall x F(x)\})$ is consistent. Also observe that for a proper choice of variable $y$, $A \rightarrow (\vee \Theta \vee B \vee (\forall x) F(x))$ is equivalent to $F' = \forall y(A \rightarrow \vee \Theta \vee B \vee \forall x F(x) \vee F(y))$. Now since $(\Delta, \Theta \cup \{F'\})$ is consistent, there exists a $u$ such that $(\Delta, \Theta \cup \{A \rightarrow \vee \Theta \vee B \vee \forall x F(x) \vee F(u)\})$ is consistent. But then we must have that $(\Delta \cup \{A\}, \Theta \cup \{B, \forall x F(x), F(u)\})$ is consistent.

LEMMA 3. *Let* $(\Delta, \Theta)$ *be a saturated, consistent and complete theory in a certain language with constant domains and let* $(A \rightarrow B) \in \Theta$. *Then* $(\Delta \cup \{A\}, \{B\})$ *is a consistent theory with constant domains in the same language.*

*Proof.* Clearly $(\Delta \cup \{A\}, \{B\})$ is consistent (by the deduction theorem). Assume now that $(\Delta \cup \{A\}, \{B, \forall x F(x)\})$ is consistent, and assume that for all $u$ of the language,

$$\Delta \cup \{A\} \vdash B \vee F(u) \vee \forall x F(x).$$

Then

$$(\#) \qquad \Delta \vdash A \to B \vee \forall x F(x) \vee F(u).$$

Now for properly chosen $y$, look at $F' = (\forall y)(A \to B \vee \forall x F(x) \vee F(y))$. Since $(\Delta, \Theta)$ is complete, $F' \in \Delta$ or $F' \in \Theta$. If $F' \in \Theta$, then by the constant domains property of $(\Delta, \Theta)$, for some $u$, $A \to B \vee \forall x F(x) \vee F(u) \in \Theta$ which together with $(\#)$ would contradict the consistency of $(\Delta, \Theta)$. Therefore, $F' \in \Delta$, i.e. $\Delta \vdash A \to B \vee \forall x F(x)$ (Since $y$ is not free in $A$ or $B$). But this contradicts the consistency of $(\Delta \cup \{A\}, \{B, \forall x F(x)\})$. We therefore conclude that for some $u$, $(\Delta \cup \{A\}, \{B, \forall x F(x), F(u)\})$ is consistent.

LEMMA 4. *Let $(\Delta, \Theta)$ be a saturated, consistent and complete theory of constant domains in a certain language. Assume that $F = (A \to B) \in \Theta$. Then there exists a complete, saturated and consistent theory of constant domains $(\Delta_F, \Theta_F)$ in the same language such that $A \in \Delta_F$, $B \in \Theta_F$, and $\Delta \subseteq \Delta_F$.*

*Proof.* By Lemma 3, $(\Delta \cup \{A\}, \{B\})$ is a consistent theory of constant domains in the same language. We now define an increasing sequence of consistent theories of constant domains

$$(\Delta_n^*, \Theta_n^*), \quad \text{with} \quad \Delta \subseteq \Delta_n^*, \ \Delta_n^* \subseteq \Delta_{n+1}^* \ \{B\} \subseteq \Theta_n^* \subseteq \Theta_{n+1}^*.$$

Let $B_1, B_2, B_3, \ldots$ be an enumeration of all formulas of the language. We proceed by induction:

$$0: \qquad \Delta_0^* = \Delta \cup \{A\}, \ \Theta_0^* = \{B\}.$$

Suppose $(\Delta_n^*, \Theta_n^*)$ was defined and further assume that $\Theta_n^*$ and $(\Delta_n^* - \Delta)$ are finite and that this theory is consistent and of constant domains. To define $(\Delta_{n+1}^*, \Theta_{n+1}^*)$ we distinguish several cases. We work parallel to the cases of Lemma 1.

*Case* 1. $(\Delta_n^*, \Theta_n^* \cup \{B_n\})$ is inconsistent. In this case, $(\Delta_n^* \cup \{B_n\}, \Theta_n^*)$ must be consistent. We distinguish several subcases.

(1a), (1b), (1c) Proceed as in the proof of Lemma 1.

(1d) $B_n = \exists x A(x)$. In this case we claim that for some $u$ of the language, $(\Delta_n^* \cup \{A(u), \exists x A(x)\}, \Theta_n^*)$ must be consistent. For otherwise, since $(\Delta_n^* - \Delta)$ is finite, we get

$$(\#) \qquad \Delta \vdash \wedge A_i^* \wedge \exists x A(x) \wedge A(u) \to \vee \Theta_n^*$$

with $\{A_i^*\} = \Delta_n^* - \Delta$, and $u$ any variable of the language.

Now look at $F^* = (\forall y)(\wedge A_i^* \wedge \exists x A(x) \wedge A(y) \to \vee \Theta_n^*)$. Since $(\Delta, \Theta)$ is complete, $F^* \in \Theta$ or $F^* \in \Delta$. If $F^* \in \Theta$, then by the constant domain property of $(\Delta, \Theta)$, for some $u$, $(\wedge A_i^* \wedge \exists x A(x) \wedge A(u) \to \vee \Theta_n^*)$ is in $\Theta$, which is impossible, in view of $(\#)$ and the consistency of $(\Delta, \Theta)$. Therefore $F^* \in \Delta$. But then $\Delta \vdash \wedge A_i^* \wedge \exists x A(x) \to \vee \Theta_n^*$, which contradicts the consistency of $(\Delta_n^* \cup \{\exists x A(x)\}, \Theta_n^*)$. Thus for some $u$, $(\Delta_n^* \cup \{\exists x A(x), A(u)\}, \Theta_n^*)$ must be consistent. Let then $(\Delta_{n+1}^*, \Theta_{n+1}^*)$ be just this theory.

(1e) Proceed as in Lemma 1.

*Case* 2. $(\Delta_n^*, \Theta_n^* \cup \{B_n\})$ is consistent.

(2a), (2b), (2c), (2d), proceed as in the case of Lemma 1.

(2e) $B_n = \forall x A(x)$. Since the theory $(\Delta_n^*, \Theta_n^* \cup \{B_n\})$ is consistent and since $(\Delta_n^*, \Theta_n^*)$ has the constant domain property, there must exist a $u$, in the same language, such that $(\Delta_n^*, \Theta_n^* \cup \{B_n, A(u)\})$ is consistent. Let $(\Delta_{n+1}^*, \Theta_{n+1}^*)$ be this theory. Thus all the cases are checked. By Lemma 2 $(\Delta_{n+1}^*, \Theta_{n+1}^*)$ is of constant domains and also fulfills all the other conditions of the sequence. Now let $\Delta_F = \bigcup \Delta_n^*$, $\Theta_F = \bigcup \Theta_n^*$. Clearly $(\Delta_F, \Theta_F)$ is consistent, saturated, complete and of constant domains in the same language.

*Proof of Theorem A Section* 1. The reader can verify that whatever is provable is valid in the interpretation. To show the other direction let $(\Delta, \Theta)$ be a consistent $2h$ theory; by Lemma 1, we can assume that a consistent and complete and saturated theory $(\Delta, \Theta)$ of constant domains is given. Let $S$ be the set of all consistent, and complete and saturated theories $(\Delta', \Theta')$ of the same language of constant domains such that $\Delta \subseteq \Delta'$. Let $(\Delta', \Theta')R(\Delta'', \Theta'')$ iff $\Delta' \subseteq \Delta''$. Let $0 = (\Delta, \Theta)$. Let $\mathbf{D} = \{X | X = \{(\Delta', \Theta') | A \in \Delta'\}$ for some *wff* $A\}$. Let $g(x) = \{(\Delta', \Theta') | x \in \Delta'\}$.

We claim, for all $A$,

$$(*): \qquad \|A\|_g = \{(\Delta', \Theta') | A \in \Delta'\}.$$

*Proof.* For $x$ atomic proposition or a propositional variable $(*)$

holds by definition. The cases of $\wedge$, $\vee$ present no difficulties. Since the theories are saturated the case of $\exists$ presents no difficulties. Since the theories are of constant domains the case of $\forall$ presents no difficulties.

Assume $(\Delta', \Theta') \in \|A \to B\|_g$.

Iff $(A \to B) \notin \Delta'$, then $(A \to B) \in \Theta'$. Then $(\Delta'_{A \to B}, \Theta'_{A \to B}) \in S$ and $(\Delta', \Theta') R (\Delta'_{A \to B}, \Theta'_{A \to B})$ and $A \in \Delta'_{A \to B}$, $B \notin \Delta'_{A \to B}$, which contradicts the definition of $\|A \to B\|_g$.

The other direction is easy.

*Proof of Theorem B Section* 1. Now to get Theorem B, first verify that all theorems are valid in the interpretation. Then for the other direction proceed as in the proof of theorem A, and observe that since $\exists x(x \leftrightarrow A)$ is an axiom it must be a member of $\Delta$ and so since $(\Delta, \Theta)$ is saturated, $\|A\|_g = g(x)$, for some $x$.

*Remark.* Although our proof of the completeness theorem used the fact that the language is countable, the theorem is true for any language, as the language of any one $A$ is countable.

**Exercise D.** Show that $C2h$ has the disjunction property.

## 3. Examples of connectives in $2h$

In $2h$ one can define many $n$-place connectives (wffs with $n$ propositional variables) that are not definable in $h$. For example the 0-place wff $\forall p(p \vee \neg p)$ is neither equivalent, in $2h$, to truth nor to falsity and is therefore not definable in $h$.

In this section we define a wff with one variable $F(x)$, such that $2h \nvdash \forall x F(x)$ and for some consistent extension $2h^*$ of $2h$, $2h^* \vdash \neg \forall x(F(x) \leftrightarrow A(x))$ for all first order one place wffs $A(x)$. This shows of course that $F$ is not definable in $h$. We also define in $2h$ some of the new operators and connectives introduced (for $h$) in Chapter 7.

Let $(S, R, 0, D)$ be a propositional structure for $h$. It is possible to turn this structure into a structure for $C2h$, called the full second order propositional structure $(S, R, 0, D)$. We give the following definition.

DEFINITION 1. Let $(S, R, 0)$ be a partially ordered set with ordering $R$ and first element $0$. Let $g$ be a function on $P \cup V$ (of $C2h$) such that $g(x) \subseteq S$ for all $x \in P \cup V$ and all $t, s \in S$, $t \in g(x) \wedge tRs \to s \in g(x)$. We require that $g(f) = \emptyset$. We write, for $x \in V$, $g =_x g'$ iff for all $y \neq x$, $y \in P \cup V$, $g(y) = g'(y)$. We define by induction the notion $[A]_t^g$, called the 2nd order truth value of $A$ in $(S, R, 0, g)$.

(a) $[A]_t^g = 1$ iff $t \in g(A)$, for $A \in P \cup V$

(b) $[A \wedge B]_t^g = 1$ iff $[A]_t^g = 1$ and $[B]_t^g = 1$.

(c) $[A \vee B]_t^g = 1$ iff $[A]_t^g = 1$ or $[B]_t^g = 1$.

(d) $[A \to B]_t^g = 1$ iff for all $s$, $tRs$ and $[A]_s^g = 1$ imply $[B]_s^g = 1$.

(e) $[\forall x A(x)]_t^g = 1$ iff for all $g' =_x g$ $[A(x)]_t^{g'} = 1$.

(f) $[\exists x A(x)]_t^g = 1$ iff for some $g' =_x g$ $(A(x)]_t^{g'} = 1$.

$A$ is said to be valid in $(S, R, 0, g)$ iff $[A]_0^g = 1$. $A$ is said to be valid in $(S, R, 0)$ iff $A$ is valid in $(S, R, 0, g)$ for all $g$.

The reader can verify that all theorems of $C2h$ are valid in any $(S, R, 0, g)$.

*Remark* 2. Given a propositional first order structure for $h$, $(S, R, 0, D)$, we can turn it into a full 2nd order structure for $C2h$, provided we identify the sets $P$ and $V$, i.e., identify the atomic propositions and propositional variables. We can let $g_D(x) = \{t \mid D(t, x) = 1\}$. In this case we can write $[A]_t^D$ instead of $[A]_t^{g_D}$.

Let us now continue with the problem of defining the wff $F(x)$ in $C2h$.

DEFINITION 3. We define the following formulas of $2h$.

(a) $G(x) = \forall y((y \to x) \vee ((y \to x) \to x))$

(b) $H(x) = \forall y(y \vee (y \to x)) \to x$

(c) $F(p) = \forall u(H((p \vee u)) \wedge G(u) \wedge ((p \to u) \to u) \to (p \vee u))$

DEFINITION 4. Let $\omega^*$ be the set of all finite sequences of natural numbers. Let $R \subseteq \omega^{*2}$ be the relation defined by $xRy$ iff $x$ is an initial segment of $y$ or $x = y$. Let $\Lambda$ denote the empty sequence. Let $S_0 = \{0\} \cup (I \times \omega^*)$, where $I = 2 \times \omega$. Let $\rho$ be the lexicographic ordering on $I$.

Define $\leqslant_0$ on $S_0$ *by*:

$$\leqslant_0 = \{0\} \times S \cup \{((i, \Lambda), (j, y)) \mid i, j \in I, i = j \vee i\rho j\} \cup \cup \{((i, x), (i, y)) \mid xRy\}.$$

LEMMA 5. (a) $\forall p F(p)$ *is valid in* $(\omega^*, R, \Lambda)$.

(b) $\forall p F(p)$ *is not valid in* $(S_0, \leqslant_0, 0)$.

*Proof*. First show that in any $(S, \leqslant, 0)$ and any $D$,

(c) $[G(x)]_t^D = 1$ and $t \leqslant s \wedge t \leqslant s'$ and $\sim s \leqslant s' \wedge \sim s' \leqslant s$ imply $[x]_s^D = 1$ or $[x]_{s'}^D = 1$.

To show (c) assume that $[x]_s^D = [x]_{s'}^D = 0$. Take a $q \neq x$, and a $D' \equiv_q D$ with $[q]_r^{D'} = 1$ iff $s \leqslant r$. Then clearly $[y \to x]_s^{D'} = 0$ and $[y \to x]_{s'}^{D'} = 1$ and therefore $[G(x)]_t^D = 0$, a contradiction.

(c) Means that whenever $G(x)$ holds at $t$, then the following set; $\{s \mid t \leqslant s \text{ and } [x]_s = 0\}$ is either empty or linearly ordered by $\leqslant$.

We now prove that $\forall p F(p)$ is valid in $(\omega^*, R, \Lambda)$. Let $D$ be arbitrary, and show $[F(p)]_\Lambda^D = 1$. Let $t$, $u$ be such that $[H((p \vee u)) \wedge G(u) \wedge ((p \to u) \to u)]_t^D = 1$.

From (c) above, $[G(u)]_t^D = 1$ implies that $\{s \mid tRs \wedge [u]_s^D = 0\}$ is either empty, in which case $[u]_t = 1$ and we are finished, or non-empty and is therefore linearly ordered by $R$. Assume then that it is not empty and is therefore of order type $\omega_0$. Let $t = s_0, s_1, s_2, \ldots$ be all the elements of this set in increasing $R$ order. Since $[(p \to u) \to u]_t^D = 1$ and $[u]_t^D = 0$, there must be an $s$, such that $tRs$ and $[p]_s = 1$ and $[u]_s = 0$. Thus $s = s_n$ for some $n$. Look at $s_{n-1}$. We claim that if $s_{n-1}Rr \wedge s_{n-1} \neq r$ then $[p \vee u]_{s_{n-1}}^D = 1$. For if $s_nRr$, then $[p]_r^D = 1$ and if $\sim s_nRr$, then $[u]_r^D = 1$. We have thus shown that $[\forall y(y \vee (y \to (p \vee x)))]_{s_{n-1}}^D = 1$. Therefore, since $H((p \vee u))$ holds at $t$, we get that $[p \vee x]_{s_{n-1}} = 1$, and therefore $[p]_{s_{n-1}}^D = 1$. Continuing in this manner we get that $[p]_t^D = 1$.

We have thus shown that if the antecedent of $F(p)$ holds at $t$, so does the conclusion.

This proves (5a).

To prove (5b) we must exhibit a $D$ such that $F(p)$ is not valid in $(S_0, \leqslant_0, 0, D)$. Let $u$, $p$, $D$ be such that:

(1) $D(t, u) = 0$ iff $t = (i, \Lambda) \vee t = 0$.

Clearly $\{t \mid [u]_t^D = 0\}$ is linearly ordered by $\leqslant_0$ and therefore $[G(u)]_0^D = 0$.

(2) $D(t, p) = 1$ iff $t \neq 0 \wedge t \neq ((0, m), x))$ for $m \in \omega, x \in \omega^*$.

We now claim that $[H((p \vee u)) \wedge G(u) \wedge ((p \to u) \to u)]_0^D = 1$ and $[p \vee u]_0^D = 0$. First let $t$ be such that $[u]_t^D = 0$. Then $t$ must be of the form $t = 0$ or $t = (i, \Lambda)$. Then clearly at a point of the form $s = ((i, m), \Lambda)$, $m$ large enough, we have $t \leq_0 s$ and $[p]_s^D = 1$, $[u]_s^D = 0$. Thus $[(p \to u) \to u]_0^D = 1$. Now assume $[\forall y(y \vee (y \to p \vee u))]_t^D = 1$. For this to hold we must have that for all $s$, $t \neq s$, $t \leq_0 s$ we have $[p \vee u]_s^D = 1$. This implies that $t$ cannot be $0$ or the form $((0, m), \Lambda)$ because at $s = ((0, m+1), \Lambda)$, $[p \vee u]_s^D = 0$. But if $t$ is not of this form then certainly $[p \vee u]_t^D = 1$. Thus this shows that $[H(p \vee u))]_0^D = 1$.

We thus conclude that $[F(p)]_0^D = 0$ and (5b) is proved.

LEMMA 6. *Let $(S, \leq, 0)$ be such that for any $t \in S$, there exists $s \in S$ such that $t \leq s$ and $(\omega^*, R, \Lambda)$ is isomorphic to $(\{r \mid s \leq r\}, \leq, s)$.*

*Let $A(p_1, \ldots, p_n)$ be a formula of $h$ such that $h \nvdash A$. Then $\neg\forall p_1, \ldots, p_n A$ is valid in $(S, \leq, 0)$.*

*Proof.* Since any non-provable $A$ of $h$ can be falsified in a $(\omega^*, R, \Lambda, D)$, a suitable $D'$ can falsify $A$ at $s$.

COROLLARY 7. *Let $T$ be $T = \{\neg\forall p_1, \ldots, p_n\, A(p_1, \ldots, p_n),\ h \nvdash A\}$ then $T$ is valid in $(\omega^*, R, \Lambda)$ and in $(S_0, \leq_0, 0)$ of Definition 4. $\neg\neg\forall pF(p)$ is also valid (for $F$ of Definition 3).*

*Proof.* $(S_0, \leq_0, 0)$ has the property of Lemma 6.

LEMMA 8. *Let $C2h^*$ be the system obtained by adjoining to $C2h$ the axioms of $T$ and $\neg\neg\forall pF(p)$. Then*

(a) $C2h^* \nvdash \forall pF(p)$

(b) *For any $A(p)$ of $h$ such that $\nvdash_h A(p)$ we have that $C2h^* \vdash \neg\forall p(F(p) \leftrightarrow A(p))$.*

*Proof.* In $(S_0, \leq_0, 0)$, $C2h^*$ is valid and $\forall pA(p)$ is not. Thus (a) follows. To show (b) assume $\forall p(A(p) \leftrightarrow F(p))$. Since $C2h^* \vdash \neg\forall pA(p)$, we get $C2h^* \vdash \neg\forall pF(p)$, but also $C2h^* \vdash \neg\neg\forall pA(p)$ a contradiction.

Thus $C2h^* \vdash \neg\forall p(A(p) \leftrightarrow F(p))$.

*Remark* 9. The connective $\#(x_1, \ldots, x_n)$ of Chapter 7, Section 4 is definable in $2h$ by $\#(x_1, \ldots, x_n) =_{\text{def}} \forall x(x \vee x \to B(x_1, \ldots, x_n))$.

$\#$ will have the same truth table via this interpretation as the one given in Chapter 7 Section 4. Of course, since the structures involved are not the same, the interpretation is not faithful, but if we add appropriate axioms to $2h$, it can be made faithful.

**Exercise 10.** Show that in $C2h$, $f$, $\wedge$, $\vee$, $\exists$ are definable in terms of $\rightarrow$, $\forall$.

$$A \wedge B =_{\text{def}} \forall x((A \rightarrow (B \rightarrow x)) \rightarrow x)$$

$$A \vee B =_{\text{def}} \forall x((A \rightarrow x) \wedge (B \rightarrow x) \rightarrow x)$$

$$\exists x A(x) =_{\text{def}} \forall y(\forall x(A(x) \rightarrow y) \rightarrow y)$$

$$f =_{\text{def}} (\forall x)x.$$

THEOREM 11. *The connective $F(x)$ is not definable in $h$.*
*Proof.* Follows from Lemma 8.

**Exercise 12.** A connective of M. Kaminski, which is related to our connective $\#$, provides an example of a finitely axiomatizable logic $J$ with a connective $\#$ that is not expressible in $2h$. The semantic condition for the connective is

$$[\#p]_t = 1 \quad \text{iff} \quad \{s \mid [p]_s = 1\} \quad \text{bars the point } t,$$

where a set $S_0$ bars a point $t$ iff every maximal linearly ordered set passing through $t$ intersects $S_0$. (This definition is like in Beth interpretation.) The axiomatization for this logic is with the following axioms

$$\#(p \rightarrow q) \rightarrow (\#p \rightarrow \#q)$$

$$\#p \rightarrow \neg\neg p$$

$$p \rightarrow \#p$$

$$\#\#p \rightarrow \#p$$

## NOTES

The material in this chapter is based on the author's papers.

Kreisel has proposed and studied various connectives defined in 2nd order intuitionistic propositional logic. See his paper and the paper of C.A. Goad.

CHAPTER 10

# MODIFIED KRIPKE INTERPRETATION

## 1. Theory of Equality

We begin this chapter with an example. Consider a language that contains the symbol =. We want to characterize it as the symbol for equality. One approach – the syntactical point of view – is to take the axioms of equality $T_=$

(1) $$x = x$$

$$x = y \to y = x$$

$$x = y \to (y = z \to x = z)$$

$$x = y \to (A(x) \to A(y)), \qquad A \text{ any wff.}$$

The semantical approach, in the Kripke interpretation, is to introduce the truth condition for $x = y$, for an assignment $g$,

(2) $$[x = y]_t^g = 1 \quad \text{iff} \quad g(x) = g(y).$$

These two approaches are not equivalent. We must add one more axiom to $T_=$ to obtain completeness for the condition (4.2): The axiom is

(3) $$x = y \vee \neg x = y.$$

We leave verification to the reader.

Thus it seems that from the point of view of the interpretation, the satisfactory theory of equality is $T_= + (3)$. Intuitionistic mathematics, however, takes $T_=$ as the theory of equality and calls $T_= + (3)$ the theory of decided equality. The difference is significant. We shall later see that the pure theory of equality $T_=$ is undecidable, while $T_= + (3)$ is decidable.

Rather than leave the situation as it is, we find it more useful to change our point of view of the interpretation. Given a Kripke structure $\mathbf{M} = (S, R, 0, D, U)$, we can regard it as a set of classical structures, $\mathbf{A}_t$, $t \in S$, where the domain of $\mathbf{A}_t$ is $U_t$ and the atomic relations are interpreted via $D_t(-) = D(t, -)$. Our conditions on $D$, $U$ are that $tRs$ implies $U_t \subseteq U_s$ and $D(t, A) \subseteq D(s, A)$. We change these requirements to the slightly weaker requirements that $\mathbf{A}_t$ is homomorphic into $\mathbf{A}_s$ and that if $tRs \wedge sRr$ then the following diagram is commutative

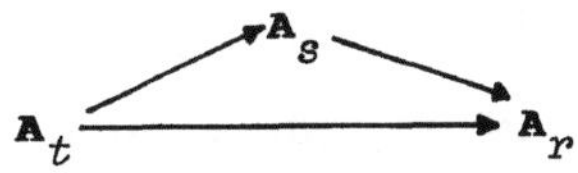


This is slightly weaker requirement because the mapping *homomorphism into* does not require that the mapping of $U_t \to U_s$ is one to one, HPC is still complete for this interpretation. This interpretation is useful for other purposes as well, as we shall see in the next chapter.

## 2. Function symbols and =

Consider a language $L$ with function symbols and equality symbol $=$, as well as relation symbols.

DEFINITION 1. (a) A modified Kripke structure for $L$ has the form $(S, R, 0, \mathbf{A}_t, \nu)$, $t \in S$, where $(S, R, 0)$ is a partially ordered set with first element $0 \in S$, and $\mathbf{A}_t$, $t \in S$, is classical structure for the language. $\nu$ is a function such that for each $t, s \in S$ such that $tRs$, $\nu_{t,s}$ is a homomorphism of $\mathbf{A}_t$ into $\mathbf{A}_s$, with the property that the following diagram is commutative for $tRs \wedge sRr$

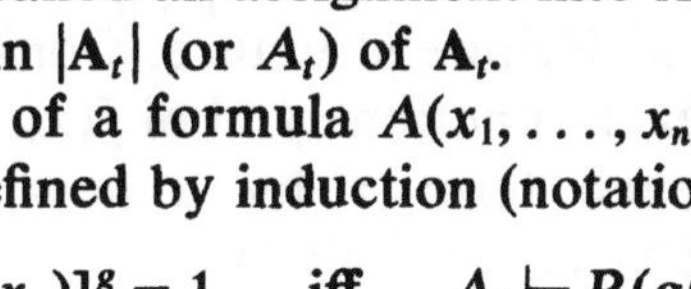


(b) A function $g$ is called an assignment into $\mathbf{A}_t$ iff for each variable $x$, $g(x)$ is in the domain $|\mathbf{A}_t|$ (or $A_t$) of $\mathbf{A}_t$.

(c) The truth value of a formula $A(x_1, \ldots, x_n)$ at a world $t$ under the assignment $g$ is defined by induction (notation $[A]^g_t$).

(a) $[P(x_1, \ldots, x_n)]^g_t = 1$ iff $A_t \models P(g(x_1), \ldots, g(x_n))$, for atomic $P$ or equation between terms $t_1 = t_2$

(b) $[A \wedge B]^g_t = 1$ iff $[A]^g_t = 1$ and $[B]^g_t = 1$

(c) $[A \vee B]^g_t = 1$ iff $[A]^g_t = 1$ or $[B]^g_t = 1$

(d) $[f]^g_t = 0$

(e) $[A \to B]_t^g = 1$ iff for all $s$, $tRs \wedge [A]_s^{g_s} = 1$ implies $[B]_s^{g_s} = 1$ where $g_s(x) = \nu_{t,s}(g(x))$

(f) $[\exists x A(x)]_t^g = 1$ iff for some $g^1 = {}_x g$, $[A]_t^{g^1} = 1$

(g) $[\forall x A(x)]_t^g = 1$ iff for all $s$ such that $tRs$ and all $g^1 = {}_x g_s$, $[A]_s^{g^1} = 1$.

(d) A modified Kripke structure is a tree modified Kripke structure if $(S, R, 0)$ is a tree. See (3.1.2).

We can give a semantic condition which corresponds to the axioms of equality $T_=$

(2) $[x = y]_t^g = 1$ iff $\mathbf{A}_t \models g(x) = g(y)$.

The condition which corresponds to the axiom

$$x = y \wedge \neg x = y$$

is

(3) $g(x) \neq g(y) \to \nu_{t,s} \neq \nu_{t,s} g(y)$

for any $t$ and any assignment $g$ into $\mathbf{A}_t$.

THEOREM 4. *Consider the system $H^=$ in a language equality and function symbols and the additional axioms $T_=$. Then this system is complete for the modified Kripke structure interpretation. In fact we can assume that the Kripke structures involved are tree Kripke structures* (See 3.12).

*Proof.* First verify that all axioms and rules are valid in this interpretation.

Let $(\Delta, \Theta)$ be a consistent theory of $\Vdash_{H'}$. We want to show it is valid in a some modified Kripke structure. Consider a new language obtained by replacing each function symbol $f(x_1, \ldots, x_n)$ by a relation $R_f(x_1, \ldots, x_n, y)$ and each term $\tau$ by a new variable $x_\tau$. Let $\Delta_0$ contain the axioms $\forall x_1, \ldots, x_n \exists y R_f(x_1, \ldots, x_n, y)$, $(\forall x_1 \ldots x_n, y, y')$ $[R_f(x_1, \ldots, x_n, y) \wedge R_f(x_1, \ldots, x_n, y') \to y = y']$, for every function symbol $f(x_1, \ldots, x_n)$. Also if $\tau = f(\tau_1, \ldots, \tau_n) \in \Delta$, $\tau$, $\tau_i$ terms, we add the axiom $R_f(x_{\tau_1}, \ldots, x_{\tau_n}, x_\tau)$ to $\Delta_0$.

Translate each $A$ of the language with function symbols into an $A^*$ of the language without function symbols by eliminating all function symbols from $A$. I.e., replace any term $\tau$ by $x_\tau$ and any equation

$\tau = f(\tau_1, \dots, \tau_n)$ by $R_f(\tau_1, \dots, \tau_n, \tau)$. Let $\Delta^1 = \Delta_0 \cup \{A^* \mid A \in \Delta\}$, $\Theta^1 = \{A^* \mid A \in \Theta\}$. Then $(\Delta^1, \Theta^1)$ is a theory in the new language with $=$ without function symbols with new relations $R_f$ and variables $x_\tau$. By regarding $=$ as just a binary atomic relation, we can apply the completeness theorem of (3.1.6) and obtain a structure and the identity assignment $g$ into the structure that validates $\Delta^1$ and in which all members of $\Theta^1$ are not valid. Let this structure be $(S, R, 0, D, U)$. Note that by the proof of the completeness theorem (3.1.6), the structure can be assumed to be a tree Kripke structure (see 3.1.2e).

For $t \in S$, $\equiv_t = D(t, =)$ is a congruence relation on $U_t$, relative to $D(t, A)$, $A$ any wff. This is guaranteed by the axioms of equality. Let a classical structure $\mathbf{A}_t$ for the original language be defined as follows:

The domain of $\mathbf{A}_t$ is $U_t/D(t, =)$. For any atomic $P$, $\mathbf{A}_t \models P(\dots x_i/\equiv_t \dots)$ iff $[P(\dots x_i \dots)]_t = 1$

$$\mathbf{A}_t \models f(\dots x_i/\equiv_t \dots) = y/\equiv_t \quad \text{iff} \quad [R_f(\dots x_i \dots, y)]_t = 1.$$

If $tRs$, let $\nu(x/\equiv_t) = x/\equiv_s$. It is easy to verify that $(S, R, 0, \mathbf{A}_t, \nu)$ is a modified Kripke structure and that under $g(x) = x/\equiv_t$ all members of $\Delta$ are valid and all members of $\Theta$ are not valid. This completes the proof of Theorem 4.

*Remark* 5. (a) We have already mentioned that the axiom $x = y \vee \neg x = y$ corresponds to the condition that $\nu_{t,s}$ in a modified Kripke structure is a 1–1 map.

(b) A modified Kripke structure is said to be a decided modified Kripke structure iff for any atomic $A$ and any $g$, $[A]_t^g = 0$ implies $[A]_s^{g_s} = 0$, for any $t, s$ such that $tRs$. It is easy to verify that $\mathbf{M}$ is a decided modified structure iff $P \vee \neg P$ is valid for all atomic $P$.

(c) A modified Kripke structure is said to be of constant domains iff $\nu_{t,s}$ are all onto.

(d) It follows from our method of proof that $CD$ is complete for the class of modified Kripke structures with constant domains.

## 3. Some model theory

DEFINITION 1. (a) Let $\mathbf{A} = (S, R, 0, \mathbf{A}_t, \nu)$ and $A' = (S', R', 0', A'_t, \nu')$ be two modified Kripke structures then we say $\mathbf{A} \equiv \mathbf{A}'$ ($\mathbf{A}$ and $\mathbf{A}'$ are elementarily equivalent) iff for any closed wff $A$ of the language $[A]_0 = [A]_{0'}$.

(b) $\mathbf{A} \subset \mathbf{A}'$ iff there exists a function $\alpha : S \to S'$ and for each $t$, a function $\beta_t : A_t \to A'_t$ such that

(1) $tRs \to \alpha(t)R\alpha(s)$.

(2) $\beta_t(x) = \beta_s(x)$, for $tRs$ and $x \in A_t$.

(c) We write $\mathbf{A} < \mathbf{A}'$ iff $\mathbf{A} \subset \mathbf{A}'$ and for each $t \in S$ and each $A(x_1, \ldots, x_n)$ and $g$ with $g(x_i) \in \mathbf{A}_t$, $[A(x_1, \ldots, x_n)]_t^g = [A(\beta_t(g)(x_1), \ldots \beta_t(g)(x_n))]_{\alpha(t)}$ where $\beta_t(g)(x) = \beta_t(g(x))$.

(d) The truncation of $\mathbf{A}$ at $t_0 \in S$, denoted by $\mathbf{A}^{t_0}$, is the modified structure with $S_{t_0} = \{t \mid t_0 R t\}$, $R \restriction S_{t_0}$, $\mathbf{A}_t$, $t \in S_{t_0}$ and $\nu_{t_0} = \nu \restriction S_{t_0}$.

*Remark* 2. $\mathbf{A} < \mathbf{A}'$ iff $\mathbf{A} \subset \mathbf{A}'$ and for any $t \in S$, $\mathbf{A}^t \equiv \mathbf{A}'^{\alpha(t)}$, in a language with names for each $x \in \mathbf{A}_t$.

DEFINITION 2. Let $\mathbf{A}(i)$, $i \in I$ be a family of structures $(S(i), R(i), 0(i), \mathbf{A}_t(i), \nu(i))$ and let $F$ be an ultrafilter over $I$. Define a structure $\mathbf{A} = (S, R, 0, \mathbf{A}_t, \nu)$ called the ultraproduct of $\mathbf{A}(i)$ over $F$ denoted by $\Pi\mathbf{A}(i)/_F$.

Let

$$S = \Pi S(i)/_F$$

$$R = \Pi R(i)/_F$$

$$0 = \{(i, 0(i) \mid i \in I\}/_F$$

(so far we took the ultraproducts in the usual sense of $(S(i), R(i), 0(i))$. For $\gamma/_F \in S$ let $\mathbf{A}_{\gamma/F} = \Pi\mathbf{A}_{\gamma(i)/F}$.

For $\gamma/_F$, $\delta/_F$, define $\nu : \mathbf{A}_{\gamma/_F} \to \mathbf{A}_{\delta/_F}$ by $\nu(x/_F) = \{i, \nu(i)(x(i))\}/_F$, that is $\nu = \Pi\nu(i)/_F$.

It is easy to verify that these definitions are invariant under $F$ and, indeed, define a modified Kripke structure.

In case all $\mathbf{A}(i) = \mathbf{B}$ we write $\mathbf{A} = \mathbf{B}^I/_F$.

THEOREM 3. *Let $A(x_1, \ldots, x_n)$ be any wff and let $\bar{a}_{1/_F}, \ldots, \bar{a}_{n/_F} \in A_{\gamma/_F}$ let $g$ be an assignment with $g(x_i) = \bar{a}_{i/_F}$ and let $g_i$ be an assignment into $\mathbf{A}(i)$ with $g_j(x_i) = \bar{a}_i(j)$ then $A(g(x_1), \ldots, g(x_n)_{\gamma/_F} = 1$ iff*

$$\{j \mid [A(g_j(x_1), \ldots, g_j(x_n))]_{\gamma(j)} = 1\} \in F.$$

*Proof.* By induction on $A$.

*Remark* 4. We write $g = \Pi g_i/_F$ for a situation like this of Theorem 3.

THEOREM 5 (*Compactness*). *Let* $(\Gamma, \Theta)$ *be a theory such that every theory* $(\Gamma', \Theta')$ *with* $\Gamma', \Theta'$ *finite and* $\Gamma' \subseteq \Gamma$, $\Theta' \subseteq \Theta$, *is validated in* $\mathbf{A}^{(\Gamma', \Theta')}$. *Then* $(\Gamma, \Theta)$ *is validated in some structure.*

*Proof.* Let $I =$ set of all pairs of finite $\Gamma' \subseteq \Gamma'$, $\Theta' \subseteq \Theta$. Define

$$J_{(\Gamma', \Theta')} = \{(\Gamma^*, \Theta^*) \mid \mathbf{A}^{(\Gamma', \Theta')} \text{ validates } (\Gamma^*, \Theta^*)\}.$$

Then $J_{(\Gamma', \Theta')} \cap J_{(\Gamma'', \Theta'')} = J_{(\Gamma' \cup \Gamma'', \Theta'' \cup \Theta'')} \neq \emptyset$. So we can enlarge $\{J_{(\Gamma', \Theta')}\}$ to an ultrafilter $F$ over $I$. Now $\mathbf{A} = \Pi \mathbf{A}^{(\Gamma', \Theta')}/F$ validates $(\Gamma, \Theta)$.

In classical logic, we have a nice algebraic characterization for the classes **K** that have the property: '**K** is the class of all models of a (certain) theory Γ'; namely, **K** is such a class iff **K** is closed under ultraproducts and elementary equivalence, (classes **K** fulfilling this are said to be $EC_\Delta$-classes). In our case this is not true, and so the definitions below should be compared with the classical ones, as there are several alternative notions.

DEFINITION 6. A class **K** of intuitionistic structures is said to be an $EC^+_\Delta$-*class* iff it is the class of all structures validating some theory $(\Gamma, \emptyset)$. **K** is an EC-*class* iff it is the class of all structures validating some theory $(\{F\}, \{E\})$ or a complement of such a class. **K** is said to be $EC_\Delta$ iff **K** is closed with respect to ultraproducts and elementary equivalence.

THEOREM 7. $\mathbf{K} \in EC^+_\Delta$ *iff* **K** *is closed with respect to ultraproducts, elementary equivalence and truncations.*

*Proof.* (⇒) is easy. To prove (⇐) let Γ be the set of sentences true in all structures of **K**. Let **A** validate Γ. We want to show $\mathbf{A} \in \mathbf{K}$. Let $(\Gamma', \Theta')$ be a finite theory of which **A** validate. We now find a structure $\mathbf{A}^{(\Gamma', \Theta')} \in \mathbf{K}$ which validates $(\Gamma', \Theta')$. Observe that

$$\bigwedge_{F \in \Gamma'} F \to \bigvee_{E \in \Theta'} E$$

is false in **A** so it cannot be true in all structures of **K**. Let $\mathbf{B} \in \mathbf{K}$ be a structure in which the sentence is false. Then for some $s \in S_\mathbf{B}$

$$\Big[\bigwedge_{F \in \Gamma'} F\Big]_s = 1, \qquad \Big[\bigvee_{E \in \Theta'} E\Big]_s = 0.$$

Let $\mathbf{A}^{(\Gamma',\Theta')} = \mathbf{B}^s$. Let $I$ = set of all finite $(\Gamma', \Theta')$ which $\mathbf{A}$ validates. $J_{(\Gamma',\Theta')} = \{(\Gamma'', \Theta'') \mid \mathbf{A}^{(\Gamma'',\Theta'')}$ validates $(\Gamma', \Theta')\}$. $\{J_{(\Gamma',\Theta')}\}$ may be enlarged to an ultrafilter $F$ and $\mathbf{A} \equiv \Pi \mathbf{A}^{(\Gamma',\Theta')}/F \in \mathbf{K}$.

THEOREM 8. *There exists a* $\mathbf{K} \in \mathrm{EC}_\Delta$ *such that* $\mathbf{K}$ *is not the class of all structures validating any theory* $(\Gamma, \Theta)$.

*Proof.* Let $p, q$ be two propositional variables. Let $\mathbf{K}$ be the class of all structures that not validating $(\{p\}, \{q\})$. We have that $\mathbf{K} \in \mathrm{EC}_\Delta$. Assume now that $\mathbf{K}$ is the class of all structures of some $(\Gamma, \Theta)$. We show that if $F \in \Gamma$ then $F$ is provable. Otherwise there exists a structure $\mathbf{A}$ such that $[F]_0 = 0$, $\mathbf{A} \notin \mathbf{K}$ so $[p]_0 = 1$ and $[q]_0 = 0$.

Form the following structures $(S \cup \{0'\}, R', 0', \mathbf{A}_t)$ where $0'$ is the base point of $S \cup 0'$, $0' \notin S$; $B_t = A_t$ for $t \in S$; $R = R \cup \{0'\} \times (S \cup \{0'\})$.

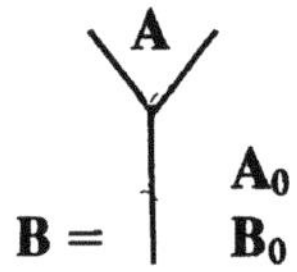


$\mathbf{B}_0$ is like $\mathbf{A}_0$ except that $[p]_{\mathbf{B}_0} = 0$. Thus $\mathbf{B} \in \mathbf{K}$, hence $[F]_{\mathbf{B}_0} = 1$ and therefore $[F]_{\mathbf{A}_0} = 1$, a contradiction.

Now if $\Gamma$ is a set of tautologies we conclude that: $\mathbf{A} \notin \mathbf{K}$ iff some $E \in \Theta$ holds in $\mathbf{A}$. But if $E$ holds in $\mathbf{A}$ then $E$ holds in any truncation of $\mathbf{A}$, i.e., $\mathbf{A} \notin \mathbf{K} \Rightarrow \mathbf{A}^t \notin \mathbf{K}$ for all $t \in S$. To get a contradiction take the structure given by

$$\begin{array}{l|ll} \mathbf{A}_1 & p = 1, & q = 1 \\ \mathbf{A}_0 & p = 1, & q = 0. \end{array}$$

Clearly $\mathbf{A}_1 \in \mathbf{K}$.

THEOREM 9. *If* $\mathbf{K} \in \mathrm{EC}_\Delta$ *then* $\mathbf{K}$ *is the intersection of* EC*-classes.*

*Proof.* Let $A$ be the set of all pairs $(\{E\}, \{F\})$ that hold in all structures of $\mathbf{K}$ and let $B$ be the class of pairs that do not hold in every structure of $\mathbf{K}$.

Let $\mathbf{M}$ be a structure in which all pairs in $A$ hold and all pairs in $B$ do not hold. We want to show that $\mathbf{M} \in \mathbf{K}$. Since $\mathbf{K}$ is closed under elementary equivalence it is enough to show that $\mathbf{M}$ is elementarily equivalent to a structure in $\mathbf{K}$.

Let $(\Gamma, \Theta)$ be the theory of $\mathbf{M}$. For every finite $E_1, \ldots, E_n \in \Gamma$,

$F_1, \ldots, F_n \in \Theta$ we claim that there exists a model **N** in **K** at which all $\delta_i$'s hold and all $\vartheta_i$'s do not hold. If it is not the case then the pair $(\{\wedge E_i\}, \{\vee F_i\})$ is in $A$ and therefore holds in **M** which is impossible. We continue now as in the proof of Theorem 7 to obtain a structure in **K** elementarily equivalent to **M**.

## NOTE

Model theory, and ultra-products were given independently by several people, J. Cleave, P. Aczel, S. Gornemann and the author.

CHAPTER 11

# THEORIES IN HPC 1

$H$ is taken as the underlying logic of intuitionistic mathematics. Thus intuitionistic theories are often presented as theories in $H$. This is particularly useful since there does not exist a large body of intuitionistic mathematics and thus we must take recourse, in many cases, to axiomatic formulations.

Not all theories formulated in $H$ are meaningful or acceptable from the point of view of intuitionistic mathematics. In this chapter, we study the general properties of theories in $H$.

Let $\Vdash$ be the consequence relation arising from the Kripke interpretation. It is also the maximal consequence relation that agrees with $\vdash_H$. A theory in $\Vdash$ has the form $(\Delta, \Theta)$. If we consider theories of the form $(T, \emptyset)$ (written just as $T$) we get a more restricted class of theories; as we have not got classical negation, we cannot reduce $(\Delta, \Theta)$ to an equivalent theory of the form $(T, \emptyset)$. Still, when we are dealing with $H$, and not $\Vdash$, it is natural to take as a theory a set of axioms $T$ and see what can be derived from it using the axioms and rules of $H$. This is the form of all the theories considered in the literature.

Given a set of axioms $T$, $T$ gives rise to two theories, $T^C$ and $T^H$. $T^C$ is the classical theory obtained by taking the logic $C$ as the underlying logic and $T^H$ is the theory in $H$, obtained by taking $H$ as the underlying logic. If $T^H$ is valid in a Kripke structure $\mathbf{M}$, $\mathbf{M}$ is regarded as a partially ordered set of classical structures $A_t$, and we can ask, what is the connection between the $\mathbf{A}_t$'s and $T^C$. Such relations are studied in this chapter.

Notice that several formulations of a classical theory, e.g., $T_1$, $T_2$, may give rise to different theories in $H$; i.e., $T_1^C = T_2^C$ but $T_1^H$ and $T_2^H$ are not equivalent. So when we speak about e.g. 'the intuitionistic theory of fields' we should be careful which formulation to take in $H$ if any at all. We may also have that $T^H$ is an acceptable theory of intuitionistic mathematics while $T^C$ is inconsistent.

## 1. Some intuitionistic theories

We begin with an example. Consider the theory $G$ with the axioms below, in a language with $+$, 0, $=$.

$$\forall xyz((x+(y+z))=((x+y)+z)) \tag{1}$$

$$\forall x(x+0=x) \tag{2}$$

$$\forall x\exists y(x+y=0) \tag{3}$$

Let $(S, R, 0, \mathbf{A}_t, \nu)$ be a modified Kripke structure validating $G$. It is easy to see that each $\mathbf{A}_t$ is a group and for $tRs$, $\nu_{t,s}$ is a homomorphism of $\mathbf{A}_t$ into $\mathbf{A}_s$.

Conversely, suppose we are given a structure $(S, R, 0, \mathbf{A}_t, \nu)$ such that each $\mathbf{A}_t$ is a group. Is $G$ valid in the structure? The answer is yes, as is easy to see. Thus, from the point of view of the Kripke interpretation, groups in $H$ are partially ordered sets of groups, with certain properties as above.

The axiom $\delta = \forall xy(x=y \vee \neg x=y)$ added to $G$ yields the semantic condition that the embeddings $\nu_{t,s}$ are isomorphism into.

THEOREM 5. *In the theory $G^H$ of groups in $H$, if $G^H \vdash A \vee B$ then $G^H \vdash A$ or $G^H \vdash B$, where $A \vee B$ contains no free variables or constants.*

*Proof*. Assume that $G^H \nvdash A$ and $G^H \nvdash B$, then there are structures $(S_i, R_i, 0_i, \mathbf{A}_t^i, \nu_i)$ $i=1,2$ such that $A$ is not valid in $0_1$ and $B$ is not valid in $0_2$. Let $\mathbf{A}_0$ be the group with one element only. Form the following modified Kripke structure $S = S_1 \cup S_2 \cup \{0\}$. $R = R_1 \cup R_2 \cup \{0\} \times S$. $\mathbf{A}_t$ is $\mathbf{A}_t^i$ for $t \in S_i$, $\mathbf{A}_0$ is as above. For the unit of $\mathbf{A}_0$ let $\nu_{0,t}$ take the unit to the unit of $\mathbf{A}_t$. $\nu_{t,s} = \nu_{t,s}^i$ for $t, s \in S_i$. In this structure neither $A$ nor $B$ can hold.

Theorem 5 is a special case of a more general theorem which we shall prove in the next section.

The theory $CG^H$ of commutative groups has the following additional axiom

$$\forall yx(y+x=x+y) \tag{6}$$

It is easy to verify that $CG^H$ is complete for the class of modified Kripke structures of the form $(S, R, 0, \mathbf{A}_t, \nu)$, where $\mathbf{A}_t$ is a commutative group.

**Exercise 7.** Show that the theory $G^{CD} + \forall xy(x = y \vee \neg x = y) = G^C$.

The theory $R$ of rings with unity, in a language with $+, \cdot, 0, 1$ and axioms: ($\cdot$ is usually omitted).

(8) axioms of $G$ for $+, 0$.

(9) $\forall xyz((xy)z = x(yz))$

(10) $\forall xyz((x + y)z = xz + yz)$

$\forall xyz(z(x + y) = zx + zy)$

(11) $\forall x(x1 = 1x = x)$

can be treated similarly. It is complete for the class of all modified Kripke structures in which $A_t$ are rings with unity.

We can add

(12) $\neg 0 = 1$

and the respective condition is that $0 \neq 1$ in all the rings $\mathbf{A}_t$.

Difficulty for this correspondence begins when we consider the axioms for inverse. There are at least three (equivalent in $C$) axioms

(13) $\forall x(\neg x = 0 \rightarrow \exists y(xy = 1)$

(14) $\forall x \exists y(\neg x = 0 \rightarrow xy = 1)$

(15) $\forall x \exists y(x = 0 \vee xy = 1)$

Of these axioms, only (15) has the property that if a modified structure for $R$ satisfies (15) then each $\mathbf{A}_t$ is a ring with division.

Conversely, if each $\mathbf{A}_t$ is a division ring, then (15) is valid.

The axiom (16) of commutativity

(16) $\forall yx(yx = xy)$

has this property as well.

Let $F_1$ be the set of axioms of $R$ with (12), (15) and (16). Then $F_1^H$ is complete for the class of modified Kripke structures in each $\mathbf{A}_t$ is a field. In fact since $\vdash \neg 0 = 1$ and $\vdash x = 0 \vee \exists y(xy = 1)$ we can get $x = 0 \vee \neg x = 0$, for all $x$, and so $x = y \vee \neg x = y$ for all $x$, $y$.

Consider now $F_2$, obtained from $F_1$ by replacing (15) by (14). The following is a structure satisfying $F_2^H$, where $\mathbf{A}$ is any field.

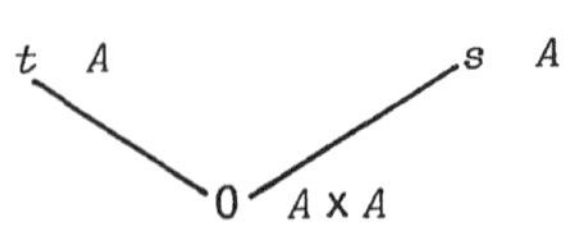

where $\nu_{0,t}(x, y) = x$, $\nu_{0,s}(x, y) = y$. 0 is interpreted as $(0, 0)$ and 1 as $(1, 1)$. Here we see that it is not necessary for each $\mathbf{A}_t$ to be a field, in order that $F_2^H$ be valid.

Actually the intuitionistic theory of fields $F$ is formulated using the apartness relation $x \# y$. We shall discuss $\#$ in a later section.

*Remark.* Although we have shown that the theory of groups $G^H$ is complete for partially ordered sets of groups, this is only an interpretation. This is not the intuitionistic notion of group. The theory of intuitionistic groups is formulated axiomatically in $H$, but the intended interpretation are groups constructed in intuitionistic mathematics.

## 2. Some general theorems

Let $\mathbf{M} = (S, R, 0, \mathbf{A}_t, \nu)$ be a modified Kripke structure.

THEOREM 1. **M** *validates every instance of* $P \vee \neg P$, *P atomic iff* $\nu_{t,s}$ *are all isomorphisms into.*

*Proof.* Immediate.

THEOREM 2. *Suppose for each t, s in S, tRs implies* $\mathbf{A}_t$ *is an elementary substructure of* $\mathbf{A}_s$, *then for all t and any assignment g and wff A*

$$[A]_t^g = 1 \quad \text{iff} \quad \mathbf{A}_t \models^g A$$

*(i.e., A holds classically at* $\mathbf{A}_t$ *iff* $[A]_t = 1$*).*

*Proof.* By induction on $A$. The cases of $A$ atomic and the induction steps for $f$, $\wedge$, $\vee$, $\exists$ present no difficulties. Assume $[A \to B]_t^g = 0$, then for some $s$ such that $tRs$, $[A]_s^{g_s} = 1$, $[B]_s^{g_s} = 0$. By the induction hypothesis $\mathbf{A}_s \models^{g_s} A$ and $A_s \not\models^{g_s} B$. Since $\mathbf{A}_t$ is an elementary substructure of $\mathbf{A}_s$, $\mathbf{A}_t \models^g A$ and $\mathbf{A}_t \not\models^g B$, i.e., $\mathbf{A}_t \not\models^g A \to B$.

The other direction is easy. If $[\forall x A(x)]_t^g = 0$, then for some $s$ such that $tRs$ and $g' =_x g_s'$. $[A(x)]_s^{g'} = 0$ and hence $\mathbf{A}_s \not\models^{g'} A(x)$. Therefore since $\mathbf{A}_t$ is an elementary substructure of $\mathbf{A}_s$, for some $g'' =_x g$, $\mathbf{A}_s \not\models^{g''} A(x)$, i.e., $\mathbf{A}_s \not\models \forall x A(x)$. The other direction is easy.

Thus Theorem 2 is proved.

DEFINITION 3. Let the class of regular wff be defined as the smallest class of wff containing any wff of the form

$$(*) \qquad (\mathcal{Q}\ldots) \bigwedge_i \left( \bigwedge_j A_{ij} \rightarrow \bigvee_j B_{ij} \right)$$

where $(\mathcal{Q}\ldots)$ is a string of quantifiers, $B_{ij}$, $A_{ij}$ are atomic and the empty conjunction is truth and the empty disjunction is falsity.

The class of regular wff is closed under the operation that if $C_{ij}$ are in the class then

$$(\mathcal{Q}\ldots) \bigwedge_i \left( \bigwedge_j A_{ij} \rightarrow \bigvee_j (B_{ij} \vee C_{ij}) \right)$$

where $(\mathcal{Q}\ldots)$, $A_{ij}$, $B_{ij}$ are as before.

THEOREM 4. *Let $A$ be a regular wff such that $[A]_t^g = 1$, then* $\mathbf{A}_t \models^g A$.

*Proof.* By checking the closure conditions of the class of regular wffs.

COROLLARY 4. *Let $T$ be a set of regular axioms such that $T^C$ is model complete, then $T^H + \{P \vee \neg P \mid P$ atomic$\}$ equals $T^C$.*

*Proof.* By Theorems 1 and 4 any structure for $T^H + \{P \vee \neg P \mid P$ atomic$\}$ is a partially ordered set of classical structures of $T^C$ with $\nu$ being isomorphism into. By model completeness of $T^C$, $\nu$ is an elementary embedding and hence by Theorem 2 we conclude that for any $A$, $A \vee \neg A$ is valid in the structure.

DEFINITION 5. (a) Let $T$ be a set of axioms. $T^C$ is said to be prime if there exists a classical structure $\mathbf{M}$ validating $T^C$ such that for any other structure $\mathbf{M}'$ validating $T^C$ we have that $\mathbf{M}$ is isomorphic into $\mathbf{M}'$.

(b) $T^H$ is said to have the disjunction property iff for any closed wff $A$ and $B$, $T^H \vdash_H A \vee B$ iff $T^H \vdash_H A$ or $T^H \vdash_H B$. $T^H$ is said to have the existence property iff whenever $T^H \vdash_H \exists x A(x)$, ($\exists x A(x)$ closed) then for some variable or constant $u$, $T^H \vdash_H A(u)$.

(c) $T^H$ is said to be a semiregular theory iff for all regular $A$, $T^C \vdash_C A$ implies $T^H \vdash_H A$.

(d) $T^H$ is said to be a regular theory iff all the members of $T$ are regular wffs.

(e) Let $\mathbf{M}_i = (S_i, R_i, 0_i, \mathbf{A}_t^i, \nu^i)$, $t \in S_i$ $i \in I$ be a family of modified Kripke structures with $S_i$ pairwise disjoint sets and let $\mathbf{M} = (S, R, 0, \mathbf{A}_t, \nu)$ be a modified Kripke structure such that

$$S = \{0\} \cup \bigcup S_i$$

$$R = \{0\} \times S \cup \bigcup R_i$$

$$\mathbf{A}_t = \mathbf{A}_t^i, \quad \text{for} \quad t \in S_i$$

$$\nu \upharpoonright S_i = \nu^i.$$

Then $\mathbf{M}$ is said to be a result of adjunction of $\mathbf{A}_0$ to the structures $\mathbf{M}_i$. It will be schematically shown as

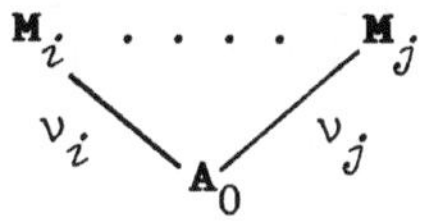


where $\nu_i = \nu_{0,0_i}$.

If $\nu_i$ are all isomorphism into, we say that $\mathbf{M}$ is the result of the decided adjunction of $\mathbf{A}_0$ to $\mathbf{M}_i$.

THEOREM 6. *Let $T$ be a set of axioms of the form $(\forall x_1, \ldots) \wedge_i (\wedge_j B_{ij} \rightarrow (\exists y_1, \ldots) \wedge_k \vee_j C_{ijk})$ where $B_{ij}$ and $C_{ijk}$ are atomic. Let $M$ be a modified Kripke structure. Then $T^H$ is valid in* $\mathbf{M}$ *iff $T^C$ is classically valid in each $\mathbf{A}_t$ of* $\mathbf{M}$.

*Proof.* Since the above wffs are regular, if they are valid in $\mathbf{M}$ then they are classically valid in each $\mathbf{A}_t$ of $\mathbf{M}$.

Assume that a wff of the above form is not valid in $\mathbf{M}$. Then for some $t$ and an assignment into $\mathbf{A}_t$ and some $i$, $[\wedge_j B_{ij}]_t^g = 1$ and $[(\exists y_1, \ldots) \wedge_k \vee_j C_{ijk}]_t^g = 0$. Now it is clear that

$$\mathbf{A}_t \models^g \bigwedge_j B_{ij} \quad \text{and} \quad \mathbf{A}_t \models^g \sim (\exists y_1, \ldots) \bigwedge_k \bigvee_j C_{ijk}.$$

THEOREM 7. *Let $T$ be a set of axioms of the form $(\mathcal{Q} \ldots) \wedge \vee B_{ij}$, where $B_{ij}$ are atomic. Let* $\mathbf{M}$ *be a modified Kripke structure of constant domains. Then $T^H$ is valid in* $\mathbf{M}$ *iff $T^C$ is classically valid in each $\mathbf{A}_t$ of* $\mathbf{M}$.

*Proof.* One direction follows from the fact that $T$ is a set of regular axioms. If a wff of the above form is not valid in $\mathbf{A}_t$ then it is not valid classically in $\mathbf{A}_t$, as the quantifiers range over $\mathbf{A}_t$ as a domain.

THEOREM 8. *Let $T$ be a set of axioms of the form*

$$(\forall x_1, \ldots)(\exists y_1, \ldots) \bigwedge_i \left( \bigwedge_j B_{ij} \rightarrow (\exists z_1, \ldots) \bigwedge_k \bigvee_j C_{ijk} \right)$$

*where $B_{ij}$ and $C_{ijk}$ are atomic and let* **M** *be a decided modified Kripke structure. Then $T^H$ is valid in* **M** *iff each* $\mathbf{A}_t$ *of* **M** *classically validates $T^C$.*

*Proof.* Since the axioms of $T$ are regular, one direction is immediate. Assume a wff of the above form is not valid in **M**. Then for some $t$ and some $g$ into $A_t$, we have $(\exists y_1, \ldots) \wedge_i (\wedge B_{ij} \rightarrow (\exists z_1, \ldots) \wedge \vee C_{ijk}))$ is not intuitionistically valid at $t$. So for each $g'$ such that $g' =_{y_1} g$, $g' =_{y_2} g, \ldots$ there exists an $i$ and $s$ such $tRs$ such that $[\wedge B_{ij}]_s^{g'} = 1$, $[(\exists z_1 \ldots) \wedge \vee C_{ijk}]_s^{g'} = 0$. Therefore, since the structure is decided, $[\wedge B_{ij}]_t^{g'} = 1$, $[(\exists z_1, \ldots) \wedge \vee C_{ijk}]_t^{g'} = 0$. Therefore for some $g$ and every $g' =_{y_m} g$, $(m = 1, 2, \ldots)$.

$$\mathbf{A}_t \models^{g'} \sim \left( \bigwedge B_{ij} \rightarrow (\exists z_1, \ldots) \bigwedge \bigvee C_{ijk} \right)$$

that is $\mathbf{A}_t \models \sim (\forall x_1, \ldots)(\exists y_1, \ldots) \wedge (\wedge B_{ij} \rightarrow (\exists z_1, \ldots) \wedge \vee C_{ijk}))$.

THEOREM 9. *Let $T$ be a set of axioms of the form of Theorem 6 and let* **M** *be a result of the adjunction of* $\mathbf{A}_0$ *to* $\mathbf{M}_i$*, where* $\mathbf{M}_i$ *intuitionistically validate $T^H$ and* $\mathbf{A}_0$ *validates classically $T^C$, then* **M** *intuitionistically validate $T^H$.*

*Proof.* Immediate, from Theorem 6.

THEOREM 10. *Let $T$ be a set of axioms of the form of Theorem 8 and let* **M** *be a result of the decided adjunction of* $\mathbf{A}_0$ *to* $\mathbf{M}_i$*, where* $\mathbf{M}_i$ *intuitionistically validate $T^H$ and every instance of $P \vee \neg P$, $P$ atomic and* $\mathbf{A}_0$ *classically validates* $\mathbf{A}_0$*, then* **M** *intuitionistically validate $T^H$ and every instance of $P \vee \neg P$, $P$ atomic.*

*Proof.* Immediate, from Theorem 8.

THEOREM 11. (a) *Let $T$ be a set of axioms of the form of Theorem 6 such that $T^C$ is prime, then $T^H$ has the disjunction property and the existence property.*

(b) *Let $T$ be a set of axioms of the form of Theorem 8 such that $T^C$ is prime then $T^H + \{P \vee \neg P \mid \text{for } P \text{ atomic}\}$, has the disjunction property and the existence property.*

*Proof.* (a) Suppose $T^H \not\vdash A$, $T^H \not\vdash B$, $A$, $B$ closed, then there exist structures $\mathbf{M}^1$, $\mathbf{M}^2$, in which respectively $A$, $B$ are false. Let $\mathbf{M}$ be the result of the adjunction of the prime structure of $T^C$ to $\mathbf{M}^1$, $\mathbf{M}^2$. In $\mathbf{M}$, $A \vee B$ is not valid, by the previous theorems of this section. The existence property is proved similarly.

(b) Is proved similarly.

THEOREM 12. *Let $\mathbf{M}^i$, be structures validating a regular theory $T^H$ and let $\mathbf{M}$ be an adjunction of $\mathbf{M}^i$ with a structure validating $T^C$, then $\mathbf{M}$ validates $T^H$.*

*Proof.* Let $A$ be any regular axiom of $T^H$. If $[A]_0 = 0$ in $\mathbf{M}$ then show that $\mathbf{A}_0 \models \neg A$, which is a contradiction.

DEFINITION 13. Let $x$ be an intermediate propositional logic and let $T$ be a set of axioms for a theory. Let $H^x$ be the extension of $H(=HPC)$ with all the theorems of $x$ as axioms and let $T^x$ be the theory $T$ with $H^x$ as the underlying logic. Thus relating this notation to the previous one in this section, $T^h$ is $T^H$ and $T^c$ is $T^C$.

We say that a set of axioms $T$ has the deJongh property iff for all $x$, $x \underset{\neq}{\supset} h \rightarrow T^x \underset{\neq}{\supset} T^h$.

THEOREM 14. *Let $T$ be a set of axioms that has the following properties*:

(a) *The axioms of $T$ are of the form of Theorem 6.*

(b) *For each $n$, there exist wff $A_1, \ldots, A_n$ such that for each $1 \leq i \leq n$, $A_i \wedge \bigwedge_{j \neq i} \neg A_j$ is $T^H$ consistent.*

(c) *$T^C$ has a prime classical structure.*

*Then $T$ has the deJongh property.*

*Proof.* Let $\alpha(p_1, \ldots, p_n)$ be a wff of $h$ such that $h \not\vdash \alpha$. We show that for some $E_1, \ldots, E_n$ of $T^h$, $T^h \not\vdash \alpha(E_1, \ldots, E_n)$ where $p_1, \ldots, p_n$ are all the atoms of $\alpha$.

Since $h \not\vdash \alpha(p_1, \ldots, p_n)$, there exists a tree $J_n$ (of Definition 4.2.5) and an assignment $D$ such that $[\alpha]_{J_n} = 0$. (See Chapter 4, Section 2). Let $S$ be the set of endpoints of $J_n$. Suppose $\bar{\bar{S}} = m$. Let $P_1, \ldots, P_m$ be wffs promised by (b) of the theorem. For each $1 \leq i \leq n$ let $\mathbf{M}_i$ be modified Kripke structure of $T^H \cup \{P_i \wedge \bigwedge_{j \neq i} \neg P_j\}$. We can assume $S_i$ are pairwise disjoint. Let $\mathbf{B}$ be the prime model of $T^C$. We can regard

**B** as a submodel of each $\mathbf{A}_{0_i}$. Now construct the following modified Kripke structure $\bar{\mathbf{M}}$ based on $J_n$. Enumerate the points of $S$ as $\{s_1, \ldots, s_m\} = S$. Let $\bar{S} = J_n \cup \bigcup_{i=1} S_i$. Let $\bar{R}$ be the transitive and reflexive closure of $\bigcup_i R_i \cup R_{J_n} \cup \{(s_i, 0_i \mid i = 1, \ldots, m\}$.

Let $\bar{\mathbf{A}}_t$ be

$$\mathbf{A}_t^i \quad \text{for} \quad t \in S_i,$$

$$\mathbf{A}_{0_i} \quad \text{for} \quad t = s_i \in S$$

$$\mathbf{B} \quad \text{for} \quad t \in J_n - S.$$

Define $\bar{\nu}_{t,s}$ in the obvious way, since $B$ can be regarded a substructure of each $\mathbf{A}_{0_i}$. We claim

LEMMA. (a) *If $C \in T$ then $C$ is valid in the structure.*

(b) *If $S_t$ is the set of all endpoints of $J_n$ above $t \in J_n$ then $[\neg\neg(\bigvee_{s_i \in S_t} P_i)]_s = 1$ iff $t$ is below $s$.*

*Proof.* (a) By Theorem 6.

(b) Clear.

Let $p$ be an atom of $\alpha$ and let $S_p = \{t \in J_n \mid [p]_t = 1\}$. Let $E_p = \bigvee_{t \in S_p}(\neg\neg\bigvee_{s_i \in S_t} P_i)$. Then clearly $[E_p]_t = 1$ iff $[p]_t = 1$ where the values of [ ] are taken in the respective structures. It therefore follows by induction on the subformulas $\beta$ of $\alpha$ that

$$(*) \qquad [\beta(p_1, \ldots, p_n)]_t = [\beta(E_{p_1}, \ldots, E_{p_n})]_t$$

The theorem follows from (*) and the lemma.

## 3. End extensions

DEFINITION 1. Let $T^H$ be a theory in a certain language and let $<$ be a binary relation definable in $T^H$. Let $\mathbf{M} = (S, R, 0, \mathbf{A}_t, \nu)$ be a structure for $T^H$. We say that $\mathbf{M}$ is an $<$-end extension structure iff for any $t$, $s$ such $t \neq s \wedge tRs$ and any $y \in (|\mathbf{A}_s| - \nu_{t,s}|\mathbf{A}_t|)$ and $x \in |\mathbf{A}_t|$, there *does not* exist an $n$ such that $y <^n \nu_{t,s}(x)$, where $u <^n z$ abbreviates $(\exists x_1, \ldots, x_{n-1})(u < x_1 \wedge x_1 < x_2 \wedge \cdots \wedge x_{n-1} < z)$.

THEOREM 2. *$T^H$ is complete for a class of $<$-end extension structures iff for all $n$, $T^H \vdash \forall x^{a,n}(A(x) \vee B) \to \forall^{a,n} x\, A(x) \vee B$, $x$ not free in $B$, and $\forall^{a,n} x\, C(x)$ is $\forall x(\bigvee_{m=1}^{n} x <^m a \to C(x))$, for $n \geqslant 1$ and $\forall^{a,0} x$ is $\forall a$.*

*Proof*. One direction is easy to verify, namely, that the axiom schema above is valid in all $<$-end extension structures.

For the other direction we need a series of definitions and lemmas.

DEFINITION 3. (a) Let $t = (\Delta, \Theta)$ be a saturated $T^H$ theory. Let $x$, $y$ be variables in the language of $(\Delta, \Theta)$. We write $x <_t y$ iff $x < y \in \Delta$.

(b) Let $U$, $V$ be two sets of variables in the language of $t$, we say $U \ll_t V$ iff

(1) $U \cap V = \emptyset$.

(2) For *no* $u \in U$, $v \in V$, $n$ do we have $v <_t^n u$.

(c) We say $U \leqslant_t V$ iff for any $x \in U$ there exists $v \in V$ such that $x = v \in \Delta$ or $x <_t^n v$ for some $n$.

(d) Let $t = (\Delta, \Theta)$ be a saturated theory in a language with variables $U$ and let $s = (\Delta', \Theta')$ be a saturated theory in a language with variables $U' \supseteq U$. We say that $s$ is an end-extension of $t$ iff (a) $\Delta \subseteq \Delta'$ and (b) $U \ll_s (U' - U)$.

(e) We shall use the following abbreviations.

(1) If $\Delta$ is in a language with variables $U$, we shall say that $\Delta$ is in a language $U$.

(2) If $U_0 \subseteq U$ and $\Delta_0$ is the set of all wff $A$ such that $A$ is in the language $U_0$ and $A \in \Delta$, we shall simply write $\Delta_0 \equiv \Delta \cap U_0$

We can now proceed with some useful lemmas, for theories satisfying the condition of Theorem 2.

LEMMA 4. *Let $t_0 = (\Delta_0, \Theta_0)$ be a saturated theory in a language $U_0$. Let $t_1^* = (\Delta_1^*, \Theta_1^*)$ be a saturated theory in the language $U_1^*$. Assume that $\Delta_0 \subseteq \Delta_1^*$. Then $t_0$ can be extended to a saturated theory $t$ in a language $U$ and $t_1^*$ can be extended to a saturated theory $t^*$ in a language $U^* \supseteq U$ such that $t^*$ is an end extension of $t$.*

*Proof*. Let $U_1^* = U_1 \cup V_1$ with the property that $U_1 \leqslant_{t_1^*} U_0$ and $U_0 \ll_{t_1^*} V_1$.

*Claim* 1. *$(\Delta_0, \Theta_0 \cup \Theta_1^* \cap U_1)$ is a consistent theory of the language $U_0 \cup V_1$.*

*Proof*. Otherwise for some $A_0 \in \Delta_0$, $B_0 \in \Theta_0$, $B_1^* \in \Theta_1^* \cap U_1$ we get: $\vdash A_0 \to B_0 \vee B_1^*$. Let $x_i$ be the variables of $U_1 - U_0$ that appear free in

$B_1^*$. Since $U_1 \leqslant_{t_1^*} U_0$, there exists $n_i$, $a_i \in U_0$ such that $x_i < n_i$, $a_i \in \Delta_1^*$. We can therefore continue:

$$\vdash A_0 \to \forall \begin{matrix} & n_i & \\ \cdots & a_i & \cdots \\ \cdots & x_i & \cdots \end{matrix} (B_0 \vee B_1^*), \qquad x_i \neq a_i$$

$$\vdash A_0 \to B_0 \vee \forall \begin{matrix} & n_i & \\ \cdots & a_i & \cdots \\ \cdots & x_i & \cdots \end{matrix} B_1^*,$$

where

$$\left(\forall \begin{matrix} a_1, n_1 \\ x_1 \end{matrix}\right)\left(\forall \begin{matrix} a_2, n_2 \\ x_2 \end{matrix}\right)$$

is abbreviated by

$$\left(\begin{matrix} & n_1 & n_2 \\ \forall & a_1 & a_2 \\ & x_1 & x_2 \end{matrix}\right)$$

This sentence is in the language $U_0$.

Since $A_0 \in \Delta_0$, $B_0 \in \Theta_0$, we get

$$\forall \begin{matrix} & n_i \\ \cdots & a_i \\ \cdots & x_i \end{matrix} B_1^* \in \Delta_0 \subseteq \Delta_1^*,$$

a contradiction to the consistency of $(\Delta_1^*, \Theta_1^*)$.

Now since $(\Delta_0, \Theta_0 \cup \Theta_1^* \cap U_1)$ is consistent, it can be extended to a theory $t_2 = (\Delta_2, \Theta_2)$, saturated in the language $U_2 \supseteq U_0 \cup U_1$.

*Claim* 2. *The theory* $(\Delta_1^* \cup \Delta_2, \Theta_1^*)$ *is consistent in the language* $U_2 \cup U_1^*$.

*Proof*. Otherwise for some $A_1^* \in \Delta_1^*$, $A_2 \in \Delta_2$, $B_1^* \in \Theta_1^*$ we have

$$\vdash A_2 \to (A_1^* \to B_1^*).$$

Let $x_i$ be the variables in $A_2$ from $U_2 - (U_0 \cup U_1)$ and let $y_j$ be the variables from $V_1$ that appear in $A_1^* \to B_1^*$. Then

$$\vdash \exists x_i A_2 \to \forall y_j (A_1^* \to B_1^*).$$

This wff is in the language if $U_0 \cup U_1$ and since $\exists x_i A_2 \in \Delta_2$, we get $\forall y_j (A_1^* \to B_1^*) \in \Delta_2$. On the other hand, since $A_1^* \in \Delta_1^*$, $B_1^* \in \Theta_1^*$,

$\forall y_j(A_1^* \to B_1^*) \in \Theta_1^* \cap U_1$, and therefore $\forall y_j(A_1^* \to B_1^*) \in \Theta_2$, a contradiction.

Now since the theory of Claim 2 is consistent, extend it to a saturated theory $t_3^* = (\Delta_3^*, \Theta_3^*)$ in a language $U_3^* \supseteq U_1^* \cup U_2$. We can write $U_3^* = U_3 \cup V_3$, with $U_3 \leqslant_{t_3^*} U_2$ and $U_2 \lll_{t_3^*} V_3$.

We can now continue, as we are in the situation of Claim 1 and construct the extension $t_4$ of $t_2$ and then using the situation of Claim 2 again, construct $t_5^*$ etc.

We can thus assume that we constructed the sequences of theories $t_{2n}$ and $t_{2n+1}^*$ with the properties:

(1) $t_{2n} = (\Delta_{2n}, \Theta_{2n})$ is in the language $U_{2n}$.

(2) $t_{2n+1}^* = (\Delta_{2n+1}^*, \Theta_{2n+1}^*)$ is in the language $U_{2n+1}^* = U_{2n+1} \cup V_{2n+1}$.

(3) (a) $t_{2n+2}$ extends $t_{2n}$, $U_{2n} \subseteq U_{2n+2}$

(b) $t_{2n+3}^*$ extends $t_{2n+1}^*$, $U_{2n+1}^* \subseteq U_{2n+3}^*$.

(4) (a) $U_{2n+1} \leqslant_{t_{2n+1}^*} U_{2n}$.

(b) $U_{2n+1} \lll_{t_{2n+1}^*} V_{2n+1}$.

(c) $U_{2n} \subseteq U_{2n+1}^*$, $U_{2n+1} \subseteq U_{2n+2}$.

(5) $\Delta_{2n} \subseteq \Delta_{2n+1}^*$, $\Theta_{2n+1}^* \cap U_{2n+1} \subseteq \Theta_{2n+2}$.

We claim that

$$t = \left(\bigcup \Delta_{2n}, \bigcup \Theta_{2n}\right), \qquad t^* = \left(\bigcup \Delta_{2n+1}^*, \bigcup \Theta_{2n+1}^*\right)$$

fulfill the properties of Lemma 4.

To show that $t^*$ is an end extension of $t$, let $x \in \bigcup U_{2n+1}^* - \bigcup U_{2n}$. Then by (4c), $x \notin \bigcup U_{2n+1}$. Let $y \in U_{2n}$, then since $x \in V_{2n+1}$ we get $y <_{t^*} x$.

LEMMA 5. *Let $t_1^* = (\Delta_1^*, \Theta_1^*)$, $t_2 = (\Delta_2, \Theta_2)$ be two saturated theories such that $\Theta_1^* \cap U \subseteq \Theta_2$, where $U$ is the common language of these two theories and $\Delta_2 \cap U \subseteq \Delta_1^*$, then $t_1^*$ can be extended to a $t^*$ and $t_2$ to a $t$ such that $t^*$ is an end extension of $t$.*

*Proof.* Follows from the process given in the proof of Lemma 4 especially beginning with Claim 2.

DEFINITION 6. Let $\omega^*$ be the set of all finite sequences of natural numbers including the empty sequence $\Lambda$. Let $\rho$ be the relation of being an initial segment of.

LEMMA 7. *Let* $\mathbf{T} \subseteq \{x | \Lambda \bar{\rho}_1 x\} \subseteq \omega^*$. *Let* $\alpha$ *be a function associating with each* $x \in \mathbf{T}$ *a theory of linear order* $\alpha(x) = (\Delta_x, \Theta_x)$ *saturated in a language (denumerable)* $U_x$. *Assume that* $x\rho y$ *implies* $\Delta_x \subseteq \Delta_y$, $U_x \subseteq U_y$. *Then there exists a function* $\beta$, *associating with each* $x \in \mathbf{T}$ *a theory* $(\Delta_x^*, \Theta_x^*)$ *with properties as above and the additional properties that:*

(a) $\beta(x)$ *is an extension of* $\alpha(x)$.

(b) $x\rho y$ *implies* $\beta(y)$ *is an end extension of* $\beta(x)$.

(c) *All languages involved are countable.*

*Proof.* Assume $\mathbf{T}$, $\alpha$ are given. Let $f$ be a function from $\omega$ into $\mathbf{T}$ such that $f(0) \neq \Lambda$ and each element of $\mathbf{T}$ is obtained $\aleph_0$ times, and $\Lambda$ is always the highest element in $\mathbf{T}$ below $f(n)$, $f(n+1)$; for any $n$. We shall define a sequence of functions $\beta_{m,n}$, $m$, $n \in \omega$, such that $\beta_{m,n}(x)$, for $x \in \mathbf{T}$ is a saturated theory. The required $\beta$ shall be the limit of these $\beta_{m,n}(x)$.

*Stage* 0, 0. Let $\beta_{0,0} = \alpha$.

*Stage* 0, 1. If $f(0) = x \neq \Lambda$, let $\Lambda = y_0, \ldots, y_n = x$ be the sequence ($\rho$ sequence) of $\mathbf{T}$ leading to $x$. Look at the pair $\alpha(y_{n-1})$, $\alpha(y_n)$. This pair fulfills the conditions of Lemma 4 and therefore there exist extensions in a richer language, with properties as in Lemma 4. Call these theories $\beta_{0,1}(y_n)$, $\beta_{0,1}(y_{n-1})$. Now look at $\alpha(y_{n-2})$, $\beta_{0,1}(y_{n-1})$. This pair again fulfills the conditions of Lemma 4. Let $\beta_{0,2}(y_{n-2})$, $\beta_{0,2}(y_{n-1})$ give this pair. Define also $\beta_{0,2}(y_n) = \beta_{0,1}(y_n)$. Continue in this manner until you reach $\Lambda = y_0$, and at that stage $\beta_{0,n}(\Lambda), \ldots, \beta_{0,n}(y_n)$ are defined. Each time we extend our theories according to Lemma 4, we use completely new variables. We also know from the construction that $\beta_{0,n}(y_i) = \beta_{0,m}(y_i)$ if $m \geqslant i+1$.

Now we observe that $\beta_{0,n}(\Lambda)$ and $\beta_{0,n}(y_1)$ fulfill the conditions of Lemma 5. Let therefore $\beta_{0,n+1}(\Lambda)$, $\beta_{0,n+1}(y_1)$ be two extensions, in a richer language fulfilling the conditions promised in Lemma 5. Let $\beta_{0,n+1}(y_i) = \beta_{0,n}(y_i)$ for $i > 1$. Now $\beta_{0,n+1}(y_1)$, $\beta_{0,n+1}(y_2)$ fulfill the assumption of Lemma 5. Therefore there exist extensions $\beta_{0,n+2}(y_1)$, $\beta_{0,n+2}(y_2)$ fulfilling the properties promised by Lemma 5. Let $\beta_{0,n+2}(\Lambda) = \beta_{0,n+1}(\Lambda)$, $\beta_{0,n+2}(y_i) = \beta_{0,n+1}(y_i)$, for $i > 3$.

We can thus go on until we reach $y_n = x$ and then start going back again etc., etc.

Let $\beta_{1,0}$ be defined by:

$$\beta_{1,0}(z) = \alpha(z), \qquad \text{for } z \notin \{y_0, \ldots, y_n\}$$

$$\beta_{1,0}(z) = \left(\bigcup_n \Delta_{\beta_{0,n}(z)}, \bigcup_n \Theta_{\beta_{0,n}(z)}\right), \qquad \text{otherwise.}$$

It is easy to see that $\beta_{1,0}(y_i)$ is an end extension of $\beta_{1,0}(y_{i-1})$.

*Stage m*. Suppose $\beta_{m,0}$ is given. Assume $f(m) = x$. Let $y_0, \ldots, y_n$ be as before. Notice that since in the process of construction, whenever we extended a theory, we used completely new variables, then if $\Lambda \rho v$, then the condition of Lemma 5 hold for $\beta_{m,0}(\Lambda)$, $\beta_{m,0}(v)$. We can then again run up and down $y_1, \ldots, y_{n-1}, x$ and define $\beta_{m+1,0}$.

Now define $\beta$ by

$$\beta(x) = \left(\bigcup_n \Delta_{\beta_{n,0}(x)}, \bigcup_n \Theta_{\beta_{n,0}(x)}\right).$$

It is easy to verify that $\beta$ has the required properties.

We can now turn to the

*Proof of Theorem* 3. Let $(\Delta, \Theta)$ be a consistent $T^H$ theory of $L$. We shall construct a tree structure $(\Delta, \Theta)$ with all the required properties.

Recall the way a structure $(\Delta, \Theta)$ is constructed in Chapter 3. Two lemmas are needed:

(1) *Let $(\Delta', \Theta')$ be a saturated theory in a language U. Let $F = A \to B \in \Theta'$, then there exists a saturated theory $(\Delta'_F, \Theta'_F)$ in a language with possibly more variables V such that*

$$\Delta' \cup \{A\} \subseteq \Delta'_F$$

$$B \in \Theta'_F$$

(2) *Similarly, if $F = \forall x A(x) \in \Theta'$, then there exists $(\Delta'_{(F)}, \Theta'_{(F)})$ such that $\Delta' \subseteq \Delta'_{(F)}$ and for some x of the language of $(\Delta'_{(F)}, \Theta'_{(F)})$ we have $A(x) \in \Theta'_F$. As before, we have that $(\Delta'_{(F)}, \Theta'_{(F)})$ is a theory in a possibly richer language.*

Now, to construct a structure validating $(\Delta, \Theta)$, let $S$ be a set of saturated $T^H$-theories such that

(1) $\qquad (\Delta, \Theta) \in S$.

(2) If $(\Delta', \Theta') \in S$ and $F = A \to B$ (or $F = \forall x A(x)$) is in $\Theta'$ then a theory $(\Delta'_F, \Theta'_F)$ (or $\Delta'_{(F)}, \Theta'_{(F)}$) is in $S$.

Defining $R$ on $S$ as $\subseteq$ (i.e., $(\Delta_1, \Theta_1) R (\Delta_2, \Theta_2)$ iff $\Delta_1 \subseteq \Delta_2$) and letting $[P(x_1, \dots, x_n)]_{(\Delta', \Theta')} = 1$ iff $P(x_1, \dots, x_n) \in \Delta'$, for $P$ atomic, we obtain a structure validating $(\Delta, \Theta)$, as we can show that for any $A$

$$[A]_{(\Delta', \Theta')} = 1 \qquad \text{iff } A \in \Delta'.$$

Although $(S, R)$ is not a tree, we can always arrange for an $S$ to be such that, $S = \mathbf{T} \subseteq \{x \mid \Lambda \rho x\}$, and for a function $\alpha$, such that $\alpha(x)$, $x \in S$ is a saturated theory $(\Delta_{\alpha(x)}, \Theta_{\alpha(x)})$ and the required properties hold for $R = \rho$ and $x \rho y \to \Delta_{\alpha(x)} \subseteq \Delta_{\alpha(y)}$, and a structure is obtained from the theories by $[A]_x = 1$ iff $A \in \Delta_{\alpha(x)}$.

So from now on when we speak about Henkin type structures we mean something like the above.

Following Chapter 3 let $\mathbf{T}_1$ be a Henkin structure validating $(\Delta, \Theta)$. We have no assurance that it has the required properties. Let $\alpha_1$ be the function giving the respective Henkin type saturated theories $(\Delta_{\alpha_1(x)}, \Theta_{\alpha_1(x)})$ associated with each $x \in \mathbf{T}_1$.

Let $\beta_1$ be a function promised in Lemma 7, with the properties of Lemma 7. Since the theories given by $\beta_1$ are in a richer language, we have no assurance that $\mathbf{T}_1$, $\beta_1$ is a good structure, i.e., we may have a theory $(\Delta_{\beta_1(x)}, \Theta_{\beta_1(x)})$ with $A \to B \in \Theta_{\beta_1(x)}$ but for no $y$ such that $x \rho y$ can we have $A \in \Delta_{\beta_1(y)}$, $B \in \Theta_{\beta_1(y)}$. This may happen because $A \to B$ may be in the new language. However there is no obstacle in following the process of Chapter 3 (or the basic lemmas of the completeness proof) and taking $\mathbf{T}_2 \supseteq \mathbf{T}_1$, $\alpha_2 \supseteq \beta_1$ such that $\mathbf{T}_2$, $\alpha_2$ is a proper structure. But now it is not clear that this structure has the end extension property required by Theorem 3. Of course we know that the property holds for $\alpha_2(x)$, $\alpha_2(y)$, $x$, $y \in \mathbf{T}_1$, but it may fail for $\mathbf{T}_2$. However, we can always define $\beta_2$ as in Lemma 7. We can go on getting $\alpha_3$, $\mathbf{T}_3$, $\beta_3$, $\alpha_4, \dots$.

Now let $\mathbf{T} = \bigcup_n \mathbf{T}_n$; for $x \in \mathbf{T}$, let $\alpha(x) = \lim \alpha_n(x)$ where $\lim \alpha_n(x) = (\Delta_x, \Theta_x)$ with $\Delta_x = \bigcup_{i \geq m} \Delta_{\alpha_i(x)}$, $\Theta_x = \bigcup_{i \geq m} \Theta_{\alpha_i(x)}$, and $x \in \mathbf{T}_m$.

One can verify that this is a Henkin type structure validating $(\Delta, \Theta)$ fulfilling the requirements of Theorem 3. It is important to remember that whenever we use additional variables, in any of the processes, we always use completely new ones.

CHAPTER 12

# THEORIES IN HPC 2

## 1. Equality

The $H$ theory of equality $E^H$ has the axioms:

(a) $x = x.$

(b) $x = y \to y = x.$

(c) $x = y \to (y = z \to x = z).$

(d) For any wff $A$ $\quad x = y \to (A(x) \to A(y)).$

Considerations in the two previous chapters show that this theory is complete for the class of all modified Kripke structures of the form $(S, R, 0, \mathbf{A}_t, \nu)$, where $\mathbf{A}_t$ is a classical model for equality and $\nu_{t,s}$ are homomorphism into.

If we add the axiom

(e) $x = y \vee \neg x = y$

we get the theory $E^H_D$ of decided equality. It is complete for the class of all structures for equality in which $\nu_{t,s}$ are 1–1 embeddings.

The reader can easily observe that $E^C$ is a prime theory (see 11.2.5)) and hence by (11.2.11) $E^C$ has the disjunction property and the existence property.

To obtain the deJongh property for $E^H$, we must show property (b) of (11.2.14). Let $A_i(x_1, \ldots, x_n)$, $i = 1, \ldots, n$ be

$$\forall y(y = x_i \vee \neg y = x_i) \wedge \bigwedge_{i \neq j} \neg x_i = x_j.$$

**Exercise.** Consider the theory of stable equality, $E^H$ + the axiom $\neg\neg(x = y) \to x = y$. Examine it for the disjunction, existence and deJongh properties.

## 2. The apartness relation

This is a basic relation that appears in intuitionistic mathematics. The language contains the binary relation # with equality =. The following axioms are taken besides the axioms for equality:

(1) $\neg x \# y \to x = y.$

(2) $x \# y \to x \# z \vee y \# z.$

From (1) and (2) one easily concludes

(3) $x \# y \to y \# x.$

(4) $x \# y \to \neg(x = y).$

(5) $\neg\neg(x = y) \to x = y.$

There is no special condition on the structures that characterises the apartness relation, except that the axioms hold.

**Exercise.** Does the theory of apartness relation have the disjunction, existence or deJongh properties?

## 3. Linear order

The language contains $<$ and =. The axioms are:

(1) axioms for equality.

(2) $\neg(x < x).$

(3) $x < y \vee x = y \vee y < x.$

(4) $x < y \to (y < z \to x < z).$

The decided theory of linear order has the additional axioms:

(5) $x < y \vee \neg(x < y).$

(6) $x = y \vee \neg(x = y).$

By results of the previous chapter the theory of linear order is complete for the class of all modified structures of the form $(S, R, 0, \mathbf{A}_t, \nu)$ where each $\mathbf{A}_t$ is a classical model of linear order. The decided theory of linear order is complete for the additional condition that $tRs$ implies that $\mathbf{A}_t$ is a substructure of $\mathbf{A}_s$.

Consider the additional schema (added to the decided theory of linear order)

(7) $\forall y \leqslant u(A(y) \vee B) \rightarrow \forall y \leqslant uA(y) \vee B$, $y$ is not free in $B$, $y \neq u$.

We remark that if we add the axiom schema (8) below to the decided theory of linear order we obtain the theory of linear order of classical logic

(8) $\forall y(A(y) \vee B) \rightarrow \forall yA(y) \vee B$, $y$ not free in $B$.

This is so because (8) is the axiom schema for the logic CD of constant domains.

THEOREM 8. *The decided theory of linear order with* (7) *is complete for the class M of all modified Kripke structures of the form* $(T, \rho, \Lambda, U_t, <_t)$ *where:*

(a) $T \subseteq \{x \in \omega^* \mid \Lambda \rho x\}$

(b) $y \in T \wedge x\rho y \rightarrow x \in T$

(c) $(U_t, <_t)$, $t \in T$ *is a totally linearly ordered countable set.*

(d) *For* $t\rho_1 s$, *the structure* $(U_s, <_s)$ *is an* end extension *of* $(U_t, <_t)$. *(That is* $(U_t, <_t)$ *is a substructure of* $(U_s, <_s)$ *with* $a \in U_t \wedge b \in U_s - U_t \rightarrow a <_s b$.)

*Proof.* Follows from the previous chapter about end extensions.

Theories of dense linear order are obtained by adding the axioms:

(9) $x < y \rightarrow \exists z(x < z \wedge z < y)$.

In the presence of (1)–(6), (9) is equivalent to:

(10) $\exists z(x < y \rightarrow x < z \wedge z < y)$.

The theory of dense linear order (with (9)) is complete for the class of all modified Kripke structures in which each $\mathbf{A}_t$ is a classical model of dense linear order.

Other formulations of this theory include the axiom (3*) instead of (3). This gives rise to a different theory, for which (9) and (10) are no longer equivalent.

(3*) $\neg x = y \rightarrow x < y \vee y < x$.

**Exercise.** (a) Which of the theories of linear order have the disjunction and existence properties? Use (11.2.11).

(b) Which of the theories of linear order has the deJongh property? Use (11.2.14) and try for condition (b) of (11.2.14):

$$A_i(x_1, \ldots, x_n) = \bigwedge_{i \neq j} \neg x_i = x_j \wedge \forall y(x < y \vee y < x).$$

## 4. One unary function

Consider a theory with one unary function symbol $f$ and equality $=$. If we add no other axioms, we get the theory of one unary function. We can add the axiom, $\forall xy(x = y \vee \neg(x = y))$ and get decided equality. Consider the additional schemas $(*)_n$:

$$(*)_n \qquad \forall_x^{a,n}(A(x) \vee B) \rightarrow (\forall_x^{a,n} A(x) \vee B), \quad x \text{ not free in } B;$$

where $\forall_x^{a,n} A(x)$ is $\forall x(\bigvee_{0 \leq m \leq n} f^m(x) = a \rightarrow A(x))$.

THEOREM. *The above theory is complete for the class of all modified Kripke structures of the form* $(T, \rho, \Lambda, U_t, f_t)$ *such that*:

(a) $(\Lambda \in T) \wedge (s \in T \wedge t\rho s \rightarrow t \in T)$.

(b) $(U_t, f_t)$ *is a classical model of unary function.*

(c) *If* $x \in U_t$ *and* $t\rho s$ *and* $y \in U_s - U_t$ *then not* $f^n(y) = x$ *for any natural number* $n$.

(d) *If* $t\rho s$, *then* $(U_t, f_t)$ *is a substructure of* $(U_s, f_s)$.

*Proof.* Follows from the results of the previous chapter about end extensions.

Another possible axiom is $(\#)_m$.

$$(\#)_m \qquad \forall x \exists y_1, \ldots, y_m\Big(\Big(\bigwedge_i f(y_i) = x\Big) \wedge \bigwedge_{j \neq i} \neg y_i = y_j$$
$$\wedge \forall z(f(z) = x \rightarrow \vee z = y_i)\Big)$$

which says that there are exactly $m$ 'predecessors' to any $x$.

**Exercise.** Examine the various theories of unary function for the disjunction, existence and deJongh properties.

## 5. Groups

The theory of groups $G$ is in a language with $+, 0, =$. The axioms are:

(a) axioms for equality.

(b) $x+(y+z)=(x+y)+z$.

(c) $x+0=0+x=x$.

(d) $\forall x\exists y(x+y=0)$.

As we have seen in the previous chapters it is complete for the class of all modified structures of the form $(S, R, 0, \mathbf{A}_t, \nu)$ where $\mathbf{A}_t$ are all groups and $\nu_{t,s}$ is homomorphism into.

If we add the axiom

(e) $\forall xy(x=y \vee \neg x=y)$

we get completeness for the additional condition that $\nu_{t,s}$ are all isomorphisms into.

The axiom

(f) $\forall xy(x+y=y+x)$

corresponds to the condition that all the groups $\mathbf{A}_t$ are commutative. The theory $G+(f)$ is denoted by $CG$.

Since for the case of axiom (e) we get that $\nu_{t,s}$ is an isomorphism into, we can identify $\mathbf{A}_t$ as a subgroup of $\mathbf{A}_s$.

DEFINITION 1. Let $G_m$ be the theory with axioms (a)–(f) and the addition axiom $(g_m)$

$(g_m)$ $mx=0$.

Clearly, by the results of the previous chapter, $G_m$ is complete for the class of all structures with $\mathbf{A}_t$ being a commutative group satisfying $\forall x(mx=0)$ and that $tRs$ implies $\mathbf{A}_t$ is a subgroup of $\mathbf{A}_s$.

DEFINITION 2. Let $G_m^*$ be the extension of $G_m$ with the axiom

$(s_m)$ $$\bigwedge_{1\leqslant n\leqslant m}(\exists y(x=ny)\vee\neg\exists y(x=ny)).$$

Then the following holds.

THEOREM 3. *$G_m$ is complete for the class $\mathbf{M}_m^*$ of all modified Kripke structures of the form $(S, R, 0, \mathbf{G}_t, I_t)$ where $(S, R, 0)$ is a partially ordered set. For each $t \in S$, $\mathbf{G}_t = \Sigma_{\beta \in I_t} \mathbf{G}_\beta^t$, where $\Sigma$ is a direct sum, $\mathbf{G}_\beta^t$ are cyclic and $tRs \to I_t \subseteq I_s$ and $I_0 \neq \emptyset$.*

*Proof.* By Chapter 10, we can assume that $G_m$ is complete for a class of Kripke structures in which $R$ is well founded and all the other properties mentioned at the beginning of this section also hold. Now each $\mathbf{G}_t$, $t \in S$ is a classical abelian group such that $\mathbf{G}_t \models \forall s(ms = 0)$. By results of classical group theory, $\mathbf{G}_t = \Sigma_{\beta \in I_t} \mathbf{G}_\beta^t$, a direct sum of cyclic groups. Since $tRs$ implies $G_t$ is a subgroup of $G_s$, we get by axiom $(s_m)$ that $G_t$ is a pure subgroup of $\mathbf{G}_s$ (i.e., since $\exists y \in |\mathbf{G}_s|(x = ny) \to \exists y \in |\mathbf{G}_t|(x = ny)$ holds as well as $\forall x(mx = 0)$ we get that $\mathbf{G}_t$ is a pure subgroup of $\mathbf{G}_s$) and hence $\mathbf{G}_t$ is a direct summand of $\mathbf{G}_s$. Since $R$ is well founded $(S, R, 0, \mathbf{A}_t)$ can be described as in the theorem.

Notice that if $m$ is the product of pairwise distinct primes then $G_m^* = G_m$.

The theory $DCG$ of divisible abelian groups is obtained by adding to $CG$, for each $m$, the axiom $\forall x \exists y(x = my)$.

Theories of linearly ordered abelian groups are obtained by combining the theory of abelian group with theories of linear order. That is, take, for $+$, $<$, $=$:

(a) axioms for commutative groups

(b) axioms for linear order

(c) $x < y \to x + z < y + z$.

Depending on the formulation of the linear order, we may get a nice characterisation of the Kripke interpretation, using Chapter 10.

**Exercise.** (a) Examine the various theories of groups for the disjunction and existence properties. Apply Theorem 11.2.11.

(b) Examine the various theories of groups for the deJongh property. Apply Theorem 11.2.14. For condition (b) of (11.2.14) you can sometimes use $A_n = \forall x(p_n x = 0)$, where $p_n$ is the $n$th prime. What about the theory $\mathbf{G}_m$?

## 6. Fields

The language contains 0, 1, +, ·, =. The axioms are:

(1) axioms for commutative group for + and 0 (see Section 5).

(2) axioms for equality

(3) $(xy)z = x(yz)$

$x(y+z) = xy + xz$

$xy = yx.$

(4) $\neg 0 = 1$

For the axiom for inverse we have two possibilities.

(5a) $\forall x \exists y(x = 0 \vee xy = 1)$

(5b) $\forall x(x \neq 0 \rightarrow \exists y(xy = 1)).$

The following axiom may be added for algebraic closure

(6) For each $n \geqslant 1$, the following:

$$(\forall x_0, \ldots, x_n)\exists z\left(\sum_{i=0}^{n} x_i z^i + z^{n+1} = 0\right).$$

Observe that in the formulation with (5a), call it $F_1$, $\forall xy(x = y \vee \neg(x = y))$ is provable. Thus by the previous chapter this theory is complete for the class of all modified Kripke structures $(S, R, 0, \mathbf{A}_t, \nu)$ with the property that each $\mathbf{A}_t$ is a field and $\nu$ is isomorphism into. If we add axiom 6 we get a theory that is identical with the classical theory of algebraically closed fields, since this classical version is model complete.

If we take (5b) instead of (5a) we get a non-classical theory, even with the addition of (6). There is no simple condition on structures that characterises this theory with (5b).

A more acceptable theory from the point of view of intuitionistic mathematics is formulated with the apartness relation. The language contains 0, 1, +, ·, #, =. The axioms are (1)–(4), together with the axioms for apartness and

(5c) $x \# 0 \rightarrow \exists y(xy = 1).$

(6) may be added for algebraically closed fields.

Theories of real closed fields are formulated in a language with an additional unary predicate $P(x)$. The axioms are (1)–(5) (different versions of (5) give different intuitionistic versions) and

(7) $\neg P(0)$

$P(x) \wedge P(y) \to P(x+y)$

$P(x) \wedge P(y) \to P(xy)$

(8) $\forall xy \exists z(x+y=0 \to x=z^2 \vee y=z^2)$

(9) $\forall x_0, \ldots, x_{2n} \exists y \left(\sum x_i y^i + y^{2n+1} = 0\right)$

(10a) $x+y=0 \to x=0 \vee P(x) \vee P(y)$

(10b) $x+y=0 \wedge \neg P(x) \wedge \neg P(y) \to x=0.$

Again, the version with axioms (1)–(4), (5a), (7)–(9), (10a) yields a theory that is identical with the classical theory.

**Exercise 11.** Examine the various theories of fields for the disjunction, existence and deJongh properties.

## 7. A Fragment of Arithmetic

DEFINITION 1. Heyting's arithmetic, HA, is formulated in a language with $+, \cdot, 0, ', =$ and the following axioms (a), (b), (c1), (c2), (d), (e), (f):

(a) axioms for equality

(b) $\neg(x'=0)$

$x'=y' \leftrightarrow x=y$

(c1) $\neg x=0 \to \exists y(y'=x)$

(c2) $x=0 \vee \exists y(y'=x)$

(d) $x+0=x$

$x+y'=(x+y)'$

(e) $x \cdot 0 = 0$

$x \cdot y' = x \cdot y + x$

(f) For each $A$, the induction axiom

$$A(0) \wedge \forall x(A(x) \rightarrow A(x')) \rightarrow \forall x A(x).$$

**Exercise 2.** (a) Show by induction that in HA

$$x = y \vee \neg x = y.$$

(b) Show that (c2) follows from (c1).

DEFINITION 3. (a) Let $A$ be the subsystem of $HA$ without axioms (c1), (c2) and (f).

(b) Let $RHA$ be the subsystem of $HA$ with axioms of $A$ and (c2).

(c) Let $RHA^{\omega}$ be the extension of $A$ with axiom (c1).

(d) Let $RHA^{st}$ be the extension of $RHA$ with the axiom

$$\neg\neg(x = y) \rightarrow (x = y).$$

**Exercise 4.** Show that $\forall xy(x = y \vee \neg(x = y))$ is not provable in $RHA^{st}$.

*Solution.* Let N be the classical model of natural numbers with the usual $+$, $\cdot$, 0. Consider $\mathbf{N} \times \mathbf{N}$. Let $+$ and $\cdot$, $'$, be defined coordinatewise, and let $O$ be $(0, 0)$.

Consider the Kripke structure

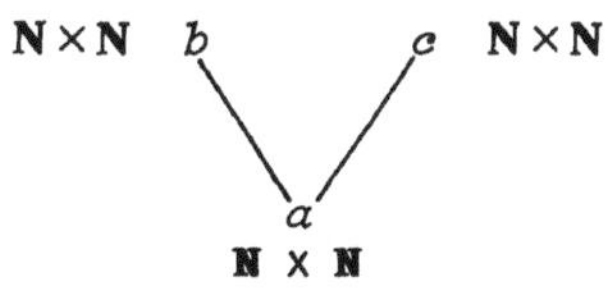


with $\nu$ being the identity mapping and $=$ is interpreted as equality at $a$ and at $b$ we take $=$ as the congruence relation $(x, y) \sim (u, v)$ iff $x = u$, and at $c$ $=$ is taken as the congruence relation $(x, y) \sim (u, v)$ iff $y = v$.

Show that all axioms of $RHA^{st}$ hold in the structure but $\forall xy(x = y \vee \neg(x = y)$ does not hold.

**Exercise 5.** Show that $\exists x \neg \forall y(x = y \vee \neg(x = y))$ is a consistent sentence of $RHA^{st}$.

THEOREM 6. *HA has the disjunction property.*

*Proof.* Suppose $HA \nvdash A_1$ and $HA \nvdash A_2$. Then we show $HA \nvdash A_1 \vee A_2$, $A_i$ closed formulas.

Let $\mathbf{M}_1, \mathbf{M}_2$ be two modified Kripke structures validating *HA*, such that $A_i$ is false in $\mathbf{M}_i$. Let $\mathbf{N}$ be the standard classical model of arithmetic and let $M$ be the result of the disjunction

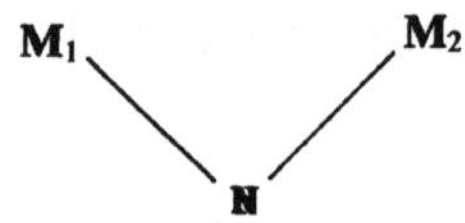


$\mathbf{N}$ is embedded into $M_i$ via the identity map since all the standard natural numbers have names in *HA* and = is decided. $\mathbf{N}$ is already a substructure structure of any classical structure of $\mathbf{M}_i$. It is easy to verify that one gets a structure validating *HA* in which $A_1 \vee A_2$ is not valid.

**Exercise 7.** Show *HA* has the existence property.

**Exercise 8.** Show that for any wff $A$ of classical arithmetic $P$ (i.e., $HA + CPC$),

$$P \vdash A \quad \text{iff} \quad HA \vdash \neg\neg A.$$

**Exercise 9.** Show the deJongh property for *HA*.

## NOTE

The section on end-extension is new. The rest is pretty standard. The results on fragments of arithmetic are based on a paper of K. Gordon and D. deJongh.

## CHAPTER 13

# COMPLETENESS OF HPC WITH RESPECT TO RE AND POST STRUCTURES

Validity in recursive structures has been investigated by several authors. Kreisel has shown that there exists a consistent sentence of classical predicate calculus (CPC) that does not possess a recursive model. The sentence is a conjunction of the axioms of a variant of Bernay's set theory, including the axiom of infinity. The language contains additional constants besides $\epsilon$. Later Kreisel and Mostowski presented a sentence (not possessing recursive models) which was a conjunction of axioms of a variant of Bernay's set theory without the axiom of infinity but still with additional constants besides $\epsilon$. Later Mostowski improved the result by giving a sentence which can be demonstrated in Heyting arithmetic to be consistent and to have no recursive models. Rabin obtained a simple proof that some sentence of set theory with the single nonlogical constant $\epsilon$ does not have any recursively enumerable models.

More generally, Mostowski has shown that the set of all sentences valid in all RE models is not arithmetical and Vaught improved it by showing that it holds for a language of one binary relation. In fact, Vaught gives a way of translating $n$-place relations to 2-place ones that preserves the RE characteristic of the model.

The work described here is related to two interpretations of Heyting's predicate calculus (HPC) in terms of recursive structures (both different from the intended interpretation of Brouwer–Heyting: as we shall see below) by Prawitz and Kreisel.

DEFINITION 0. (a) Let $(N, S, R, 0, D, U)$ be a Kripke or a Beth structure. For $t \in S$, $(D_t, U_t)$ can be regarded as a classical model for the language of HPC. We denote this model by $\mathbf{A}_t$, with domain $|\mathbf{A}_t| = U_t$. For atomic $P$ and $a_1, \dots, a_n \in |\mathbf{A}_t|$ write $\mathbf{A}_t \models P(a_1, \dots, a_n)$ just in case $(a_1, \dots, a_n) \in D_t(P)$. Notice that for a Kripke structure, $\mathbf{A}_t \models P(a_1, \dots, a_n)$ iff $[P(a_1, \dots, a_n)]_t = 1$. Not so for Beth structures.

(b) We say that a wff $B(P_1, \dots, P_n)$ is $s$-valid in a structure, where $P_1, \dots, P_n$ are all the atomic wff appearing in $B$ iff all substitution instances $B^* = B(C_1, \ldots, C_n)$ of $B$ are valid in the structure (Beth or Kripke). We can also speak of completeness with respect to $s$-validity in a class of structures.

(c) A Kripke structure is said to be locally $RE(LRE)$ iff for each $t \in S$, the relation

$$\mathbf{A}_t \models \mathbf{P}(a_1, \ldots, a_n), \qquad P \text{ atomic, } a_i \in |\mathbf{A}|_t$$

is recursively enumerable. A Kripke structure is said to $RE$ iff the relation

$$t \in S \wedge s \in S \wedge tRs \wedge (\mathbf{A}_t \models P(a_1, \ldots, a_n))$$

is recursively enumerable in $t, s, P, a_1, \ldots, a_n$; (for all atomic $P$, $a_i \in |\mathbf{A}|_t$). A Kripke structure is said to be recursive ($R$) iff the relation $t \in S \wedge s \in S \wedge tRs \wedge (\mathbf{A}_t \Vdash P(a_1, \ldots, a_n))$, $P$ atomic, $a_i \in |\mathbf{A}|_t$, is recursive. We similarly define these notions for Beth structures.

## 1. The results

First let us state a general positive result. The other results appear in Table I (see opposite).

THEOREM A. *If T is a decidable, finitely axiomatizable theory of HPC, then T is complete for validity in RE Kripke structures.*

## 2. Positive results

We begin with the proofs of several positive results. Some of the theorems below are stronger than those in Table I.

THEOREM 1. *Let T be a decidable and finitely axiomatizable HPC theory; then T is complete for a class of RE-Kripke structures.*

*Proof.* Let $\tau$ be an axiom for $T$ and let $E$ be a non-provable sentence (i.e., $HPC \nVdash \tau \to E$), we shall construct an $RE$-Kripke structure in which $\tau$ holds and $E$ does not hold. The idea of the proof is the following: Kripke used a modification of the semantic tableaux method to prove completeness of HPC for the class of Kripke structures. Thus, Kripke's proof allows one to construct, for any wffs $E$ such that $HPC \nVdash E$, a countermodel for $E$. Now, suppose $T$ is an $HPC$ theory which is decidable and finitely axiomatizable. Let $\tau$ be

TABLE I

Validity of recursive structures

| kind of validity / fragment | 1. Validity<br>2. *s*-validity<br>for LRE<br>Kripke<br>structures | 1. Validity<br>2. *s*-validity<br>for LRE<br>Beth<br>structures | 1. Validity<br>2. *s*-validity<br>for RE<br>Kripke<br>structures | 1. Validity<br>2. *s*-validity<br>for RE<br>Beth<br>structures | 1. Validity<br>2. *s*-validity<br>in R Kripke<br>or Beth<br>structures |
|---|---|---|---|---|---|
| HPC | 1. Set of valid statements not arithmetical<br>2. complete | 1. complete<br>2. complete | 1. not arithmetical<br>2. ? | ? | 1. Incomplete (Dyson–Kreisel)<br>2. ? |
| Monadic HPC or pure equality | 1. complete<br>2. complete | 1. complete<br>2. complete | 1. not arithmetical<br>2. ? | ? | 1. incomplete<br>2. ? |
| Fragment without $\exists, \vee, \neg$ | 1. complete<br>2. complete | 1. complete<br>2. complete | 1. complete<br>2. complete | 1. complete<br>2. complete | ? |
| Fragment where all atomic predicates are negated | 1. complete<br>2. complete | 1. complete<br>2. complete | 1. not arithmetical<br>2. ? | ? | 1. incomplete (not arithmetical)<br>2. ? |

an axiom for $T$. Then $T \vdash_{HPC} E$ if $HPC \vdash \tau \to E$. We can therefore follow Kripke and construct a countermodel for $\tau \to E$. This countermodel, however, need not be *RE*. The idea of the proof is to modify Kripke's construction in such a way as to obtain a recursively enumerable countermodel. How do we do that? Well, why doesn't the Kripke construction yield an *RE* structure? An examination of the semantic tableaux method used by Kripke shows that there are stages in the construction of the tableaux where we have a choice like the following:

> The wff AVB should appear in the left (truth) side of the tableau, therefore it is consistent to add either $A$ or $B$. We don't know which, so we split the tableau into two tableaux, one for each possibility.

When we construct the structure we take an infinite path which is free from contradiction. The choice of this path is not *RE*.

If, however, the theory $T$ is decidable, we do not have to take two possibilities. Since the tableaux involved at stage $n$ are all finite, we use the decision procedure to decide which of $\{A, B\}$ to add and proceed to the next stage. Thus a path free from contradiction is constructed in an *RE* manner. This is precisely what we do in what follows. It is a construction which is like that of Kripke except that, whenever we have a choice, we use the decision procedure to make that choice. The final structure is obtained from the union of the left (truth) hand side of the tableaux and is *RE*.

We begin with some definitions. Assume that $V$ is the set of variables of the language, and let $U \subseteq V$ and $\beta(x_1, \ldots, x_n)$ be given a formula, $\alpha$ is said to be a $U$ subformula of $\beta$ iff: (a) all free variables of $\alpha$ are from $U$ and (b) the free variables of $\alpha$ can be changed in such a way that the resulting $\alpha'$ is a subformula of $\beta$. Let $f$ be a total recursive function such that: (a) for each $n$, $f(n)$ is an ordered pair $(t, \beta)$, where $t$ is an ordered sequence of natural numbers, with 0 as a first element, and $\beta$ is a formula of *HPC*, and (b) each pair $(t, \beta)$ is obtained as value of $f$ an infinite number of times.

Towards the construction of the desired countermodel, we shall define an auxiliary recursive sequence of finite functions that are actually approximations of the structure we have in mind. This sequence of functions, $g_0, g_1, \ldots$ is such that each $g_k$ has the following properties:

(a) $g_k$ is a finite function with domain comprised of finite sequences of natural numbers taken from $\{0\} \cup \{i \in \omega \mid i < k\}$ with 0 as the first element. The value of $g_k(t)$ is a triple $(\Delta_t^k, \Theta_t^k, V_t^k)$. Where $\Delta_t^k$ and $\Theta_t^k$ are finite sets of formulas and where $V_t^k$ is a finite set of variables containing all the free variables of the formulas of $\Delta_t^k \cup \Theta_t^k$.

Once $g_k$ has been defined, a function $\gamma_k$ can also be defined from $g_k$ with the properties that dom $\gamma_k$ = dom $g_k$ and that $\gamma_k(t)$, $t \in$ dom $\gamma_k$ is a wff. If $t \leqslant s$, for $t$, $s \in$ dom $g_k$ means that $t$ is a (not necessarily proper) initial segment of $s$, then the inductive (on $\leqslant$) definition of $\gamma_k$ is the following:

(1) $$\gamma_k(t) = \wedge \Delta_t^k \to \vee \Theta_t^k,$$

if $t$ is a top point of dom $g_k$ (i.e., $\forall s \in \text{dom}\, g_k(t \leqslant s \to t = s)$).

(2) If $s_1, \ldots, s_n$ are the immediate $\leqslant$ successors of $t$ in dom $g_k$, then $\gamma_k(t) = \wedge \Delta_t^k \to \vee \Theta_t^k \vee \forall_i \gamma_k^*(s_i)$, where $\gamma_k^*(s_i)$ is obtained from $\gamma_k(s_i)$ by universally quantifying over the variables of $V_{s_i}^k - V_t^k$.

*Abbreviation.* Let $\bar{\Theta}_t^k$ mean:

(1) $\vee \Theta_t^k$, if $t$ is a top point of dom $g_k$.

(2) $\vee \Theta_t^k \vee \vee_i \gamma_k^*(s_i)$, if $s_i$ are the $\leqslant$ immediate successors of $t$ in dom $g_k$.

(b) If $g_k(t) = (\Delta_t^k, \Theta_t^k, V_t^k)$, then $HPC \not\vdash \wedge \Delta_t^k \to \vee \bar{\Theta}_t^k$.

(c) If for $t$, $g_k(t)$ is defined then $g_k(s)$ is defined for all initial segments $s$ of $t$, and $V_s^k \subseteq V_t^k$ and $\Delta_s^k \subseteq \Delta_t^k$ hold.

(d) If $g_k(t)$ is defined, then $g_{k+1}(t)$ is defined and $\Delta_t^k \subseteq \Delta_t^{k+1}$, $\Theta_t^k \subseteq \Theta_t^{k+1}$, $V_t^k \subseteq V_t^{k+1}$.

(e) If $t'$ is a successor of $t$ and $g_{k'}(t')$ is defined, then there exists a $k \leqslant k'$ and a $\beta$ of the form (1) $\beta_1 \to \beta_2$ or of the form (2) $\forall x \beta_3(x)$ such that $HPC \not\vdash \wedge \Delta_t^k \to \beta \vee \bar{\Theta}_t^k$ and either

(1) $$\Delta_{t'}^k = \Delta_t^k \cup \{\beta_1\}, \qquad \Theta_{t'}^k = \{\beta_2\}, \qquad V_{t'}^k = V_t^k \qquad \text{or}$$

(2) $$\Delta_{t'}^k = \Delta_t^k, \qquad \Theta_{t'}^k = \Theta_t^k \cup \{\beta_3(u)\} \quad \text{and} \quad V_{t'}^k = V_t^k \cup \{u\}$$

where $u$ is a variable that does not appear in $\bigcup_{s \leqslant t} V_s^k$.
($s \leqslant t$ means that $s$ is an initial segment of $t$, or $s = t$.)

We now turn to the definition of the sequence $g_n$, $n \in \omega$. Let $g_0$ have domain $\{(0)\}$ and let $g_0((0)) = (\{\tau\}, \{E\}, \{v_0\})$. Assume that $g_k$ has been defined and proceed to define $g_{k+1}$. We look at $f(k) = (t, \beta)$ and distinguish several cases according to what $(t, \beta)$ is. Let $s_i$ be the immediate successors of $t$ in dom $g_k$ (if they exist).

*Case* 1. If either $g_k(t)$ is not defined or $\beta$ is not a $V_t^k$ subformula of any member of $\Delta_t^k \cup \Theta_t^k$, then let $g_{k+1} = g_k$.

*Case* 2. Case 1 does not hold and $HPC \vdash \wedge \Delta_t^k \to \bar{\Theta}_t^k \vee \beta$. We distinguish several subcases.

(2a) $\beta = \beta_1 \wedge \beta_2$. Let dom $g_{k+1}$ = dom $g_k$ and

$$g_{k+1}(s) = \begin{cases} g_k(s), & \text{if not} \quad s \geq t \\ (\Delta_s^k \cup \{\beta_1, \beta_2\}, \Theta_s^k, V_s^k) & \text{if} \quad s \geq t, \quad \text{and} \\ s \in \text{dom}\, g_{k+1}. \end{cases}$$

(2b) $\beta = \beta_1 \vee \beta_2$. In this case we cannot have that both $HPC \vdash \beta_i \wedge \wedge \Delta_t^k \to \bar{\Theta}_t^k$, $i = 1, 2$, because that would imply that

$$HPC \vdash \wedge \Delta_t^k \to \bar{\Theta}_t^k.$$

So let $\beta_*$ be the first of the $\beta_i$ (i.e., $\beta_1$ and otherwise $\beta_2$) such that $HPC \nvdash \beta_* \wedge \wedge \Delta_t^k \to \bar{\Theta}_t^k$. Define domain $g_{k+1}$ = domain $g_k$ and let

$$g_{k+1}(s) = \begin{cases} g_k(s) & \text{if not} \quad s \geq t, \\ (\Delta_s^k \cup \{\beta_*\}, \Theta_s^k, V_s^k) & \text{if} \quad s \geq t, \quad \text{and} \\ s \in \text{dom}\, g_{k+1}. \end{cases}$$

(2c) $\beta = \exists x \beta_3(x)$. In this case let $u$ be a new variable such that $u \notin \bigcup_{s \in \text{dom}\, g_k} V_s^k$, and let dom $g_{k+1}$ = dom $g_k$ and let for $s \in \text{dom}\, g_k$

$$g_{k+1}(s) = \begin{cases} g_k(s), & \text{if not} \quad s \geq t \\ (\Delta_s^k \cup \{\beta_3(u)\}, \Theta_s^k, V_s^k \cup \{u\}) & \text{if} \quad s \geq t. \end{cases}$$

(2d) $\beta = \beta_1 \to \beta_2$ or $\beta = \forall x \beta_1(x)$. In this case let dom $g_{k+1}$ = dom $g_k$, and let for $s \in \text{dom}\, g_{k+1}$

$$g_{k+1}(s) = \begin{cases} g_k(s) & \text{if not} \quad s \geq t \\ (\Delta_s^k \cup \{\beta\}, \Theta_s^k, V_s^k) & \text{if} \quad s \geq t. \end{cases}$$

(2e) $\beta$ is atomic then $\beta \neq f$ for otherwise the inductive hypothesis (b) on $g_k$ cannot hold. In this case let dom $g_{k+1}$ = dom $g_k$ and let for $s \in \text{dom}\, g_{k+1}$:

$$g_{k+1}(s) = \begin{cases} g_k(s) & \text{if not} \quad s \geq t \\ (\Delta_s^k \cup \{\beta\}, \Theta_s^k, V_s^k) & \text{if} \quad s \geq t. \end{cases}$$

*Case* 3. Case 1 does not hold and $HPC \nvdash \wedge \Delta_t^k \to \bar{\Theta}_t^k \vee \beta$.

(3a) $\beta = \forall x \beta_3(x)$.

Let $u$ be a new variable (i.e., $u \notin \bigcup_s V_s^k$, $s \in \operatorname{dom} g_k$) and define $g_{k+1}$ as follows:

dom $g_{k+1} = \operatorname{dom} g_k \cup \{t * (k)\}$, (where $*$ is concatenation) and

$$g_{k+1}(s) = \begin{cases} g_k(s) & \text{if} \quad s \neq t * (k) \\ (\Delta_t^k, \{\beta_3(u)\}, V_t^k \cup \{u\}) & \text{for} \quad s = t * (k). \end{cases}$$

(3b) $\qquad \beta = \beta_1 \to \beta_2$.

Let dom $g_{k+1} = \operatorname{dom} g_k \cup \{t * (k)\}$, and let

$$g_{k+1}(s) = \begin{cases} g_k(s) & \text{if} \quad s \neq t * (k) \\ (\Delta_t^k \cup \{\beta_1\}, \{\beta_2\}, V_t^k) & \text{if} \quad s = t * (k). \end{cases}$$

(3c) If $\beta$ is not of the form of (3a) or (3b) let $g_{k+1} = g_k$.

*Claim #.* $\not\vdash \wedge \Delta_t^k \to \bar{\Theta}_t^k$, for $t \in \operatorname{dom} g_k$.

*Proof.* The proof is by induction on $k$.

Assume that # holds for $g_k$ and show that it holds for $g_{k+1}$. We check all the cases of constructing $g_{kH}$ from $g_k$.

(a) $g_{k+1}$ was obtained from $g_k$ as in case 2 of the construction: We need some lemmas.

LEMMA 1. *If $t \leqslant s$, $t$, $s \in \operatorname{dom} g_k$ then $\vdash \gamma_k(s) \to \gamma_k(t)$.*

LEMMA 2. *Assume that $t$, $g_k$ and $g_{k+1}$ are such that:*

(a) $t \in \operatorname{dom} g_k$

(b) $\operatorname{dom} g_k = \operatorname{dom} g_{k+1}$

(c) *for all $s \geqslant t$, $s \in \operatorname{dom} g_k$ we have that $\Theta_s^{k+1} = \Theta_s^k$ and $\Delta_s^{k+1} = \Delta_s^k \cup \{\beta\}$, where $\beta$ is some fixed wff.*

*Then*

$$\beta \vdash \gamma_{k+1}(t) \leftrightarrow \gamma_k(t).$$

*Proof.* By induction on dom $g_k$.

(1) If $t$ is a top point then clearly

$$\beta \vdash [\beta \wedge \wedge \Delta_t^k \to \vee \Theta_t^k] \leftrightarrow [\wedge \Delta_t^k \to \vee \Theta_t^k].$$

(2) If $s_i$ are the immediate successors of $t$ in dom $g_k$ then since (by the induction hypothesis)

$$\beta \vdash \gamma_{k+1}(s_i) \leftrightarrow \gamma_k(s_i)$$

certainly

$$\beta \vdash [\beta \wedge \wedge \Delta_t^k \to \vee \Theta_t^k \vee \vee_i \gamma_{k+1}^*(s_i)] \leftrightarrow [\wedge \Delta_t^k \to \vee \Theta_t^k \vee \vee_i \gamma_k^*(s_i)],$$

where $\gamma_k^*(s_i)$ (resp. $\gamma_{k+1}^*(s_i)$) are obtained from $\gamma_k(s_i)$ (resp. $\gamma_{k+1}(s_i)$) by universally quantifying over the variables in $V_{s_i}^k - V_t^k$ (resp. $V_{s_i}^{k+1} - V_t^{k+1}$).

LEMMA 3. *Assume that* $t$, $g_k$ *and* $g_{k+1}$ *fulfill properties* (a), (b), *and* (c) *of Lemma* 2 *and let* $s_i$ *be the immediate successors of* $t$ *in* dom $g_k$ *(if they exist). Further assume that*

$$\not\vdash \wedge \Delta_t^k \wedge \beta \to \bar{\Theta}_t^k.$$

*Then also*

$$\not\vdash \wedge \Delta_t^k \wedge \beta \to \bar{\Theta}_t^{k+1}.$$

*Proof*. Otherwise,

$$\vdash \beta \to [\wedge \Delta_t^k \to \vee \Theta_t^k \vee \vee_i \gamma_{k+1}^*(s_i)].$$

Now by Lemma 2,

$$\vdash \beta \to (\gamma_{k+1}(s_i) \to \gamma_k(s_i))$$

and therefore

$$\vdash \beta \to [\wedge \Delta_t^k \to \vee \Theta_t^k \vee \vee_i \gamma_k^*(s_i)],$$

i.e., $\vdash \beta \wedge \wedge \Delta_t^k \to \bar{\Theta}_t^k$, a contradiction.

\# follows from the lemmas, as conditions (a), (b), (c) of Lemmas 2 and 3 are fulfilled.

(b) $g_{k+1}$ was obtained from $g_k$ as in case 3 of the construction.

LEMMA 4. *Let* $g_{k+1}$ *be obtained from* $g_k$ *by case* 3 *of the construction. Then for all* $s \in \operatorname{dom} g_k$

$$\gamma_k(s) \leftrightarrow \gamma_{k+1}(s).$$

*Proof*. Clearly the only case we have to check is the case that $g_{k+1} \neq g_k$. Assume that $t \in \operatorname{dom} g_k$ is such that $t * (k)$ is the new point of dom $g_{k+1}$. We first check the lemma for the case of $s = t$. We follow the cases of the construction.

(3a) In this case

$$\gamma_k(t) = \wedge \Delta_t^k \to \vee \Theta_t^k \vee \vee_i \gamma_k^*(s_i)$$

with $\beta_3(u) \in \Theta^k_{t*(k)}$, $s_i$ the immediate successors of $t$ in dom $g_k$.

$$\gamma_{k+1}(t) = \wedge \Delta^k_t \to \vee \Theta^k_t \vee \vee_i \gamma^*_k(s_i) \vee \forall u(\wedge \Delta^k_t \to \beta_3(u)),$$

with $u \notin V^k_t$. Clearly $\vdash \gamma_k(t) \leftrightarrow \gamma_{k+1}(t)$.

(3b) In this case

$$\gamma_k(t) = \wedge \Delta^k_t \to \vee \Theta^k_t \vee \vee_i \gamma^*_k(s_i),$$

with $\beta_1 \to \beta_2 \in \Theta^k_t$.

$$\gamma_{k+1}(t) = \wedge \Delta^k_t \to \vee \Theta^k_t \vee \vee_i \gamma^*_k(s_i) \vee (\wedge \Delta^k_t \wedge \beta_1 \to \beta_2).$$

Again it is clear that $\vdash \gamma_k(t) \leftrightarrow \gamma_{k+1}(t)$.

Now assume by induction that for each successor $s_i$ of $s$ the theorem holds. Then:

$$\gamma_k(s) = \wedge \Delta^k_s \to \vee \Theta^k_s \vee \vee_i \gamma^*_k(s_i).$$

$$\gamma_{k+1}(s) = \wedge \Delta^k_s \to \vee \Theta^k_s \vee \vee_i \gamma^*_{k+1}(s_i).$$

Clearly since $\vdash \gamma_k(s_i) \leftrightarrow \gamma_{k+1}(s_i)$, we get the same for $s$.

LEMMA 5. *For any $\alpha$ and $k$, $t$:*
*If* $\quad \not\vdash \wedge \Delta^k_t \to \bar{\Theta}^k_t \vee \alpha$
*then* $\quad \not\vdash \wedge \Delta^{k+1}_t \to \bar{\Theta}^{k+1}_t \vee \alpha.$

*Proof.* We check the construction of $g_{k+1}$ from $g_k$.

For case 2 the result follows from Lemma 2 and for case 3 from Lemma 4.

We have thus defined the recursive sequence of functions $g_k$, $k = 0, 1, 2, \ldots$.

Let us now use this sequence to define an *RE* Kripke structure. Let

$$S = \bigcup_k \text{domain } g_k$$

$$R = \text{'being an initial segment of'}$$

$$A_t = \bigcup_k V^k_t$$

and for atomic $P$, $[P(a_1, \ldots, a_n)]_t = 1$ iff for some $k$

$$P(a_1, \ldots, a_n) \in \Delta^k_t.$$

MAIN LEMMA. *For any $t$ and any subformula $\beta$ of $\tau \wedge E$ with* *variables in $A_t$ we have:*

$$[\beta]_t = 1 \quad \textit{iff} \quad \exists k,\ HPC \vdash \wedge\Delta_t^k \to \bar{\Theta}_t^k \vee \beta.$$

*Proof.* Clearly this holds for atomic $\beta$ by the definition of $\{g_k\}$. We proceed by induction

(1) $\beta = \beta_1 \wedge \beta_2$.

Assume that $[\beta_1 \wedge \beta_2]_t = 1$. Then by the induction hypothesis, for some $n_1 \leqslant n_2$, $\vdash \wedge\Delta_t^{n_i} \to \bar{\Theta}_t^{n_i} \vee \beta_i$, $i = 1, 2$. And therefore we get that $HPC \vdash \wedge\Delta_t^{n_2} \to \bar{\Theta}_t^{n_2} \vee \beta$. The other direction is easy.

(2) $\beta = \beta_1 \vee \beta_2$.

Assume that $\exists n$, $HPC \vdash \wedge\Delta_t^n \to \beta \vee \bar{\Theta}_t^n$. Let $m$ be such that $m \geqslant n$ and $f(m) = (t, \beta)$. Clearly, since $HPC \vdash \wedge\Delta_t^m \to \beta \vee \bar{\Theta}_t^m$, we must deal with $\beta$ in the construction of $g_{m+1}$. According to (2b) of the construction, $\Delta_t^{m+1}$ must contain either $\beta_1$ or $\beta_2$ and therefore certainly $HPC \vdash \wedge\Delta_t^{m+1} \to \bar{\Theta}_t^{m+1} \vee \beta_*$ (where $\beta_*$ is either $\beta_1$ or $\beta_2$), and so by the induction hypothesis, $[\beta_*]_t = 1$ and so $[\beta]_t = 1$. The other direction is easy.

(3) $\beta = \beta_1 \to \beta_2$.

Assume that $[\beta]_t = 1$. Let $m$ be such that $f(m) = (t, \beta)$ and Case 1 of the definition of $g_k$ does not hold. We claim that $HPC \vdash \wedge\Delta_t^m \to \vee\Theta_t^m \vee \beta$. For otherwise by construction of $g_{m+1}$, $HPC \nvdash \wedge\ \Delta_{t*(m)}^{m+1} \to \beta_2$, and $\beta_1 \in \Delta_{t*(m)}^{m+1}$. So by the induction hypothesis, $[\beta_1]_{t*(m)} = 1$, $[\beta_2]_{t*(m)} = 0$, which contradicts $[\beta]_t = 1$.

Now assume that $[\beta]_t = 0$. Then for some $s$, $t \leqslant s$ and $[\beta_1]_s = 1$ and $[\beta_2]_s = 0$. So for some $m$, $HPC \vdash \wedge\Delta_s^m \to \beta_1 \vee \bar{\Theta}_s^m$, and $HPC \nvdash \wedge\Delta_s^m \to \beta_2 \vee \bar{\Theta}_s^m$. Now assume, in order to obtain a contradiction, that for some $n$, $HPC \vdash \wedge\Delta_t^n \to \bar{\Theta}_t^n \vee (\beta_1 \to \beta_2)$. We now assume that $m \geqslant n$. In this case, $\beta \in \Delta_s^m$ since $s \geqslant t$ and this contradicts $HPC \nvdash \wedge\Delta_s^m \to \beta_2 \vee \bar{\Theta}_s^m$.

(4) The case of $\exists x\ \beta_3(x)$ is easy. It follows from (2c) of the construction.

The case of $\beta = \forall x\ \beta_3(x)$ is similar to the case of $\beta_1 \to \beta_2$.

Thus the main lemma is proved.

*Claim.* $[\tau]_{(0)} = 1$ *and* $[E]_{(0)} = 0$.

*Proof.* (a) $[\tau]_{(0)} = 1$ since $HPC \vdash \wedge\Delta_{(0)}^0 \to \tau \vee \bar{\Theta}_{(0)}^0$.

(b) $[E]_{(0)} = 0$ since otherwise by the main lemma, for some $n$

$HPC \vdash \wedge \Delta^n_{(0)} \to E \vee \bar{\Theta}^n_{(0)}$, but since $E \in \Theta^n_{(0)}$, this is impossible by the conditions on $g_k$, $k = 0, \ldots$. Thus Theorem 1 is proved.

The next theorem says that the fragment of *HPC* without $\vee$, $\exists$ is complete for *RE*-Kripke structures. This gives an example of a non-decidable fragment that is complete for *RE*-Kripke structures.

THEOREM 2. *The fragment of HPC without* $\vee, \neg$ *and* $\exists$ *is complete for the class of all RE-Kripke structures.*

*Proof.* We exhibit the *RE*-Kripke structure of constant domains in which all the non-provable formula of the fragment are false.

Let $S$ be the set of all finite sequences of wff of the fragment of the form $\alpha = (0, F_1, \ldots, F_n)$, where 0 is some fixed provable wff of the fragment and $\vdash F_{i+1} \to F_i$. Let $\bar{\alpha}$ be $F_n$, i.e., the last element of the sequence. Define $R \subseteq S \times S$ as 'being an initial segment of'. Let $A_\alpha$, for $\alpha \in S$ be the set of all variables and constants of the language, and define, for atomic $F$

$$[F]_\alpha = 1 \quad \text{iff} \quad (\text{def}), \quad HPC \vdash \bar{\alpha} \to F.$$

We now claim that for all $G$

$$[G]_\alpha = 1 \quad \text{iff} \quad HPC \vdash \bar{\alpha} \to G.$$

The proof is by induction.

Conjunction presents no difficulties since $\vdash \bar{\alpha} \to F \wedge G \Leftrightarrow \vdash (\bar{\alpha} \to F) \wedge (\bar{\alpha} \to G)$. Let $\vdash \bar{\alpha} \to (F \to G)$, then for any $\beta$ such that $\alpha R \beta$ (i.e., $\vdash \bar{\beta} \to \bar{\alpha}$), $\vdash \bar{\beta} \to (F \to G)$ and so if $\vdash \bar{\beta} \to F$, then $\vdash \bar{\beta} \to G$. Conversely, if $[F \to G]_\alpha = 1$, then since $\alpha R \alpha * (\alpha \wedge F)$ and $[F]_{\alpha * (\alpha \wedge F)} = 1$, we get $[G]_{\alpha * (\alpha \wedge F)} = 1$, $\vdash \bar{\alpha} \wedge F \to G$ or $\vdash \bar{\alpha} \to (F \to G)$. Let $\vdash \bar{\alpha} \to \forall x F(x)$, then certainly $[\forall x F(x)]_\alpha = 1$; the proof being similar to the case of $\to$. Assume $[\forall x F(x)]_\alpha = 1$, then certainly $[F(u)]_\alpha = 1$ for some $u$ that does not appear in $\bar{\alpha}$ so $\vdash \bar{\alpha} \to F(u)$ and so $\vdash \bar{\alpha} \to \forall u F(u)$. Thus the theorem is proved. Note, however, the definition of $[f]_t$ in Chapter 3. If we require $[f]_t = 0$, then the theorem is open.

THEOREM 3. *The fragment of all formulas in which all atomic predicates are negated is complete for LRE-Kripke structures* (*with* $\neg A = \text{def}(A \to f)$).

*Proof.* We begin with some definitions. Let $\mathbf{A}_t$ be a denumerable model with universe $A_t$ and diagram $D_t$. Let $D'$ be a finite subdiagram of $D_t$, (i.e., $D' \subseteq D_t$); certainly there exists a sequence $S^t_n(D') = D'$ and $\bigcup_n S^t_n(D') = D_t$. Let $S^t_n(D')$ be such a fixed sequence (for each $A_{t'}$ and each $D'$) approximating $D_t$.

Now let $(S, R, 0, \mathbf{A}_t)$ be a structure where $[G]_0 = 0$ (i.e., a counter-model for $G$) with $G$ in the fragment of Theorem 3. We shall proceed to construct a *LRE*-Kripke structure $(S^*, R^*, O^*, \mathbf{A}^*_t)$ with $[G]_{0^*} = 0$. We may assume that $S \subseteq \omega^*$ (where $\omega^*$ is the set of all finite sequences natural numbers), and that 0 is the empty sequence and that $R$ means 'being an initial segment of'. Of course, if $s \in S$ and $t \in \omega^*$ is an initial segment of $s$, then $t \in S$. This we can assume because the completeness proof of Chapter 8 yields an $(S, R, 0)$ with these properties. We define by induction the set $S^*$ and the models $A^*_a$, $a \in S^*$ and a relation $R^+ \subseteq S^{*2}$.

(1) Let $a_n = (0, S^0_n(\emptyset)) \in S^*$, for all $n$. ($\emptyset$ is the empty set). Let $A^*_a$, for $a = (0, S^0_n(\emptyset))$ be $A_0$ and let $D^*_a$ be $S^0_n(\emptyset)$. Let $a_n R^+ a_{n+1}$ hold.

(2) If $b = (t, S^t_m(D')) \in S^*$ and $tRs$ and $s$ has height $k > 0$ then let $c_n = (s, S^s_n(D' \bigcup_{t'Rt} S^{t'}_{m+k}(\emptyset)))$ for all $n$, be in $S^*$ with $A^*_{c_n} = A_s$ and $D^*_{c_n} = S^s_n(\bigcup_{t'Rt} S^{t'}_{m+k}(\emptyset)) \cup D')$. Let $bR^+c_1R^+c_2R^+\ldots$ hold. Thus $(S^*, R^+, 0^* = (0, S^0_0(\emptyset), A^*_t)$ is defined. Let $R^*$ be the transitive closure of $R^+$. Clearly if $(t, S^t_m(D'))R^*(s, S^s_n(D''))$, then $tRs$ & $D' \subseteq D''$ hold.

This construction is slightly more complicated than necessary for the proof of this theorem, but we need it for later theorems. Note that truth is defined by the diagram $D^*_a$.

*Main Claim. If* $a = (t, S^t_m(D'))$, *and if G is in the fragment, then* $[G]^*_a = [G]_t$.

*Proof.* The proof is by induction.

(1) $G = \neg P$, $P$ atomic.

Assume $[\neg P]_t = 0$. So for some $s$ such that $tRs$, $[P]_s = 1$. Since by construction the sequence (on $n$) $S^s_n(\bigcup_{t'Rt} S^{t'}_{m+k}(\emptyset) \cup D')$ approximates $A_s$, (with $k$ the height of $s$ above $t$), for some $n$ large enough $P$ holds at the model but then, since $aR^*b$, with $b = (s, S^s_n \bigcup_{t'Rt} S^{t'}_{m+k}(\emptyset) \cup D'))$, we conclude that $[\neg P]^*_a = 0$.

Conversely, if $[\neg P]^*_a = 0$, then for some $b$ such that $aR^*b[P]^*_b = 1$. But now $b$ must have the form $(s, S^s_n(\bigcup_{t'Rt} S^{t'}_{m+k}(\emptyset) \cup D'))$, i.e., $[P]_s = 1$, since $S^s_n(\ldots)$ approximates $A_s$. Thus $[\neg P]_t = 0$.

(2) The cases of $\wedge$, $\forall$, $\exists$ present no difficulties.

(3) If $[G_1 \to G_2]^*_a = 0$, then for some $b$ such that $aR^*b$, we have

$[G_1]^*_b = 1$ and $[G_2]^*_b = 0$. So if $b$ is $b = (s, (\ldots))$, we must have that $tRs$ holds and so by the induction hypothesis $[G_1]_s = 1$, $[G_2]_s = 0$ and so $[G_1 \to G_2]_t = 0$.

Assume that $[G_1 \to G_2]_t = 0$ then for some $s$, such that $tRs$, $[G_1]_s = 1$, $[G_2]_s = 0$. Then by the induction hypothesis for $b = (s, S^s_0((S^t_m(D'))))$ we have $[G_1]^*_b = 1$, $[G_2]^*_b = 0$ and since $aR^*b$ holds, we conclude that $[G_1 \to G_2]^*_a = 0$.

(4) The case of $\forall x G(x)$ is treated like the case of $G_1 \to G_2$. Clearly each $A^*_a$ is *LRE*. Thus Theorem 3 is proved.

THEOREM 4. *HPC is complete for s-validity in LRE-Kripke structure.*

*Proof*. Let $G$ be a sentence such that $\not\vdash G$. Let $(S, R, 0, A_t)$ be a structure where $G$ is false. Let $(S^*, R^*, 0^*, A^*_t)$ be the *LRE* structure constructed from $(S, R, 0, A_t)$ in the proof of Theorem 3. We cannot prove the main claim of the proof of Theorem 3 for our case now since $G$ is not necessarily in the fragment of Theorem 3. However, we may still carry the argument if we substitute for each atomic $P$, a formula $\alpha_P$, and thus we get that $[G(P_1, \ldots, P_n)]_t = [G(\alpha_{P_1}, \ldots, \alpha_{P_n})]^*_a$.

Given $P$, $\alpha_P$ is defined using $P$ and an additional unary predicate $Q$, i.e., $\alpha_P$ depends on $P$ and $Q$.

To simplify notation, for $a = (t, S^t_n(\ldots))$, let $f(a) = t$. We expect to prove, of course, that for all $a \in S^*$

$$[G(P_i)]_{f(a)} = [G(\alpha(P_i, Q))]^*_a.$$

Since we added only a *unary* $Q$, the proof yields the theorem for the monadic fragment as well.

We now proceed with the proof.

Let $Q$ be such that for every $t \in S$ there is a unique $b_t$ such that $[Q(b_t)]_a = 1 \Leftrightarrow (tRf(a)$ and $t \neq f(a))$ and let $Q$ have the property that: (for all $u$, $[Q(u)]_{a_1} = [Q(u)]_{a_2}) \Leftrightarrow (f(a_1) = f(a_2))$. Let $a(P, Q)$ be $\forall u[\forall y(Q(y) \vee (Q(y) \to Q(u))) \to (Q(u) \vee ((P \to Q(u)) \to Q(u))]$.

Let us now show that $[P]_{f(a)} = [\alpha]_a$.

Assume $[\alpha]_a$ is true. Let $b$ be the element such $[Q(b)]_{a'} = 1$ iff $f(a)Rf(a')$ and $f(a) \neq f(a')$. Certainly since $[\alpha]_a = 1$, we must have that the substitution of $b$ in $\alpha$ holds in $a$.

However since $Q(b)$ is false at $a$ and $\forall y(Q(y) \vee (Q(y) \to Q(b)))$ is true at $a$ (by the choice of $b$) we must have that $(P \to Q(b)) \to Q(b)$ is true at $a$. But $Q(b)$ is false at $a$ and so $(P \to Q(b))$ must be false at $a$.

So for some $a'$ such that $aR^*a'$ (and hence $f(a)Rf(a')$ holds), $[P]_{a'} = 1$ and $[Q(b)]_{a'} = 0$. So we must have $f(a') = f(a)$; therefore $[P]_{f(a)} = 1$. To get the other direction, suppose $[\alpha]_a = 0$. Then for some $a'$ and some $u$

$$[\forall y(Q(y) \vee (Q(y) \to Q(u))]_{a'} = 1 \quad \text{and} \quad [Q(u)]_{a'} = 0$$
$$\text{and} \quad [(P \to Q(u)) \to Q(u)]_{a'} = 0.$$

This implies that $u$ is the element $b_{f(a')}$ because otherwise $Q(b) \vee (Q(b) \to Q(u))$ would be false at $a'$. So $u = b$. Now since $[(P \to Q(u)) \to Q(u)]_{a'} = 0$, for some $a''$ such that $a'R^*a''$, $[P \to Q(u)]_{a''} = 1$ and $[Q(u)]_{a''} = 0$. $[Q(u)]_{a''} = 0$ implies that $f(a'') = f(a')$ (since $u = b_{f(a)}$) and since $[(P \to Q(u)]_{a''} = 1$, for no $a^*$ such that $a''R^*a^*$ and such that $f(a^*) = f(a')$ can have $[P]_{a^*} = 1$ (since then $[Q(u)]_{a^*} = 1$, which is impossible). We therefore conclude that $[P]_{f(a')} = 0$ (for otherwise such $a^*$ will exist).

Now to conclude the proof observe that since $aR^*a'$, $f(a)Rf(a')$ and so $[P]_{f(a)} = 0$. Thus the proof that $[P]_{f(a)} = [\alpha]^*_a$ is concluded. Continuing as in the case of Theorem 3 we can now prove by induction that

$$[G(P_i)]_{f(a)} = [G(\alpha(P_i, Q)]^*_a$$

for all $G$. Thus Theorem 4 is proved.

THEOREM 5. *Let $T$ be an HPC theory with axioms in the form $(Q\ldots) \bigwedge_k (\bigwedge_i \alpha^k_i \to \bigvee_j \beta^k_j)$, ($\alpha_i$, $\beta_j$ atomic) such that every classical model of $T$ is RE, then $T$ is complete for validity in LRE-Kripke structures.*

*Proof.* Let $(S, R, 0, \mathbf{A}_t)$ be a structure of $T$ where $G$ is false. For each $s \in S$, all axioms of $T$ hold in $S$. We claim that from the structure of the axioms of $T$ it follows that $A_s$ is a classical model of $T$ and is therefore *RE*.

This is easy to verify. Thus Theorem 5 is proved.

THEOREM 6 (*Dyson–Kreisel*). *HPC is complete for validity in LRE Beth structures.*

*Proof.* Let $G$ be any nonprovable formula and let $(S, R, 0, \mathbf{A}_t)$ be a Kripke structure whose $G$ is false. Let $(S^*, R^*, 0^*, \mathbf{A}^*_a)$ be the *LRE*-Kripke structure constructed in the proof of Theorem 3. We regard this structure as a Beth structure, we claim that for any atomic $P(x_1, \ldots)$ and any $a = (t, S_n^t(D'))$ (with $x_i \in |\mathbf{A}|_t$) we have

$$[P]_t = \|P\|_a.$$

(1) Assume that $[P]_t = 0$, then certainly for no $n'$ can we have that $A^*_{a'} \models P$, where $a' = (t, S^t_{n'}(D'))$. Now since $(t, S^t_{n'}(D'))$ form a strictly $R^*$ increasing sequence of points above $a$ which $P$ does not (classically) hold then by the definition of $\|P\|$, we have $\|P\|_a = 0$.

(2) Assume that $\|P\|_a = 0$ and show that $[P]_t = 0$. We prove this by showing that $[P]_t = 1$ leads to a contradiction.

Since $\|P\|_a = 0$, there must exist an $R^+$ increasing sequence ($R^+$ was defined in the construction in the proof of Theorem 3) $b_0 = a, b_1, b_2, \ldots$ such that $A^*_{b_k} \not\models P$. Clearly, if $b_k = (s_k, S^{s_k}_{n_k}(D'_k))$ then $s_k R s_{k+1}$ and $D'_k \subseteq D'_{k+1}$ hold. We distinguish two cases:

(a) for some $k_0$, $s_k = s_{k_0}$, for $k \geq k_0$.

In this case by definition of $R^+$ we must have that $D'_{k+1} = S^{s_{k_0}}_{s_{k_0+k-k_0}}(D'_{k_0})$. Now since $[P]_t = 1$, we have that $[P]_{s_{k_0}} = 1$ and since $D'_k$, $k \geq k_0$ approximates $D_{s_{k_0}}$, we get that for $k$ large enough $D'_k = P$. Which is a contradiction.

(b) $s_k$ is $R$-unbounded in $k$. Since $[P]_t = 1$, there exists an $n$ large enough such that $S^t_n(\emptyset) \models P$. Let $s_k$ be such that $s_k \neq s_{k-1}$, and $s_k$ has height $\geq n$. Then by definition, since $b_k = (s_k, S^{s_k}_{n_k}(D'_k))$ we must have that $D'_k \supseteq \bigcup_{t' R s_k} S^{t'}_n(\emptyset)$ and therefore $D'_k = P$, and again, a contradiction.

To complete the proof of Theorem 6, we now proceed by induction to show that

$$[F]_t = \|F\|_{(t, S^t_n(\ldots))}$$

for all $F$ of the language.

The cases of $\wedge$, $\forall$, $\rightarrow$ present no difficulties.

Assume that $[F \vee G]_t = 1$. Then, e.g., assume that $[F]_t = 1$. By the induction hypothesis $\|F\|_{(t, S^t_n(\ldots))} = 1$ and so $\|F \vee G\|_{(t, S^t_n(\ldots))} = 1$.

Assume $\|F \vee G\|_{(t, S^t_n(\ldots))} = 1$. Look at the sequence $(t, S^t_{n'} \ldots)$ $n' \geq n$. This is strictly $R^*$ increasing sequence and is therefore barred by a point $a' = (t, S^t_{n'} \ldots)$ $n' \geq n$ where either $\|F\|_{a'} = 1$ or $\|G\|_{a'} = 1$. By the induction hypothesis, either $[F]_t = 1$ or $[G]_t = 1$, i.e., $[F \vee G]_t = 1$.

The proof for the case of $\exists$ is similar to that of $\vee$.

## 3. Negative results

THEOREM 7. (a) *The set of sentences valid in LRE Kripke structures is not arithmetical.*

(b) *The set of sentences of the fragment where all atomic formula are negated that are valid in all RE Kripke structures, is not arithmetical.*

(c) *The monadic fragment or theory of equality is not complete for validity in RE Kripke structure.*

*Proof.* The method is by reduction. For cases (a) and (b) we use the fact (Mostowski) that the set of sentences of *CPC* valid in *CPC–RE* models is not arithmetical. We simply exhibit a faithful interpretation of *CPC* into *HPC*, that yields completeness of *CPC* for *RE* structures from completeness of *HPC* (or fragment of (b)) for *LRE* (or *RE* resp.) Kripke structures.

For the case (c) we carry a similar reduction for the classical theory of a reflexive and symmetrical relation.

Let us now proceed with the proofs of the various parts of Theorem 7. To prove (a) notice that every formula of classical logic can be written in the form: $(Q\ldots) \wedge (F_i \rightarrow G_i)$.

Where $Q$ is a string of quantifiers, $F_i$ is a conjunction of atomic formulae and $G_i$ is a disjunction of atomic formulae (the empty conjunction is truth and the empty disjunction is falsity).

LEMMA. *If* $(Q\ldots) \wedge (F_i \rightarrow G_i)$ *has a locally RE Kripke structure, then it has a classical RE model.*

*Proof.* Assume $(S, R, 0)$ is a model of the formula. Then

$$[(Q\ldots) \wedge (F_i \rightarrow G_i)]_0 = 1$$

therefore

$$(Q^{A_0}) \wedge ([F_i \rightarrow G_i]_0 = 1)$$

where $Q^{A_0}$ is the same string of quantifiers relativized to $A_0$. Therefore

$$(Q^{A_0}) \wedge ([F_i]_0 = 1 \rightarrow [G_i]_0 = 1).$$

The last sentence means that $(Q\ldots) \wedge (F_i \rightarrow G_i)$ is valid classically at $A_0$. $A_0$ is, of course, an *RE* model.

Since both a sentence and its negation can be classically written in the form $(Q\ldots)\wedge(F_i\to G_i)$ we get a reduction of the classical case to the intuitionistic. Since in the classical case the set of valid formula is not arithmetical, we get (a).

To prove (b), note that every sentence can classically be written as $(Q\ldots)\wedge(F_i\to G_i)$ where $F_i$ is a conjunction of negated atomic formulae and $G_i$ is a disjunction of negated atomic formulae. By reasoning as before, we obtain that if the sentence is true in an *RE* Kripke structure, then

$$(Q^{A_0}\ldots)\wedge([F_i]_0=1\Rightarrow[G_i]_0=1)\Rightarrow(Q^{A_0}\ldots)\wedge([G_i]_0=0\Rightarrow[F_i]_0=0).$$

So if $F=\wedge\neg P_j$, $G=\vee\neg P_j^*$ we get

$$(Q^{A_0}\ldots)\bigwedge\Big(\bigwedge_j[\neg P_j^*]_0=0\Rightarrow\bigvee_j[\neg P_j]_0=0\Big)$$

but since $[\neg P_j]_0=0$ means: for some $t\in S[P_j]_t=1$ and $[f]_t=0$, we get that $[\neg P_j]=0$ is *RE*. This completes the proof of (b).

(c) To obtain the result for the monadic fragment with 1 letter and for equality, observe the following.

The classical theory of a reflexive and symmetrical relation $<$ is faithfully interpretable in the monadic fragment. In fact, if the monadic fragment is complete for *RE* Kripke structures, the classical theory of a reflexive and symmetrical relation is complete for *RE* models. Since the classical theory of reflexive and symmetrical relation is not complete for *RE* models, we get the result for the monadic fragment, and similarly for equality.

All we have to do now is to present the interpretation. Let $S$ be a sentence of the reflexive and symmetric relation. Write $S$ in the form

$$S=(Q\ldots)\bigwedge_k(\wedge F_i^k\to\vee G_j^k)$$

where $(Q\ldots)$ is a string of quantifiers and $F_i$, $G_j$ are atomic formulas. If $F$ is atomic of the form $x<y$, let $F^*$ be the translation of $F$ and let it be $\neg(P(x)\wedge P(y))$ in the case of the monadic fragment (and $x\neq y$ in the case of equality). Translate $S$ into $S^*$ where $S^*=(Q\ldots)\wedge(\wedge G_j^*\to(\vee F_i^*\vee\exists xP(x))\to(\exists xP(x)\vee\exists x\neg P(x))$ in the case of the monadic fragment and $S^*=(Q\ldots)\wedge(\wedge G_j^*\to\vee F_i^*)\to\exists x(x\neq x)$ in the case of equality.

This translation is faithful (see Chapter 14) in the sense that $\neg S$ is a theorem of the classical theory of reflexive and symmetrical relation iff $S^*$ is a theorem of the fragment.

Now suppose the fragment were complete for *RE* Kripke structures and suppose $\neg S$ is not a theorem of the theory of a reflexive and symmetrical relation. We find an *RE* model where $S$ is true. Now $S^*$ is non-provable and so there exists an *RE* Kripke structure $(S, R, 0, \mathbf{A}_t)$ and a $t_0 \in S$ such that the antecedent of $S^*$ holds at $t_0$ and the consequence is false.

Define a model $(\mathbf{M}, <)$ of $S$ by $\mathbf{M} = A_{t_0}$, and $x < y$ iff $\exists t(t_0 Rt \wedge [P(x) \wedge P(y)]_t = 1)$ (or in the case of equality define $x < y$ iff $\exists t(t_0 Rt \wedge [x = y]_t = 1)$. Clearly, $<$ is *RE* on $\mathbf{M}$. Also $(\mathbf{M}, <) \models S$ (since $[S^*]_{t_0} = 1$).

We quote without proof:

THEOREM 8 (*Dyson–Kreisel*). *HPC is not complete for validity in recursive Beth structure.*

THEOREM 9 (*Kripke*). *There exists a recursive method of converting any Kripke structure into an elementary equivalent Beth structure.*

*Proof.* See Chapter 3.

## 4. Tarski's interpretation for HPC as a calculus of systems

DEFINITION 1. Let $\vdash$ be any Tarski consequence in a certain but arbitrary language. Let $\bar{P}$, for any atomic $P$ the language of *HPC*, be a wff of the language of $\vdash$, and for any variable $x$, $\bar{x}$ is a term of $\vdash$. We define by induction the notion $\vdash \Vvdash A$ for any $A$.

(1) $\vdash \Vvdash P(x_1, \ldots, x_n)$ iff $0 \vdash \{\bar{P}(\bar{x}_1, \ldots, \bar{x}_n)\}$

$P$ atomic $\neq f$.

$\vdash \Vvdash f$ iff $0 \vdash 0$

(2) $\vdash \Vvdash A \wedge B$ iff $\vdash \Vvdash A$ and $\vdash \Vvdash B$

(3) $\vdash \Vvdash A \vee B$ iff $\vdash \Vvdash A$ or $\vdash \Vvdash B$

(4) $\vdash \Vvdash A \rightarrow B$ iff for any $\vdash' \supseteq \vdash$ if $\vdash' \Vvdash A$ then $\vdash' \Vvdash B$.

(5) $\quad \vdash \Vvdash \exists x A(x)$ iff for some $y \vdash \Vvdash A(y)$

(6) $\quad \vdash \Vvdash \forall x A(x)$ iff for all $y \vdash \Vvdash A(y)$.

$A$ is said to be valid in $\vdash$ iff $\vdash \Vvdash A$. $A$ is valid if it is valid in any $\vdash$. $A$ is $s$-valid iff every substitution instance of $A$ is valid.

PROPOSITION 2. *For any A not containing* $\vee$, $\exists$, *A is a theorem of HPC iff A is* $\Vvdash$ *valid in any* $\vdash$.

*Proof.* It is easy to verify that if $HPC \vdash A$ then $A$ is $\Vvdash$ valid in any $\vdash$. To show the other direction let $\vdash_0$ be the Tarski consequence of $HPC$ itself, in the language of $HPC$, and for any atomic $p$, let $\bar{P} = P$.

LEMMA 3. *For any* $\vdash \supseteq \vdash_0$, $\vdash \Vvdash A$ *iff* $\emptyset \vdash \{A\}$. $\vdash \Vvdash f$ *iff* $\emptyset \vdash \emptyset$.

*Proof.* For atomic $A$ this holds by definition.

$\vdash \Vvdash A \wedge B$ iff $\vdash \Vvdash A$ and $\vdash \Vvdash B$ iff $\emptyset \vdash \{A\}$ and $\emptyset \vdash \{B\}$ iff $\emptyset \vdash \{A \wedge B\}$ (since $\vdash$ is an extension of $HPC$}.

If $\vdash \Vvdash A \rightarrow B$, consider $\vdash_A$ defined by $\emptyset \vdash_A \psi$ iff $\{A\} \cup \emptyset \vdash \psi$. It is easy to verify that $\vdash_A$ is a Tarski consequence such that $\vdash_A \supseteq \vdash$ and $\emptyset \vdash_A \{A\}$. Therefore $\emptyset \vdash_A \{B\}$, i.e., $\{A\} \vdash \{B\}$. Now since $\vdash$ is an extension of $HPC$ for which modus ponens is valid, the deduction theorem can be proved for $\vdash$ and so $\emptyset \vdash \{A \rightarrow B\}$.

Assume $\emptyset \vdash \{A \rightarrow B\}$, then any extension $\vdash' \supseteq \vdash$ such that $\vdash' \Vvdash A$ will satisfy by the induction hypothesis and modus ponens $\emptyset \vdash' \{B\}$ and so $\vdash' \Vvdash B$.

If $\emptyset \vdash \{A \rightarrow \forall x B(x)\}$ then clearly for any $\vdash' \supseteq \vdash_A$ and any $y \vdash' \Vvdash B(y)$ by the induction hypothesis. Also if $\vdash_A \Vvdash \forall x B(x)$ certainly $\vdash_A \Vvdash B(y)$ for $y$ not free in $\forall x B$ and so by the induction hypothesis $\emptyset \vdash \{A \rightarrow B(y)\}$ and therefore $\emptyset \vdash \{A \rightarrow \forall x B(x)\}$.

To conclude the proof of 2, notice that if $HPC \vdash_A$ then $A$ is not valid in the interpretation of Lemma 3.

## 5. Kreisel's Notion of Validity in Post Systems

We investigate completeness properties of Heyting's predicate calculus ($HPC$) with respect to several notions of validity in Post systems. Validity in Post systems was introduced by Kreisel as a generalization of Tarski's calculus of systems. Provability interpretations for $HPC$ are also mentioned in Kripke, 1965.

Kreisel's definition of validity in Post systems is the following: Let $A$ be a fixed alphabet and let $P$ be any Post system on $A$. If $x$ is a word of $A$ we write $P \mapsto x$ to mean $P$ proves $x$. Let $h$ be any function assigning words of $A$ to all the variables and predicate constants of $HPC$. For $h_1, h_2$ let $h_1 =_u h_2$ mean that for all $y \neq u$, $h_1(y) = h_2(y)$. We now define the notion of $P \Vdash\!\!\mapsto_h B$, for any wff $B$ of $HPC$ and assignment $h$.

(a) $P \Vdash\!\!\mapsto_h B(x_1, \ldots, x_n)$ iff $P \mapsto h(B)h(x_1) \ldots h(x_n)$, where $B$ is atomic.

(b) $P \Vdash\!\!\mapsto_h B \wedge E$ iff $P \Vdash\!\!\mapsto_h B$ and $P \Vdash\!\!\mapsto_h E$

(c) $P \Vdash\!\!\mapsto_h B \vee E$ iff $P \Vdash\!\!\mapsto_h B$ or $P \Vdash\!\!\mapsto_h E$

(d) $P \Vdash\!\!\mapsto_h \neg B$ iff for all consistent $P' \supseteq P$, $P' \not\Vdash\!\!\mapsto_h B$ ($P'$ is consistent iff it does not prove every word)

(e) $P \Vdash\!\!\mapsto_h B \to E$ iff for all $P' \supseteq P$, if $P' \Vdash\!\!\mapsto_h B$ then $P' \Vdash\!\!\mapsto_h E$

(f) $P \Vdash\!\!\mapsto_h \exists x B(x)$ iff for some $h_1 =_x h$, $P \Vdash\!\!\mapsto_{h_1} B(x)$

(g) $P \Vdash\!\!\mapsto_h \forall x B(x)$ iff for all $h_1 =_x h$, $P \Vdash\!\!\mapsto_{h_1} B(x)$.

$B$ is said to be valid in Post systems iff for all $A$, $P$, $h$; $P \Vdash_h B$. $B$ is said to be $s$-valid (substitution valid) in Post systems iff all the substitution instances of $B$ (in $HPC$) are valid.

Validity in Post systems (also called post structures) is related to validity in locally recursively enumerable ($LRE$) Kripke structures.

The notion of validity in Post structures, introduced by Kreisel, (and described above) is not identical with the notion of validity in $LRE$ Kripke structures. Every Post structure $P$, gives rise to a Kripke structure $(S_P, \subsetneq, P)$, where $S_P = \{Q \mid Q \supsetneq P\}$, in a natural way; however, these structures have the special property that whenever $Q \in S_P$ then all of its Post extensions are also in $S_P$.

The following holds for each Post structure $P$, and the corresponding $(S_P, \subsetneq, P)$, for $Q \in S_p$:

$$Q \Vdash\!\!\mapsto B \quad \text{iff} \quad [B]_Q = 1.$$

In general, however, the set of statements valid in all Post structures is richer (even for the propositional calculus) than the set of theorems of the logic (or the set of statements valid in all *LRE* Kripke structures).

Since the class of Post structures is essentially a subclass of the class of all *LRE* Kripke structures, we get that all negative results that hold for validity in *LRE* Kripke structures also hold for validity in Post structures. For example:

> The set of all sentences valid in all Post structures is not arithmetical.

As for positive completeness results, the case is not clear. The class of Post structures is a proper subclass of the class of all *LRE* Kripke structures, so completeness may fail to hold for the smaller class.

Of course, since Post structures have constant domains, the schema $\forall x(B \vee F(x)) \rightarrow B \vee \forall x F(x)$ $x$ not in $B$, is valid in all Post structures. So we should investigate the analogy only for systems fulfilling the above schema.

Before we continue, we have to discuss a certain point. What do we mean when we say that a Post system $P'$ is an extension of the Post system $P$ (i.e., $P \supseteq P'$)? (on the same language). There are two possibilities. Either that for all words $a$, $P \vdash a$ implies $P' \vdash a$, or that $P'$ is obtained from $P$ by the addition to $P$ of some more axioms and rules. Of course, the second definition is more restrictive than the first one.

When we talk about validity in Post structures, we consider all the Kripke structures of the form $(S_P, \subsetneq, P)$ with $S_P = \{Q \mid P \subsetneq Q\}$. So we get the different classes of structures, and possibly different theorems, if $\subsetneq$ is understood differently.

Let us use 'weak-validity' to refer to the notion of validity in Post structures, with $P \subseteq P'$, taken to mean $\forall a(P \vdash a \rightarrow P' \vdash a)$. We shall use the term 'validity' in Post structures to refer to validity in Post structures with $P \subseteq P'$, taken to mean $P'$ has more rules and axioms than $P$. We use $P \subsetneq P'$ to read $P \subseteq P'$ and $P'$ proves more words than $P$.

The notion of a Post system is that of Smullyan's book *Theory of Formal Systems.**

Kreisel introduced the notion of validity in Post structures, to generalize Tarski's calculus of systems, that used finitely axiomatized theories. The use of theories, however, (or of formal logical systems)

* The empty word $\Lambda$ is allowed and is always considered a contradiction.

gives rise to yet another different notion of validity (to be referred to as validity in formal system structures).

With any formal systems structure arising from a formal theory $T$ we associate the Kripke structure $(S_T, \subseteq, T)$ with $S_T = \{T' \mid T \subseteq T'\}$. The natural interpretation of $T \subseteq T'$ in this case is that $T'$ has more axioms and rules in addition to those of $T$. So in particular, if $T$ contains the connective $\rightarrow$ and fulfills the deduction rule, then every $T' \supseteq T$ must also fulfill this rule.

Now it is true that one can write a Post system $P(T)$ such that for all $B$, $P(T) \vdash B$ iff $T \vdash B$, but the deduction theorem is not a rule that can be adopted in a Post system and so we may have that $S_T \neq S_{P(T)}$.

Indeed as in Section 2 we can show that the fragment of *HPC* without $\exists$ and $\forall$ is complete for validity in formal system structures. It is not complete for $s$-validity in Post structures.

It is arguable whether one may allow the disjunction property as a rule for a formal system $T$. Anyway, if we allow it, then the propositional calculus is complete for validity in formal system structures; otherwise it is not.

In the sequel we investigate all three notions, weak validity and validity in Post systems.

## 6. Weak validity in Post systems

THEOREM 1. *Heyting's propositional calculus is not complete for weak validity in Post structures.*

*Proof.* We claim that the formula $((\neg\neg P \rightarrow P) \rightarrow (\neg P \vee \neg\neg P)) \rightarrow \neg P \vee \neg\neg P$ is always weakly valid.

LEMMA. *Let $(S_P, \subsetneq, P)$ be a Post structure and let $x_1, \dots, x_n$ be atomic propositions. Let $\bar{x}_1, \dots, \bar{x}_n$ be the words of $P$, assigned to $x_1, \dots, x_n$ respectively.*

*For $Q, Q' \in S_{P'}$ define:*

$$Q \approx Q' \quad \textit{iff for all } \bar{x}_i, \quad Q \vdash \bar{x}_i \quad \textit{iff} \quad Q' \vdash \bar{x}_i.$$

We now claim that for any formula $B$ with atomic propositions from among $x_1, \dots, x_n$, $Q \Vdash B$ iff $Q' \Vdash B$, (where $Q \approx Q'$).

*Proof.* By induction. For atomic $B$ this holds by definition of $\approx$.

The cases of $\wedge$ and $\vee$ present no difficulties.

Assume that $Q \not\Vdash B \to E$. Then for some $Q_1 \supseteq Q$, $Q_1 \Vdash B$ and $Q_1 \not\Vdash E$. Let $Q_1'$ be a Post system that proves all the theorems of $Q_1$ and of $Q'$ (i.e., $Q_1' = Q_1 \cup Q'$). Clearly, since $Q$, $Q'$ agree on $\bar{x}_i$, so do $Q_1$ and $Q_1'$, and so by the induction hypothesis, $Q_1' \Vdash B$, $Q_1' \Vdash E$, and therefore since $Q_1' \supseteq Q'$, $Q' \Vdash B \to E$.

The other direction of $\to$ and the case of $\neg$ are proved similarly.

To continue with the proof of Theorem 1, let $P$ be such that

$$P \vdash (\neg\neg B \to B) \to (\neg B \vee \neg\neg B)$$

and

$$P \not\Vdash \neg B \vee \neg\neg B.$$

Then for some $P_1, P_2 \supseteq P$, we have $P_1 \Vdash B$ and $P_2 \Vdash B$. In view of the lemma, we can assume that $P$ proves any word different from $\bar{x}_i$, where $x_i$ are the atomic propositions of $B$.

Let $Q = P_1 \cap P_2$, be the Post system such that $Q \vdash a$ iff $P_i \vdash a$, $i = 1, 2$. Clearly, $Q \not\Vdash \neg B \vee \neg\neg B$. So since $P$ forces the antecedent of the schema, we get that for some $Q' \supseteq Q$, $Q' \Vdash \neg\neg B$ and $Q' \not\Vdash B$. Let $w$ be a word such that $P_2 \vdash w$, and let $P_w$ be the Post system that proves every word except $w$. Clearly $P_w \supseteq P_2$, and so $P_w \Vdash \neg B$ and therefore since $Q' \Vdash \neg\neg B$. $P_w \not\supseteq Q'$, i.e., $Q' \vdash w$. This shows that $Q'$ proves all words not proved in $P_2$, and since $Q \supseteq P_1 \cap P_2$ we conclude that $Q \supseteq P_1$, but this contradicts the fact that $Q' \not\Vdash B$. Thus Theorem 1 is proved.

*Remark.* The lemma shows that actually weak validity is identical with validity in the calculus of finite problems, introduced by Medvedev. This notion of validity is decidable.

## 7. Validity in Post Systems

THEOREM 1. *Heyting's propositional calculus is not complete for validity in Post structures. In fact, the set of all sentences valid in Post structures is not closed under substitution.*

*Proof.* We claim that if $x$, $y_1$, $y_2$ are atomic propositions then

$$(x \to (y_1 \vee y_2)) \to ((x \to y_1) \vee (x \to y_2))$$

is valid in all Post structures.

Assume that for some $P$, $P \Vdash\!\!\!\!/\, (x \to y_i)$, $i = 1, 2$. Then by definition there exists $P_i \supseteq P$, $i = i, 2$ such that $P_i \Vdash x$ (and so $P_i \vdash \bar{x}$) and $P_i \Vdash\!\!\!\!/\, y_i$ (and so $P_i \vdash\!\!\!\!/\, z/\bar{y}_i$. Now look at the Post system $P'$, which is obtained from $P$ by adding the single axiom $P' \vdash \bar{x}$. Clearly $P' \subseteq P_i$, and therefore $P' \Vdash\!\!\!\!/\, y_i$, $i = 1, 2$, and since $P' \Vdash x$, we conclude that $P \Vdash\!\!\!\!/\, x \to y_1 \vee y_2$.

To show that this formula is not valid as a schema observe that as a schema, it implies $(x \to y) \vee (y \to x)$ which is clearly not valid. Thus Theorem 1 is proved.

*Conjecture* 2. $h$ is complete for $s$-validity in Post structures.

## NOTES

Based on the author's papers. In this connection the reader will profit from a survey paper by A. Troelstra.

The Dutch logicians worked on intuitionistic and *RE* models for intuitionistic logic. See the papers of de Swart, Veldman and the new Troelstra 1978. See also Van Dalen 1979.

CHAPTER 14

# UNDECIDABILITY RESULTS

Let $T$ be a set of axioms. This set gives rise to a theory $T^H$ in Heyting's predicate logic and a theory $T^C$ in the classical predicate logic. More generally, if $X$ is any intermediate logic, we get a theory $T^X$.

The decision problem we are going to study is whether $T^X$ is decidable. This chapter contains undecidability results for various theories and the method of proof is by faithfully interpreting in $T^X$ the classically undecidable theory $RS$ of a reflexive and symmetric binary relation. In the course of the proof of the faithfullness of the interpretation we shall use our knowledge of the Kripke structures validating the theory.

Let us add that with regard to intermediate propositional logics, most of the well-known ones are decidable. By a theorem of Harrop, any intermediate logic which has the finite model property (f.m.p.) and is finitely axiomatizable is decidable. There are undecidable intermediate logics (Jankov); we can obtain these from the result we proved (in Chapter 4, Section 3) concerning the existence of $\aleph_0$ mutually incomparable intermediate logics. We also remark that there exist decidable intermediate logics without the f.m.p. It is an open problem whether every finitely axiomatizable intermediate logic is decidable.

## 1. General theorems

THEOREM Ia. *Let $T$ be a finite set of axioms, then if $T^C$ is undecidable so is $T^X$.*

*Proof.* Let $A$ be an axiom such that in classical logic, $A$ and $\wedge T$ are equivalent and $A$ is regular (see (11.2.3)). Let $B$ be any regular wff. We claim

$$(*) \qquad T^X \vdash_X \neg(A \wedge B) \quad \text{iff} \quad T^C \vdash_C \neg(A \wedge B).$$

One direction is easy. Assume $T^X \nVdash_X (A \wedge B)$, then let $\Delta = \{C \mid T^X \vdash_X C\}$, then $\Delta \vdash_H \neg(A \wedge B)$ iff $T^X \vdash_X \neg(A \wedge B)$. Thus by our assumption there exists a modified Kripke structure validating $\Delta$ in which $\neg(A \wedge B)$ is false. Thus for some world $t$, $[A \wedge B]_t = 1$. Since $A$, $B$ are regular wffs $A_t \models A \wedge B$ and hence $T^C \nvdash_C \neg(A \wedge B)$.

Thus we get from (*) that $T^C \vdash_C \neg B$ iff $T^X \vdash_X \neg(A \wedge B)$, and so since $T^C$ is not decidable, $T^X$ must be undecidable.

THEOREM Ib. *If $T$ is a regular set of axioms then if $T^C$ is undecidable so is $T^X$.*

*Proof.* We show that for any regular $B$,

$$(*) \qquad T^C \vdash_C \neg B \quad \text{iff} \quad T^X \vdash_X \neg B.$$

One direction of (*) is immediate. Assume $T^X \nvdash_X \neg B$, then for some modified Kripke structure, all theorems of $T^X$ are validated but $\neg B$ is false. Thus since $T$, $B$ are regular for some world $t$, $A_t \models T \cup \{B\}$ and so $T^C \nvdash \neg B$.

From (*) we conclude that $T^X$ is undecidable if $T^C$ is undecidable.

THEOREM Ic. *For any $T$, if $T^C$ is undecidable so is $T^{MH}$.*

*Proof.* Show that for any $B$,

$$(*) \qquad T^C \vdash_C \neg B \quad \text{iff} \quad T^{MH} \vdash_{MH} \neg B.$$

One direction of (*) is immediate. For the other direction recall that $MH$ is the smallest logic $X$ such that for any $A$, $\vdash_C A$ iff $\vdash_X \neg\neg A$.

DEFINITION 2. (a) A wff is said to be in (*$AE$) form iff it has the form $(Q\ldots)\bigwedge_i (\bigwedge_j \beta_{ij} \to \bigvee_j \gamma_{ij})$, where $\beta_{ij}$, $\gamma_{ij}$ are atomic and $Q$ is a string of quantifiers of the form $\forall \ldots \forall \exists \ldots \exists$.

(b) We similarly define the (*$A$) form.

(c) Let $T$ be a set of axioms. Then by $T_D^X$ we denote the theory of $X$ with the axioms

$$T \cup \{(\text{universal closure})(B \vee \neg B) \mid B \text{ atomic}\}.$$

DEFINITION 3. Let $\mathbf{A}$ be a classical model of $\Delta$ with domain $A$. Let $L_\Delta$ be the language of $\Delta$ and let $L_{\Delta,\mathbf{A}}$ be the extension of $L_\Delta$ with names $\bar{a}$ for each $a \in A$. Let the positive diagram of $\mathbf{A}$, $D^+(\mathbf{A})$ be

$$\{\beta(\bar{a}_1, \ldots, \bar{a}_n) \mid \mathbf{A} \models \beta(a_1, \ldots, a_n) \text{ and } \beta \text{ is atomic}\}.$$

Let the diagram of $\mathbf{A}$ be $D(\mathbf{A})$:

$$D(\mathbf{A}) = \{\beta(\bar{a}_1, \ldots, \bar{a}_n) \mid \mathbf{A} \models \beta(a_1, \ldots, a_n) \text{ and } \beta \text{ is an atomic formula or a negation of an atomic formula}\}.$$

THEOREM II. *Let* $\Delta$ *be a (*AE)-set of axioms. Then* $T_D^H(\Delta)$ *is undecidable in HPC if there exists a classical model* $\mathbf{A}$ *of* $\Delta$, *and infinite subset* $A_* \subseteq A$ *and a binary formula* $\rho$ *in* $L_\Delta$ *such that* (a) *and* (b) *hold*:

(a) $\mathbf{A} \models (\forall x, y) \neg \rho(x, y)$.

(b) *For each* $(a, b) \in A_*^2$ *the following set is consistent in CPC*:

$$\Delta \cup D(\mathbf{A}) \cup \{\rho(\bar{a}, \bar{b}) \wedge \rho(\bar{a}, \bar{a}) \wedge \rho(\bar{b}, \bar{b}) \wedge \rho(\bar{b}, \bar{a})\}$$
$$\cup \{\neg \rho(\bar{x}, \bar{y}) \mid (x, y) \in A^2 - \{a, b\}^2\}.$$

THEOREM III. *Let* $\Delta$ *be a (*A)-set of axioms. Then* $T^H(\Delta)$ *is undecidable in HPC if there exists a classical model* $\mathbf{A}$ *of* $\Delta$, *an infinite subset* $A_* \subseteq A$ *and a binary formula* $\rho$ *in* $L_\Delta$ *such that* (a) *and* (b) *hold*:

(a) $\mathbf{A} \models (\forall x, y) \neg \rho(x, y)$.

(b) *For each* $(a, b) \in A_*^2$ *the following set is consistent in CPC*:

$$\Delta \cup D^+(\mathbf{A}) \cup \{\rho(\bar{a}, \bar{b}) \wedge \rho(\bar{a}, \bar{a}) \wedge \rho(\bar{b}, \bar{b}) \wedge \rho(\bar{b}, \bar{a})\}$$
$$\cup \{\neg \rho(\bar{x}, \bar{y}) \mid (x, y) \in A^2 - \{a, b\}^2\}.$$

To prove both Theorems II and III we shall faithfully interpret in the *HPC* theory (in question) the undecidable classical theory *RS* of a reflexive and symmetric relation. The proof is uniform for Theorems II and III.

We now assume that the assumptions of either Theorems II or III are valid and proceed to derive the conclusion of the theorems. We also assume that $\rho$ is in (*)-form.

DEFINITION 4. (a) Let $D(x)$ be $\neg \rho(x, x)$, and $\rho_1(x, y) = \rho(x, y) \wedge \rho(y, x)$.

(b) Let $\forall^*$ and $\exists^*$ be the following quantifiers:

$$\forall^* x F(x) \text{ is } \forall x (D(x) \vee F(x)),$$
$$\exists^* x F(x) \text{ is } \exists x [F(x) \wedge (D(x) \rightarrow \exists x \rho(x, x))].$$

(c) Let $S$ be a sentence of the classical theory *RS* written in (*)-form.

$$S = (Q \ldots) \wedge (\bigwedge x_j < y_j \rightarrow \bigvee u_i < v_i).$$

Let $S^*$, the translation of $S$, be

$$\forall u\Big\{(Q^*\ldots)\wedge\Big(\bigwedge_i\neg\rho_1(u_i,v_i)\to\Big(\bigvee_j\neg\rho_1(x_j,y_j)\vee\exists x\rho(x,x)\Big)\Big)$$
$$\to(D(u)\vee\exists x\rho(x,x))\Big\}.$$

MAIN LEMMA. (a) *Let* $\Delta$ *be the set of Theorem* II *then*

$$T_D^H(\Delta)\vdash_{HPC}S^*\Leftrightarrow RS\vdash_{CPC}\neg S.$$

(b) *Let* $\Delta$ *be the set of Theorem* III *then*

$$T^H(\Delta)\vdash_{HPC}S^*\Leftrightarrow RS\vdash_{CPC}\neg S.$$

To prove the Main Lemma we need for some more lemmas.

LEMMA 5. *Let* $\tau(x_1,\ldots,x_n)$ *be a formula in* (*)*-form and let* $(T,R,0,\mathbf{A}_t)$, $t\in T$, *be a Kripke structure then*

(#) *if* $[\tau(a_1,\ldots,a_n)]_0=1$ *then* $\mathbf{A}_0\models_{CPC}\tau(a_1,\ldots,a_n)$ *for* $a_1,\ldots,a_n\in A_0$.

*Proof.* The proof is by induction on the form of $\tau$.

For atomic $\tau$ this is clear. $\wedge$ and $\vee$ present no difficulties. If $[\wedge\beta_i\to\vee\gamma_j]_0=1$, then by definition, for all $s$ such that $0Rs$, if $[\wedge\beta_i]_s=1$ then $[\vee\gamma_j]_s=1$ (in particular for $s=0$). But this implies $\mathbf{A}_0\models(\wedge\beta_i\to\vee\gamma_i)$ since $\beta_i,\gamma_j$ are atomic. If $[\exists x\tau(x)]_0=1$, then for some $a\in A_0$, $[\tau(a)]_0=1$ and so by the induction hypothesis $\mathbf{A}_0\models\tau(a)$ and so $\mathbf{A}_0\models\exists x\tau$. If $[\forall x\tau(x)]_0=1$, then, for all $a\in A_0$, $[\tau(a)]_0=1$ so $\mathbf{A}_0\models\tau(a)$ for all $a\in A_0$ (i.e., $\mathbf{A}_0\models\forall x\tau$).

LEMMA 6. *Let* $\tau$ *be in the form* $(Q\ldots)\wedge(\wedge\beta_i\to\vee\gamma_i)$ *with* $\beta_i$ *atomic and* $\gamma_j$ *in* (*)*-form. Then* (#) *holds for* $\tau$.

*Proof.* Use induction and Lemma 5.

LEMMA 7. *If* $(T,R,0,\mathbf{A}_t)$, $t\in T$, *is a structure where* $S^*$ (*of the Main Lemma*) *is false then there exists a CPC structure of S* (*of the Main Lemma*).

*Proof.* We may assume that for some $u\in A_0$, the antecedent $S_1$ of $S^*$ holds at 0 and the consequent of $S^*$ does not hold.

We now define a *CPC* model $\mathbf{B}$ of $RS\cup\{S\}$ as follows: Let $|\mathbf{B}|=\{x\in|\mathbf{A}_0|\,|\,[Dx]_0=0\}$. $|\mathbf{B}|$ is nonempty since $[Du]_0=0$.

Define a reflexive and symmetric relation $<$ on $|\mathbf{B}|$ as follows: $x<y$ hold in $\mathbf{B}$ iff $[\neg\rho_1(x, y)]_0=0$. Since $Dx=\neg\rho(x, x)$ we get that $R$ is reflexive on $B$. We now want to show that $\mathbf{B}\models S$ classically. We need some lemmas.

LEMMA 8. *For any subformula F of* $S_1$,
(a) *if* $[\forall^* xF]_0=1$, *then* $\forall x\in B[F(x)]_0=1$;
(b) *if* $[\exists^* xF]_0=1$, *then* $\exists x\in B[F(x)]_0=1$.
*Proof.* (a) Since $\forall^* xF$ is $\forall x(Dx\vee F)$ we get that, for all $x\in A_0$,

$$[Dx\vee F(x)]_0=1.$$

But for $x\in B$, $[Dx]_0=0$ and so $[F(x)]_0=1$.

(b) Since $\exists^* xF$ is $\exists x[F\wedge(Dx\to\exists x\rho(x, x))]$ and since $[\exists x\rho(x, x)]_0=0$ (as it is a disjunct in the consequent of $S^*$) we must have $[F(x)]_0=1$ and $[D(x)]_0=0$ (i.e., $x\in|\mathbf{B}|$).

Now to complete the proof of Lemma 7 observe that since $[\exists x\rho(x, x)]_0=0$ and $[S_1]_0=1$, we get in view of Lemma 8 that the following is true:

$$(Q^B\ldots)\wedge\Big(\bigwedge_i([\neg\rho_1(u_i, v_i)]_0=1)\to\bigvee_j([\neg\rho_1(x_j, y_j)]_0=1)\Big)$$

where $(Q^B\ldots)$ is the string of quantifiers $(Q\ldots)$ relativised to $B$. Now in view of the definition of $<$ on $\mathbf{B}$ we get that

$$\mathbf{B}\models(Q\ldots)\wedge\Big(\bigwedge_j x_j<y_j\to\bigvee_i u_i<v_i\Big).$$

Thus Lemma 7 is proved.

We now proceed to prepare for the proof of Theorems II and III.

Let $(B,<)$ be a structure of $RS$ and of $S$, we shall construct two Kripke structures $\mathbf{M}_{\mathrm{II}}$ and $\mathbf{M}_{\mathrm{III}}$. $\mathbf{M}_{\mathrm{II}}$ is a structure of $T^H_D(\Delta)$ in which $S^*$ is false and $\mathbf{M}_{\mathrm{III}}$ is a structure of $T^H(\Delta)$ (of Theorem III) in which $S^*$ is false.

The construction of these Kripke structures is uniform for both Theorems II and III. Let $\mathbf{A}$ be the classical structure promised in the assumptions of Theorem II (or of Theorem III). For each $a, b\in A$ let $\mathbf{A}_{a,b}$ be a model of the consistent theory mentioned in (b) of Theorem II (or in (b) of Theorem III). $\mathbf{A}$ is homomorphic to a submodel of $\mathbf{A}_{a,b}$ in both, the case of Theorem II and of III, since $\mathbf{A}_{a,b}\models D^+(\mathbf{A})$. We shall write $\mathbf{A}\subseteq\mathbf{A}_{a,b}$. In the case of Theorem II, $\mathbf{A}_{a,b}\models D(\mathbf{A})$, and so we have actual *isomorphism into*.

We are also given a model $(B, <)$ of $S$ and of $RS$. We can assume that $B$ is countable and therefore we can say $B \subseteq A_*$. We can now define for both, the case of Theorem II and Theorem III, the structures $\mathbf{M}_{\mathrm{II}}$ and $\mathbf{M}_{\mathrm{III}}$.

Let $T = \{0\} \cup \{(a, b) \in B^2 \mid \mathbf{B} \models a < b\}$ and $R = \{0\} \times T$. Let $\mathbf{A}_0 =_{\mathrm{def}} \mathbf{A}$ and $\mathbf{A}_{(a,b)} =_{\mathrm{def}} \mathbf{A}_{a,b}$. Since in both, the case of Theorem II and III, $\mathbf{A}$ is homomorphic into $\mathbf{A}_{a,b}$ we get two well-defined Kripke structures. $\mathbf{M}_{\mathrm{II}}$ is the structure obtained by performing all these operations in the case of Theorem II and $\mathbf{M}_{\mathrm{III}}$ is the structure obtained in the case of Theorem III. Remember that, in $\mathbf{M}_{\mathrm{II}}$, $\mathbf{A}$ is isomorphic to a substructure of $\mathbf{A}_{a,b}$ and so, for any atomic $\beta$, $[\beta]_0 = 0 \Rightarrow [\beta]_{(a,b)} = 0$.

We now proceed to show that $\mathbf{M}_{\mathrm{II}}$ is a structure of $T^H_D(\Delta)$ of Theorem II and $\mathbf{M}_{\mathrm{III}}$ is a structure of $T^H(\Delta)$ (of Theorem III).

LEMMA 9. (a) *Let $\tau$ be in the (*AE)-form and let $(T, R, 0, \mathbf{A}_t)$, $t \in T$, be a Kripke structure such that for all atomic $P$ if $[P]_t = 0$ and $tRs$ then $[P]_s = 0$. Then if $[\tau]_0 = 0$ then for some $t$, $\mathbf{A}_t \models \neg\tau$ classically.*

(b) *If $\tau$ is in the (*A)-form and $(T, R, 0, \mathbf{A}_t)$, $t \in T$, is any Kripke structure, then if $[\tau]_0 = 0$, then, for some $t$, $\mathbf{A}_t \models \neg\tau$ classically.*

*Proof.* To prove (a) let $[(\forall x_1 \ldots)(\exists y_1 \ldots) \wedge (\bigwedge \beta_i \rightarrow \bigvee \gamma_j)]_0 = 0$ where $\beta_i$, $\gamma_j$ are atomic. So for some $t$ such that $0Rt$, for some $x_1, \cdots \in A_t$ and all $y_1, \cdots \in A_t$, $[\bigwedge \beta_i \rightarrow \bigvee \gamma_j]_t = 0$. So for some $s$ such that $tRs$, $[\bigwedge \beta_i]_s = 1$ and $[\bigvee \gamma_j]_s = 0$. From the condition on atomic $P$ we get that $[\bigwedge \beta_i]_t = 1$ and $[\bigvee \gamma_j]_t = 0$. So $\mathbf{A}_t \models \neg\tau$ classically. The proof of (b) is similar.

LEMMA 10. *$\mathbf{M}_{\mathrm{II}}$ is a structure of $T^H_D(\Delta)$ of Theorem* II *and $\mathbf{M}_{\mathrm{III}}$ is a structure of $T^H(\Delta)$ of Theorem* III.

*Proof.* The proof follows from Lemma 9 and the assumptions on the respective $\Delta$'s and the structures.

We now proceed to show that $S^*$ is false in $\mathbf{M}_{\mathrm{II}}$ and $\mathbf{M}_{\mathrm{III}}$.

LEMMA 11. *In $\mathbf{M}_{\mathrm{II}}$ or $\mathbf{M}_{\mathrm{III}}$ we have*

(a) $[D(u)]_0 = 0$ *iff* $u \in B$,

(b) $a < b$ *iff* $[\neg\rho_1(a, b)]_0 = 0$.

*Proof.* (a) is proved by construction.

(b) If $a < b$, then $[\rho_1(a, b)]_{(a,b)} = 1$ by construction. If $\neg a < b$, then $(a, b) \notin T$. Furthermore since $<$ is symmetric, $\neg b < a$, $(b, a) \notin \mathbf{T}$. But then $\mathbf{A}_{x,y} \models \neg\rho_1(a, b)$ and so $[\neg\rho_1(a, b)]_{(x,y)} = 1$ for all $(x, y) \neq (a, b)$.

LEMMA 12. (a) $[\exists x\rho(x, x)]_0 = 0$.

(b) $[\exists x\rho(x, x)]_{(a,b)} = 1$.

*Proof.* To prove (a) note that $\exists x\rho(x, x)$ is classically false at $\mathbf{A}$ (requirement (a) of Theorems II and III). Also note that we assume $\rho$ is in (*)-form so if it were the case that $[\exists x\rho(x, x)]_0 = 1$ then $\mathbf{A} \models_{CPC} \exists x\rho(x, x)$, by Lemma 5.

To prove (b) notice that $\mathbf{A}_{(a,b)} \models_{CPC} \rho(a, a)$ and that since $(a, b)$ is an endpoint of $T$, the classical and intuitionistic truth values coincide.

LEMMA 13. *In both* $\mathbf{M}_{\mathrm{II}}$ *and* $\mathbf{M}_{\mathrm{III}}$ *the following hold*:

(a) $[S_1]_{(a,b)} = 1$.

(b) *If* $[S_1]_0 = 0$, *then* $\mathbf{B} \models_{CPC} \neg S$.

*Proof.* (a) To see that $[S_1]_{(a,b)} = 1$ notice that $[\exists x\rho(x, x)]_{(a,b)} = 1$ and that $\exists x\rho(x, x)$ appears as an innermost disjunct in $S_1$, and compare with Lemma 12.

(b) The proof is by induction. Let $\tau_1(a_1, \ldots, a_n)$, $a_i \in B$, be a subformula of $S_1$ of the form

$$(Q_1^* \ldots) \bigwedge\Big(\bigwedge_i \neg\rho_1(u_i, v_i) \to \Big(\bigvee_j \neg\rho_1(x_j, y_j) \vee \exists x\rho(x, x)\Big)\Big),$$

where $Q_1^*$ is an initial segment of $(Q^* \ldots)$ of $S_1$. Assume that

(b′) if $[\tau_1(a_1, \ldots, a_n)]_0 = 0$ then

$$\mathbf{B} \models \neg(Q_1 \ldots) \wedge \Big(\bigwedge x_j < y_j \to \bigvee u_i < v_i\Big) \text{ (call this formula } \tau).$$

We prove (b′) for one more quantifier. Let $[\exists^* x\tau_1(x)]_0 = 0$ so

$$[\exists x\tau_1(x) \wedge (D(x) \to \exists x\rho(x, x))]_0 = 0.$$

So, for all $x \in A_0$, $[\tau_1(x) \wedge (D(x) \to \exists x\rho(x, x))]_0 = 0$ (in particular for all $x \in B$). Now for $x \in B$, $[D(x)]_0 = 0$ and $[D(x) \to \exists x\rho(x, x)]_0 = 1$ (compare Lemmas 10 and 11). We must therefore have, for all $x \in B$, $[\tau_1(x)]_0 = 0$. But now by the induction hypothesis $\mathbf{B} \models \neg\tau(x)$. Since this holds for all $x \in B$ we get $B \models \neg\exists x\tau(x)$.

Let us now assume that $[\forall^* x\tau_1(x)]_0 = 0$. So we have $[\forall x(D(x) \vee \tau_1(x)]_0 = 0$. So for some $t$ such that $0Rt$ and some $a \in A_t$ we have $[D(a) \vee \tau_1(a)]_t = 0$. By the proof of Lemma 13(a) and the fact that $\exists x\rho(x, x)$ is a disjunct in $\tau_1$, we conclude that $t = 0$. But then we have that $a \in B$

and $[\tau_1(a)]_0 = 0$. So by the induction hypothesis $\mathbf{B} \models \neg\tau(a)$, and $\mathbf{B} \models \neg\forall x\tau(x)$. The only remaining case to prove is that of no quantifiers (the basis for the induction). Suppose

$$\left[\bigwedge_i \neg\rho_1(u_i, v_i) \to \bigvee_j \neg\rho_1(x_j, y_j) \vee \exists x\rho(x, x)\right]_0 = 0.$$

Then for some $t$ such that $0Rt$ we have $\wedge\{[\neg\rho_1(u_i, v_i)]_t = 1\}$ and

$$\bigwedge\{[\neg\rho_1(x_j, y_j) \vee \exists x\rho(x, x)]_t = 0\}.$$

By Lemma 12(a), $t = 0$. By Lemma 12(b) and by Lemma 11 we get that

$$B \models \bigwedge x_j < y_j \quad \text{and} \quad B \models \bigwedge \neg u_i < v_i.$$

This concludes the proof of Lemma 13.

LEMMA 14. *If* $RS \not\vdash_{CPC} \neg S$, *then* $T_D^H(\Delta) \not\vdash_{HPC} S^*$ *and* $T^H(\Delta) \not\vdash_{HPC} S^*$, *where* $T_D^H(\Delta)$ *is the thoery of Theorem* II *and* $T^H(\Delta)$ *is the thoery of Theorem* III.

*Proof.* Assume that $\mathbf{B}$ is a structure of $RS \cup \{S\}$; construct $\mathbf{M}_{\mathrm{II}}$ and $\mathbf{M}_{\mathrm{III}}$ as above. By Lemma 10 and Lemmas 11, 12 and 13, Lemma 14 must hold.

Thus Theorems II and III are proved.

## 2. Undecidability results

THEOREM 1. *The HPC Theory of one monadic predicate is undecidable.*

*Proof.* We apply Theorem III. Let $A = A_* = \omega$. Let $P(x) = 0$ for $x \in A$ and let $\mathbf{A} = (A, P)$. Let $\rho(x, y) = P(x) \wedge P(y)$. Certainly we have that the theory of (b) of Theorem III is consistent.

THEOREM 2. *The HPC theory of a transitive and symmetric predicate R is undecidable.*

AXIOMS.

$$\forall x \forall y (xRy \to yRx),$$

$$\forall x \forall y \forall z (xRy \wedge yRz \to xRz).$$

*Proof.* We use Theorem III. The classical theory is decidable. Let $A = A_* = \omega$. Let $xRy$ be false for all $x, y \in A$. Let $\rho(x, y)$ be $xRx \wedge$

$yRy$. Clearly, for each $a$, $b \in A$, $\mathbf{A}_{a,b} = (A, R_{a,b})$ is a model of the theory of (b) of Theorem III, where $R_{a,b} = \{(a, a), (b, b)\}$.

THEOREM 3. *The HPC theory of one equivalence relation and decidable equality is undecidable.*

AXIOMS.

$$\forall xy(x = y \to x \equiv y),$$

$$\forall x(x \equiv x),$$

$$(\forall xy)(x \equiv y \to y \equiv x),$$

$$(\forall xyz)(x \equiv y \wedge y \equiv z \to x \equiv z).$$

Similar axioms for equality, and $\forall xy(x = y \vee x \neq y)$.

*Proof.* We use Theorem III. Let $A = A_* = \omega$. Let $x \equiv y$ hold in $A$ iff $x = y$. Let $\rho(x, y)$ be $x \neq y \wedge x \equiv y$.

THEOREM 4. *The HPC theory of one unary function symbol with decidable equality is undecidable.*

AXIOMS. Axioms for equality and the axiom $\forall x \forall y(x = y \vee x \neq y)$.

*Proof.* We use Theorem II. Let $A = A_* = \omega$. Let $f(m) = m + 1$ for $m \in A$. Let $P(a)$ be $\exists x \exists y(x \neq y \wedge f(x) = a \wedge f(y) = a)$ and let $\rho(a, b) = P(a) \wedge P(b)$. Clearly the theory of (b) of Theorem II is consistent.

*Remark.* The classical theory of one unary function is decidable. If we add to the theory the axiom that there are at least $n$ predecessors, the *HPC* theory is still undecidable because we can take $P$ to say that there are $n + 1$ predecessors. If we add the axiom that there exists *exactly* $n$ predecessors the theory is decidable. Notice that to say 'there are exactly $n$ predecessors' you need a (*$AEA$)-sentence!

THEOREM 5. *The HPC theory of abelian groups with a decidable equality is undecidable.*

AXIOMS. (1) Axioms for decidable equality,

(2) $\forall xy(x + y = y + x)$,

(3) $(\forall xyz)(x + (y + z) = (x + y) + z)$,

(4) $\forall x(x + 0 = x)$,

(5) $\forall x \exists y(x + y = 0)$.

*Proof.* We use Theorem II. Let $Z$ be the group with two elements $\{0, 1\}$ and let $A$ be $Z^\omega$. Addition is pointwise. Let $A_* \subseteq A$ be all $a \in Z^\omega$ that are 0 except for one coordinate. Let $P(a) = \exists y((a = y + y) \wedge y \neq 0)$. Let $\rho(a, b)$ be $P(a) \wedge P(b)$. Clearly $Z^\omega \models \forall a \neg P(a)$.

We now want to show that the theory of (b) of Theorem II is consistent. For $a, b \in A_0$, let $m, n$ respectively be the unique points such that $a(m) = 1$ and $b(n) = 1$. Assume that $m \leqslant n$. Let $Z'$ be the cyclic group of order 4, $(0, 1, 2, 3)$, and identify $Z$ with $(0, 2)$ so we can regard $Z$ as a subgroup of $Z'$. Let $\mathbf{A}(a, b)$ be the set of all functions $f$ such that $f(x) \in Z$ for $x \neq m, n$ and $f(m), f(n) \in Z'$. Clearly $\mathbf{A}(a, b)$ is an abelian group and $\mathbf{A}$ can be identified as a subgroup of $\mathbf{A}(a, b)$. Clearly $P(a) \wedge P(b)$ holds at $\mathbf{A}(a, b)$ and for $(x, y) \neq (a, b) \neg \rho(x, y)$ holds.

THEOREM 6. *The HPC theory of decidable linear order is undecidable.*

AXIOMS. Axioms for decidable equality,

$$\neg(x < x),$$
$$(x < y) \wedge (y < z) \rightarrow (x < z).$$
$$(x < y) \vee \neg(x < y),$$
$$(x < y) \vee (y < x) \vee (x = y),$$
$$\forall x \exists y (x < y),$$
$$\forall x \exists y (y < x).$$

*Proof.* We use Theorem II. Let $A_* = A =$ the set of rational numbers. Let $P(a)$ be $\exists z(z < a \wedge \neg \exists u(z < u \wedge u < a))$. Let $\rho(a, b)$ be $P(a) \wedge P(b)$.

THEOREM 7. *The HPC theory of Boolean algebras with decidable equality is undecidable.*

(The axiom system is universal with symbols for 0, complementation and intersection.)

*Proof.* Let $A = 2^\omega$. Let $A_*$ be $\{\{m\} | m \in \omega\}$. Let $x \subseteq y$ be $x \wedge y = x$ and $At(x)$ be $(\forall y \neq \emptyset)(y \subseteq x \rightarrow y = x)$. Let $P(a)$ be $(\exists x \subseteq a)[(\forall y \subseteq x)(y = \emptyset \vee \sim At(y)) \wedge x \neq \emptyset]$. Let $\rho(a, b)$ be $P(a) \wedge P(b) \wedge P(-a) \wedge P(-b)$. Clearly $\mathbf{A} \models \forall xy \sim \rho(x, y)$.

Let $a, b \in A_*$. Assume $a = \{m\}$ and $b = \{n\}$. Let $B$ be an *atomless*

Boolean algebra. Let $A_{\{m\},\{n\}}$ be $2^{\omega-a-b}\times B\times B$.

Define a function $f_{m,n}:2^{\omega}\rightarrow A_{\{m\},\{n\}}$.

$$\begin{aligned} f_{m,n}(x) &= (x,0,0) && \text{if } m\notin a \text{ and } n\notin a,\\ &= (a-\{m\},1,0) && \text{if } m\in a \text{ and } n\notin a,\\ &= (x-\{n\},0,1) && \text{if } n\in a \text{ and } m\notin a,\\ &= (x-\{m,n\},1,1) && \text{if } \{m,n\}\subseteq a. \end{aligned}$$

One can verify that $f$ commutes with the Boolean operations and therefore $f_{m,n}$: $A\rightarrow A_{\{m\},\{n\}}$ is an isomorphism into, and $A_{\{m\},\{n\}}$ is a model of the theory of (b) of Theorem II.

THEOREM 8. *The HPC theory of Brouwerian algebras with decidable equality is undecidable.*

*Proof.* The *HPC* theory of Boolean algebras is a finite extension of this theory.

## 3. Undecidability of Classically Model Complete Theories

*Algebraically Closed Fields*

(1) First formulation is in a language with $+$, $\cdot$, 0, 1 and $=$.

(a) Axioms in AE form saying that $+$ is a commutative group.

(b) $x(yz)=(xy)z$, $x(y+z)=xy+xz$, $xy=yx$.

(c) $\forall x\exists y(x=0 \vee xy=1)$.

(d) $0\neq 1$.

(e) Equality axioms: '$x=x$, $x=y\rightarrow y=x$, $x=y\wedge y=z\rightarrow x=z$'.

(f) For each $n\geqslant 0$, the following axiom holds:

$$(\forall x_0\ldots x_n)\exists z\left(\sum_{i=0}^{n} x_iz^i+z^{n+1}=0\right).$$

The system $F\dagger$ of the first formulation is decidable.

(2) The second formulation is in the same language as the first with the axiom (c′) replacing axiom (c) and the other axioms remaining unchanged.

(c′) $\forall x(x\neq 0\rightarrow\exists y(xy=1))$.

The system $F_2$ of this formulation is undecidable.

THEOREM A. *The system $F_1$ of the first formulation is decidable.*
*Proof.* To prove this first observe the following:
(3) $F_1^* \vdash_{HPC} \forall xy(x = y \vee x \neq y)$, and use the following theorem:

THEOREM M. *Let $T$ be a set of axioms for a classical theory $T^C$, written in the form $(Q\ldots) \wedge (\wedge A_i \to \vee B_j)$ where $(Q\ldots)$ is a string of quantifiers and $A_i$, $B_j$ are atomic. Then $T$ is classically model complete iff the following holds for any formula $F$. $T_1 \vdash$ intuitionistically $F$ iff $T \vdash$ classically $F$ where $T_1 = T \cup \{(\text{universal closure})(A \vee \neg A) | A$ atomic$\}$.*

Observation (3) and Theorem M show that $F_1^*$ is the same as the classical theory of algebraically closed fields.

THEOREM B. *The system $F_2^*$ of the second formulation is undecidable, in fact, a finite extension of it with the axioms*

$$\neg\neg(x = y) \to x = y \quad \textit{and} \quad \forall x \exists y(x \neq 0 \to xy = 1)$$

*is undecidable.*
*Proof.* Call this extension $F^*$. Towards showing that $F^*$ is undecidable we need various definitions, lemmas and constructions. Our purpose is to faithfully interpret in $F^*$ the classically undecidable theory of a reflexive and symmetric relation with at least five elements. The proof of faithfulness shall use Kripke structures. So let us start with definitions.
*Construction* (4). Let $(M, <)$ be a structure of a reflexive and symmetric relation $<$. Let us assume that $\bar{\bar{M}} > 4$. We construct a Kripke structure called the Kripke structure *associated* with $(M, <)$. Let

$$S_0 = \{\{x, y\} | \neg(x < y), x, y \text{ in } M\}.$$

$$S = \{0\} \cup S_0 \cup M,$$

$$R = \{0\} \times S \cup \{(x, x) | x \in S\} \cup \{(\{x, y\}, x) | \{x, y\} \in S_0\}.$$

Let $\mathbf{A}_t = K^M$, where $K$ is some *fixed* algebraically closed field. Define $+, \cdot$ on $K^M$ to be *pointwise* for all $t \in S$. Let $\bar{0} \in \mathbf{A}_t$ be the vector with $0 \in K$ in all coordinates and let $\bar{1}$ be the constant vector $1 \in K$. Define

$=$ on $\mathbf{A}_t$ as follows: For $f, g \in K^M$ let $[f=g]_0 = 1$ iff $\forall m \in M(f(m) = g(m))$. Let $[f = g]_{\{x, y\}} = 1$ iff $f(x) = g(x)$ and $f(y) = g(y)$. Let $[f = g]_x = 1$ iff $f(x) = g(x)$.

LEMMA 5. $(S, R, 0, \mathbf{A}_t)$ *thus defined validates the system* $F_2^*$.

*Proof.* Axioms (a), (b), (e) and (f) hold because $+$ and $\cdot$ were defined pointwise. $\bar{0} \neq \bar{1}$ holds by definition of $=$. To show that (c′) holds notice that $[f \neq \bar{0}]_0 = 1$ iff $\forall m \in M(f(m) \neq 0)$ and $[f \neq \bar{0}]_{(x, y)} = 1$ iff both $f(x), f(y) \neq 0$, and so the following $g$ can serve as inverse.

$$\begin{aligned} g(m) &= 0 && \text{if } f(m) = 0, \\ &= f(m)^{-1} && \text{if } f(m) \neq 0. \end{aligned}$$

The reader can also verify if $[f = g]_0 = 0$, then, for some $m \in M$, $f(m) \neq g(m)$ and so $[f = g]_m = 0$, which means that $[\neg\neg(f = g)]_0 = 0$. Similar verification for $t = \{m, n\}$ shows that $\forall xy(\neg\neg x = y \rightarrow x = y)$ must hold at 0.

Now consider the following formulas:

$$E =_{\text{def}} (\forall x, u_1, u_2, u_3) \left[\left(x^2 = x \wedge \bigwedge_{i=1}^{3} u_i^2 = u_i \wedge u_1 + u_2 + u_3 = x\right) \rightarrow \left(\bigvee_{i \neq j} (u_i + u_j = x)\right)\right]. \tag{6}$$

$$\begin{aligned} D(x) =_{\text{def}} &\langle x^2 = x \wedge (\forall u, v) \\ &[(u^2 = u \wedge v^2 = v \wedge u + v = x) \\ &\rightarrow (x = \bar{0} \vee (u = \bar{0} \wedge v = x) \vee (u = x \wedge v = \bar{0}))]\rangle \\ &\rightarrow \langle x = \bar{0} \vee x = \bar{1} \vee E\rangle. \end{aligned} \tag{7}$$

Let $M_0 \subseteq K^M$ be the set of all $f \in K^M$ such that, for some $m \in M$, we have $f = f_m$, with

$$\begin{aligned} f_m(n) &= 1 \in K && \text{if } n = m, \\ &= 0 \in K && \text{otherwise.} \end{aligned}$$

LEMMA 8. (1) $[E]_t = 0$ *iff* $t = 0$.

(2) $[D(f)]_0 = 0$ *iff* $f \in M_0$.

*Proof.* (1) is easy to see; recall that $\bar{\bar{M}} \geq 5$.

(2) If $f \in M_0$, then at 0, the antecedent of $D(f)$ is true and the consequent is false. If $f \notin M_0$, then let us show that $D(f)$ holds. Since $E$ is false only at 0, we have to check whether the antecedent can be

true at 0 and the consequent false. If $x = \bar{0}$ does not hold, then there must be $m \in M$ such that $f(m) \neq 0$. $f(m)$ must be 1 since if $f(m) \neq 1$ then $f^2 = f$ would not hold. Since $f(m) = 1$ and $f \not\in M_0$, there must be some $n \neq m$ such that $f(n) \neq 0$ and hence $f(n) = 1$. But then take $u$ and $v$ to be $f_m$ and $f_n$ and then the antecedent will not hold. Thus if $f \not\in M_0$, $D(f)$ must hold at 0.

Define $\rho(u, v)$ by

$$(9) \qquad \rho(u, v) =_{\text{def}} [E \wedge (u + v = \bar{1}) \to u = \bar{1} \vee v = \bar{1}].$$

LEMMA 10. *For $f_a, f_b \in M_0$, we have*

$$[\rho(f_a, f_b)]_0 = 1 \quad \textit{iff } a < b.$$

*Proof.* By construction.

Now let $S$ be a statement of the theory of a reflexive and symmetric relation $<$. $S$ can be written in the form

$$(Q\ldots)\bigwedge\left(\bigwedge x_i < y_i \to \bigvee u_j < v_j\right),$$

where $(Q\ldots)$ is a string of quantifiers. We want to associate with $S$ a statement $S^*$ in the theory $F$ of fields, as follows.

Let $\forall^* x\tau$ and $\exists^* x\tau$ denote, for any formula $\tau$, the formulas $\forall x(Dx \vee \tau)$ and $\exists x(\tau \wedge (Dx \to E))$ respectively. Let $S^*$ be the following formula:

$$(\forall u_1, \ldots, u_5)\Big\langle\Big[\forall x(Dx \vee \rho(x, x)) \wedge (Q^*\ldots)\bigwedge$$
$$\Big(\bigwedge \rho(x_i, y_i) \to \bigvee \rho(u_j, v_j) \vee E\Big)\Big]$$
$$\to \Big[E \vee \bigvee_{i=1}^{5} D(u_i) \vee \bigvee_{1 \leq i \neq j \leq 5} u_i = u_j\Big]\Big\rangle,$$

where $(Q^*\ldots)$ is the same string $(Q\ldots)$ of quantifiers with * added to them and interpreted as indicated above.

LEMMA 11. *Let RS be the theory of a reflexive and symmetric relation with at least 5 elements. Then*

$$RS \vdash \neg S \quad \textit{iff} \quad F \vdash S^*.$$

*Proof.* Assume that there exists a structure $(S, R, 0, \mathbf{A}_t)$ where $S^*$ is false. Then there exists a $t \in S$ and $u_i \in \mathbf{A}_t$ such that the antecedent of the main implication of $S^*$ holds and the consequent does not hold.

Define a model $(M_0, <)$ where $S$ holds as follows:

(12) Let $M_0 = \{a \in \mathbf{A}_t \mid [D(a)]_t = 0\}$.

(13) Let $a < b$ iff $[\rho(a, b)]_t = 1$.

Clearly, since $[\forall x(Dx \vee \rho(x, x))]_t = 1$ we get that $<$ is reflexive and looking at $\rho$ we see that $<$ is symmetric. We also have $[S_1^*]_t = 1$ where

$$S_1^* = (Q^* \ldots) \bigwedge (\bigwedge \rho(x_i, y_i) \rightarrow \bigvee \rho(u_j, v_j)).$$

We claim that $S$ holds at $(M_0, <)$. This follows from the following:

LEMMA 14. *Let* $\tau = \wedge(\wedge\ x_i < y_i \rightarrow \vee\ u_j < v_j)$ *and let*

$$\tau^* = \bigwedge\left(\bigwedge \rho(x_i, y_i) \rightarrow \left(\bigvee \rho(u_j, v_j) \vee E\right)\right).$$

*Further let* $(Q^* \ldots) = (Q_1^* \ldots)(Q_2^* \ldots)$. *Then for any substitution of elements* $a_i$ *of* $M_0$ *for the variables of* $Q_1^*$ *we have that* $[(Q_2^* \ldots \tau^*(a_i)]_t = 1$ *iff* $(Q_2 \ldots)\tau(a_i)$ *holds in* $(M_0, <)$.

*Proof*. By induction.

From Lemma 14 we deduce that $(M_0, <)$ is a model of $S$.

Now to prove the other direction of Lemma 11 assume that $S$ has a model $(M, <)$. We assume that $\bar{\bar{M}} \geq 5$. Construct the model $(S, R, 0, \mathbf{A}_t)$ associated with $(M, <)$. We claim

(15) $\qquad [S^*]_0 = 0.$

In fact, for any five different $u_i \in M_0 \subseteq K^M$, the antecedent of $S^*$ holds and the consequent is false. Clearly $[E]_0 = 0$ and $[\forall x(D(x) \vee \rho(x, x))]_0 = 1$, as we have verified in the course of the construction. Also since $[D(f)]_0 = 0$ iff $f \in M_0$ and since $[\rho(f_a, f_b)]_0 = 1$ iff $a < b$ then Lemma 14 holds and therefore $[S_1^*] = 1$. Thus (15) is proved.

*Real Closed Fields*

(16) The language of the first formulation contains the symbols $+$, $\cdot$, $0$, $1$, $=$ and the unary predicate $P(x)$. The axioms are axioms (a)–(e) of (1) along with the following axioms:

(g) $\qquad P(x) \wedge P(y) \rightarrow P(x + y)$, $x + y = 0 \rightarrow \neg(P(x) \wedge P(y))$, $P(x) \wedge P(y) \rightarrow P(xy)$.

(h) $\forall x \forall y \exists z(x + y = 0 \to x = z^2 \vee y = z^2)$,

$$(\forall x_0, x_1, \ldots, x_{2n}) \exists y \left( \sum x_i y^i + y^{2n+1} = 0 \right).$$

(k) $x + y = 0 \to x = 0 \vee P(x) \vee P(y)$.

(17) The language of the second formulation is the same as before. The axioms are (a), (b), (c′), (d), (e), (g), (h), and (k′) where

(k′) $x + y = 0 \wedge \neg P(x) \wedge \neg P(y) \to x = 0$.

THEOREM C. *The system of the first formulation is decidable.*

*Proof.* Observe that the formulas $\forall xy(x = y \vee x \neq y)$ and $\forall x(Px \vee \neg Px)$ are theorems of this formulation and use Theorem M. Actually this system is equal to the classical theory.

THEOREM D. *The system of the second formulation is undecidable, in fact, a finite extension G of it, with the axiom* $\neg\neg(x = y) \to (x = y)$ *and* $\forall x \exists y(x \neq 0 \to xy = 1)$ *is undecidable.*

*Proof.* We use the same interpretation that we used in the proof of Theorem B. Let $(M, <)$ be a model of a reflexive and symmetric relation. Let $K$ be a real closed field. Define the associated Kripke structure as before. We have to define the additional predicate $P$ in the structure. Let $[P(f)]_0 = 1$ iff $\forall m P(f(m))$ holds in $K$. Let $[P(f)]_{\{m,n\}} = 1$ iff $P(f(m)) \wedge P(f(n))$ holds in $K$. Let $[P(f)]_m = 1$ iff $P(f(m))$ holds in $K$. We now have to verify that axioms (g), (h) and (k′) hold. Clearly axioms (g) and (h) hold because $+$, $\cdot$ and $P$ are defined pointwise. It is easy to verify that (k′) also holds.

From now on to complete the proof of Theorem D we proceed as in the case of the proof of Theorem B and show that $G^* \vdash S^*$ iff $RS \vdash \neg S$.

**Problem.** Suppose we replace axiom (k′) of $G$ by

(k″) $x + y = 0 \wedge x \neq 0 \to Px \vee Py$.

Is the resulting system decidable?

*Divisible Abelian Groups*

Now we give a general sufficient condition for undecidability that implies the undecidability of many theories, e.g. the intuitionistic theories of divisible abelian groups, abelian groups, equality, unary function, etc. The condition also applies to the intermediate logic *CD* of constant domains, i.e. *HPC* + the axioms scheme $\forall x(A \vee B(x)) \to A \vee \forall x B(x)$, $x$ not free in $A$.

THEOREM G. *Let T be a set of axioms in a language with = such that the theory $T^C$ fulfills the conditions below, then both $T^H$ and $T^{CD}$ are undecidable.*

(a) *$T^C$ has a structure with at least two elements.*

(b) *The class of models of $T^C$ is closed under direct products* (*where the operations and relations are defined pointwise*).

(c) *The axioms of T are either equality axioms* (e) *or of the form*

(1) $(Q\ldots)\bigwedge\bigvee P_{i,j}$, *where $P_{i,j}$ are atomic, or*

(2) $(\forall x_1,\ldots)\bigwedge\left(\bigwedge F_i \to \bigvee E_j\right)$, *where $F_i$, $E_i$ are atomic.*

COROLLARY. *The following HPC or CD theories are undecidable*:

(1) *Divisible abelian groups* (*i.e. with $\forall x \exists y(x = my)$, for all natural numbers m*).

(2) *Abelian groups.*

(3) *Equality.*

(4) *Unary function* (*one*).

*etc.*

In Section 2 the *HPC* theory of abelian groups and unary function with decidable equality were shown undecidable. This result implies (2) and (4) above, for *HPC*. For the case of *CD*, however we have

THEOREM H. *Let T be a set of axioms in language with equality and function symbols only, then $T^C = T^{CD} \cup \{\forall xy(x = y \vee x \neq y)\}$.*

*Proof of Theorem G.* Let *K* be a structure of $T^C$ with at least two elements. Let $(M, <)$ be a model of a reflexive and symmetric relation

$<$ with $\bar{\bar{M}} \geqslant 5$. Define a Kripke structure as in the proof of Theorem B (i.e. Construction (4), with $K$ being now a structure of $T^C$). Define all the relations and operations in the Kripke structure pointwise and define equality as in Construction (4).

LEMMA 19. *The Kripke structure thus defined validates*

$$T^{CD} \cup \{\forall xy(\neg\neg(x = y) \rightarrow x = y)\}.$$

*Proof.* Follows from conditions (b) and (c) of Theorem G. Show that for any sentence $F$ of the form of (c) and for any $(S, R, 0, \mathbf{A}_t)$ with constant domains and for any world $s$ we have $[F]_s = 0$ implies $\mathbf{A}_{s'} \models_{CPC} \neg F$ for some $s'$ such that $sRs'$.

Now since $K$ contains at least two elements let $a, b \in K$, $a \neq b$. Let $u_1 = (a \ldots a)$, $u_2 = (b \ldots b)$, and define

$$C(x, u_1, u_2) =_{\text{def}} \forall v(\neg\neg(v = u_1 \vee v = x) \rightarrow v = x) \wedge$$
$$\neg\neg(x = u_1 \vee x = u_2).$$

Clearly, $[C(x, u_1, u_2)]_0 = 1$ iff $x$ has the form $x = (a \ldots a, b, a \ldots a)$ or $x = u_1$. From now on we write $C(x)$, for $C(x, u_1, u_2)$.

$$E =_{\text{def}} (\forall x, y, z)(C(x) \wedge C(y) \wedge C(z)$$
$$\rightarrow x = u_1 \vee y = u_1 \vee z = u_1 \vee x = y \vee y = z \vee x = z).$$

Clearly $s \neq 0$ iff $[E]_s = 1$.

$$D(x) =_{\text{def}} C(x) \rightarrow E \vee x = u_1.$$
$$\rho(x, y) = \forall t[\neg\neg(t = x \vee t = y) \wedge E \rightarrow t = x \vee t = y].$$

Clearly if $[D(x) \vee D(y)]_0 = 0$, then, by construction of the Kripke structure $[\rho(f_n, f_m)]_0 = 1$ iff $m < n$, where $x = f_m = (a \ldots a), b, a \ldots a)$, ($b$ at the $m$th place), similarly $y = f_n$.

Let $S = (Q \ldots) \wedge (\wedge\ u_i < v_i) \rightarrow \vee\ x_j < y_j)$ be a sentence, as before, of the theory $RS$. Let

$$S^* = (\forall u_1, u_2, t_1, \ldots, t_5)\Big[u_1 \neq u_2 \wedge \forall x(Dx \vee \rho(x, x))$$
$$\wedge (Q^* \ldots) \bigwedge \Big(\bigwedge \rho(u_i, v_i) \rightarrow \bigvee \rho(x_j, y_j) \vee E\Big)$$
$$\rightarrow E \vee \bigvee_{i=1}^{5} D(t_i) \vee \bigvee_{1 \leqslant i \neq j \leqslant 5} t_i = t_j\Big].$$

Again, the following holds:

$$T^H \text{ (or } T^{CD}) \vdash S^* \quad \text{iff} \quad RS \vdash \neg S.$$

## 4. Undecidability of Theories Formulated with the Apartness Relation

*Linearly Ordered Abelian Groups*

In this section we prove that the intuitionistic version $T_1$ of linearly ordered abelian groups formulated below is not decidable. (We also look at other versions.) We prove undecidability by showing that the classically undecidable theory of a reflexive and symmetric relation is faithfully interpretable in $T_1$. No version of linearly ordered abelian groups (with or without $\#$) was ever treated before.

The language of $T_1$ contains 0, +, $\#$, =, and <, (actually it will follow from the axioms that = can be defined from $\#$ by $(x = y \leftrightarrow \neg x \# y)$. $T_1$ has the following axioms:

Group E: Axioms for quality:

E1: $x = x$

E2: $x = y \rightarrow y = x$

E3: $x = y \rightarrow (y = z \rightarrow x = z)$

E4: $x = y \rightarrow (A(x) \rightarrow A(y))$

$A$ is any formula of the language.

Group S: Axioms for apartness.

S1: $\neg(x \# x)$

S2: $x \# y \rightarrow y \# x$

S3: $\neg(x \# y) \rightarrow x = y$

S4: $x \# y \rightarrow x \# z \vee y \# z$

S5: $\neg(x \# y) \vee \neg\neg(x \# y)$

Group A: Axioms for addition.

A1: $x + y = y + x$

A2: $x + (y + z) = (x + y) + z$

A3: $x+0=x$

A4: $\forall x \exists y(x+y=0)$

A5: $x \# y \to x+z \# y+z$

Group O: Axioms for order.

O1: $x<y \wedge y<z \to x<z$

O2: $x=y \vee x<y \vee y<x$

O3: $\neg(x<x)$

O4: $x<y \to x+z<y+z$

From the above axioms it follows that

O5: $x \# y \to x<y \vee y<x$

E5: $x=y \leftrightarrow \neg(x \# y)$

The following does not follow from the axioms:

S6: $x<y \to x \# y$

In fact, if we let $T_1^*$ be the theory with the axioms below, then we do not know whether $T_1^*$ is decidable. (In $T_1^*$, $x \# y$ is $x<y \vee y<x$).

Axioms for $T_1^*$: E1–E4; S1–S4; A1–A5; O1, O3, O4, O5, S6 and O6, where:

O6: $x<y \to x<z \vee z<y$.

THEOREM A. *The theory $T_1$ of linearly ordered abelian groups is undecidable.*

Towards the proof of Theorem A we need some constructions.

Let $(M, P)$ be countably infinite classical structure of a reflexive and symmetrical relation $P$. We can assume that $M$ is the set of natural numbers. We now construct a Kripke structure $(S, R, o, \mathbf{D}_t)$, called the Kripke structure associated with $(M, P)$. Let $I$ be the linearly ordered abelian group of the integers. Let

(1) $S=\{M\} \cup \{\{m, n\} \subseteq M \mid mPn\} \cup \{\infty\}$

(2) $R=\{M\} \times S \cup \{(s, s) \mid s \in S\} \cup S \times \{\infty\}$

(3) $o=\{M\}$

To define $\mathbf{D}_t$, $t \in S$, let the domain of $\mathbf{D}_t$ be $I^M$.

(4) Define $=$ on the domain as identity.

(5) Define $+$ on the domain pointwise, and let $\bar{o}$ be $(o \ldots o)$

(6) For $f, g \in I^M$, let $k(f, g)$ be the first coordinate $k$, in the ordering of $M$ (= set of natural numbers) such that $f(k) \neq g(k)$. Then define $f < g$ iff $f(k) < g(k)$, where $k = k(f, g)$. This means that $<$ is the lexicographic ordering on $I^M$.

(7) Define $\#$ on $\mathbf{D}_t$ as follows (this is the first time that the definition depends on $t$):

(7a) $\quad [f\#g]_o = 0, \quad$ for all $\quad f, g.$

(7b) $\quad [f\#g]_{\{m,n\}} = 1 \quad$ iff for some $k \notin \{m, n\}, \quad f(k) \neq g(k).$

(7c) $\quad [f\#g]_\infty = 1 \quad$ iff $\quad f \neq g.$

We have thus defined a Kripke structure for the language of $T_1$.

LEMMA 8. *All axioms of $T_1$ hold in the above structure.*

*Proof.* We check each group of axioms.

Group E: Equality was defined as identity.

Group S: S1–S3 are clear, S3 follows from (7c) and the definition of $R$. S4 follows from (7) and S5 from the definition of $R$.

Group A: $+$ was defined pointwise, A5 follows from (6).

Group O: Follows from (6).

We now turn to give some more definitions.

(8) Let $E$ be defined as: $E = \exists x(x\#\bar{o})$

(9) Let $B(x)$ be: $x \neq \bar{o} \wedge E \rightarrow x\#\bar{o}$

(10) Let $D(x)$ be $\forall y(B(y) \vee [(y\#\bar{o} \rightarrow x\#\bar{o}) \rightarrow (x\#\bar{o} \rightarrow y\#\bar{o}) \vee E)]) \rightarrow E \vee B(x)$

(11) Let $x \equiv y$ be $(x\#\bar{o} \leftrightarrow y\#\bar{o})$

(12) Let $xPy$ be $E \rightarrow x\#\bar{o} \vee y\#\bar{o}$

LEMMA 13. *In the Kripke structure defined we have that*

(14) $[E]_t = 1$ *iff* $t \neq o$.

(15) $[B(x)]_o = 0$ *iff* $x \neq \bar{o}$ *and for some* $m, n \in M$ *we have that* $x(k) = o$, *for all* $k \notin \{m, n\}$.

(16) $[D(x)]_o = 0$ *iff* $x \neq \bar{o}$ *and for some* $m \in M$ *we have that for all* $k$, $x(k) \neq o$ *iff* $k = m$.

(17) *For* $x, y$ *such that* $[x \equiv y]_o = 1$ *and* $[D(x) \vee D(y)]_o = 0$ *we have that* $x(k) \neq o$ *iff* $y(k) \neq o$ *for all* $k \in M$.

(18) *For* $x, y$ *such that* $[D(x) \vee D(y)]_o = 0$ *we have that* $[xPy]_o = 0$ *iff* $\mathbf{M} \models a(x)Pa(y)$, *where* $a(x)$ *is defined as the* $m \in M$ *such that for all* $k$, $(x(k) \neq o$ *iff* $k = m)$. (*See* 16.)

*Proof.* (14) Follows from the definition of $\#$ in (7a).

(15) Is clear from the fact that $x = y \vee x \neq y$ holds, and from (14).

(16) Assume that for some $m$ we have that $x(k) = o$ iff $k \neq m$; for all $k$. We want to show that $[D(x)]_o = 0$. Clearly $[E \vee B(x)]_o = 0$. We will show that the antecedent of $D(x)$ holds at $o$. Let $y$ be such that $[B(y)]_o = 0$, then we must show that

$$(*) \qquad [(y\#\bar{o} \to x\#\bar{o}) \to (x\#\bar{o} \to y\#\bar{o}) \vee E]_o = 1.$$

Since $[B(y)]_o = 0$, then for some $n, n^1, \in M$ we have $y(k) = o$ for all $k \notin \{n, n^1\}$. Now if it were the case that $(*)$ is false, then in view of (14) we must have that

$$[y\#\bar{o} \to x\#\bar{o}]_o = 1, \qquad [x\#\bar{o} \to y\#\bar{o}]_o = 0$$

So $[y\#\bar{o} \to x\#\bar{o}]_{\{m\}} = 1$, since $[x\#\bar{o}]_{\{m\}} = 0$, we must have $[y\#\bar{o}]_{\{m\}} = 0$, i.e., $\{m\} \supseteq \{n, n^1\}$. But this means $m = n = n^1$ ($\{n, n^1\}$ must be nonempty since $y \neq \bar{o}$) and so $[x\#\bar{o} \to y\#\bar{o}]_o = 1$, a contradiction.

To show the other direction, let $[D(x)]_o = 0$, then clearly (by 14), $[B(x)]_o = 0$, and the antecedent of $D(x)$ holds at $o$. Assume that $m$, $n \in M$ are such that for all $k$, $k \in \{m, n\} \to x(k) \neq o$. We claim that $m = n$. Otherwise let $y$ be such that $y(k) = x(k)$ for $k \neq n$ and $y(n) = 0$. Clearly $[B(y)]_o = 0$ and so we must have that $(*)$ holds for this $y$ and this $x$. Now clearly for any $t$, $[y\#\bar{o}]_t = 1$ implies $t = \infty$ or $m \notin t$, and so $[x\#\bar{o}]_t = 1$ must also hold. However, for $t = \{m\}$, $[x\#\bar{o}]_{\{m\}} = 0$. Thus we contradicted $(*)$ and therefore $m = n$. This proves the other direction of (16).

(17) Let $x$, $y$ be such that $[D(x) \vee D(y)]_o = 0$ and $[x \equiv y]_o = 1$. Assume that $x(k) = o \wedge y(k) \neq o$, for some $k \in M$. Then for $t = \{k\}$, $[y\#\bar{o}]_t = 0$, since by (16) there could be only one point $k \in M$ such that $y(k) \neq 0$. On the other hand $[x\#o]_t = 1$ since, again by (16), there must be another point $k^1 \neq k$ such that $x(k^1) \neq 0$.

(18) Suppose $m$, $n \in M$ are such $M \not\models mPn$, then $\{m, n\} \notin S$, we claim that $[E \to x\#\bar{o} \vee y\#\bar{o}]_o = 1$ for $x$, $y$ such that for all $k$,

$$x(k) \neq o \quad \text{iff} \quad k \neq m$$

$$y(k) \neq o \quad \text{iff} \quad k \neq n$$

This is clear since for all $t \in S$ $t \neq \infty, o$; $\{m, n\} \not\subseteq t$. Now assume that

$M \models mPn$, then for $t = \{m, n\} \in S$ and $x$, $y$ defined as above, $[E]_t = 1$ and $[x \# \bar{o} \vee y \# \bar{o}]_t = 0$.

Thus 18 is proved.

(19) Observe that from (17) and (18) it follows that the relation $[x \equiv y]_o = 1$ is an equivalence relation on $\{x \mid [D(x)]_o = 0\}$ and that it is a congruence relation with respect to $[xPy]_o = 0$.

(20) Thus from (17), (18) and (19) it follows that the model $(M, P)$ is isomorphic to the model $(\bar{M}, \bar{P})$ where $\bar{M}$ is the set of $\equiv$ equivalence classes of $\{x \mid [D(x)]_o = 0\}$ and $\bar{P}$ is defined by $x/{\equiv}\bar{P}y/{\equiv}$ iff $[xPy]_o = 0$.

We are now in a position to define a faithful translation from the classical theory *RS* of a reflexive and symmetric relation into the theory $T_1$ of this section. If $\alpha$ is translated into $\alpha^*$, then we shall have

$$T_1 \vdash \alpha^* \quad \text{iff} \quad RS \vdash \neg\alpha.$$

To obtain the translation, we need some more definitions.

(21) Define the following:

(a) Let the quantifiers $\forall^*$, $\exists^*$ be defined by

$$\forall^* y\alpha \quad \text{iff} \quad (\forall y)\,(D(y) \vee \alpha)$$

$$\exists^* y\alpha \quad \text{iff} \quad (\exists y)\,((D(y) \to E) \wedge \alpha)$$

(b) Let $\alpha$ be a sentence of *RS*. Write $\alpha$ in the form $(Q\ldots)\bigwedge(\wedge x_iPy_i \to \vee u_jPv_j)$, where $Q$ is a string of quantifiers and $P$ is the symbol for the relation in the language of *RS*. Let the translation $\alpha^*$ of $\alpha$, in the language of $T_1$ be the following sentence

$$\forall u\Big[(Q^*\ldots)\bigwedge(\wedge u_jPv_j \to \vee x_iPy_i \vee E) \to E \vee D(u)\Big],$$

Where $Q^*$ is the same string of quantifiers as $Q$ except that $\forall$ and $\exists$ are replaced by $\forall^*$ and $\exists^*$ of (21a), and $D$, $P$ and $E$ are as defined in (8), (10) and (12).

LEMMA 22. *For* $\alpha$, $\alpha^*$ *of* (21) *we have*: $T_1 \vdash \alpha^*$ *iff* $RS \vdash \neg\alpha$.

*Proof*. We show that $T_1 \nvdash \alpha^*$ iff $RS \nvdash \neg\alpha$. Assume that there exists a structure $(M, P)$ in which $\alpha$ holds. We will show a Kripke structure of $T_1$ in which $\alpha^*$ does not hold. Let $(S, R, o, \mathbf{D}_t)$ be the structure associated with $(M, P)$ as constructed in the beginning of this section. We want to show that $[\alpha^*]_0 = 0$. To show this, let $u$ be any element such that $[D(u)]_o = 0$, and let us show (since $[E]_o = 0$) that

$$[Q^* \ldots) \wedge (\wedge\, u_jPv_j \rightarrow \vee\, x_iPy_i \vee E]_o = 1.$$

To prove this remember that by (20), $(\bar{M}, \bar{P})$ is isomorphic to $(M, P)$. We use this to proceed and show by induction on the length of $(Q^* \ldots)$, that for any substitution of elements $f_i$ such that $[D(f_i)]_o = 0$, and any string of quantifiers $Q_1$ we have:

$$(**) \qquad [(Q_1^* \ldots)\beta^*(f_i)]_o = 1 \quad \text{iff} \quad M \models (Q_1 \ldots)\beta(a(f_i))$$

where $a(f_i)$ is as in (17) and

$$\beta = \wedge\, (\wedge\, x_iPy_i \rightarrow \vee\, u_jPv_j)$$

$$\beta^* = \wedge\, (\wedge\, u_jPv_j \rightarrow \vee\, x_iPy_i \vee E)$$

For the proof observe that the empty string of quantifiers (**) follows from (14), (18) and (20). The induction cases follow from the definitions of $\forall^*$, $\exists^*$ and (14). The definition of $\forall^*$, $\exists^*$ ensures that $\forall^*$, $\exists^*$ are $\forall$ and $\exists$ relativized to $\{x \mid [D(x)]_o = 0\}$.

Now for the other direction, let $(S, R, o, \mathbf{D}_t)$ be a model of $T_1$ where $\alpha^*$ is false. Then for some $t \in S$, $u \in \mathbf{D}_t$ we have

(23) $[(Q^* \ldots) \wedge (\wedge\, u_jPv_j \rightarrow \vee\, x_iPy_i \vee E)]_t = 1$ and $[D(u) \vee E]_t = 0$.

Let $M = \{x \in \mathbf{D}_t \mid [D(x)]_t = 0\}/\equiv$, (i.e., equivalence classes over $\equiv$ of (11)).

Clearly $\bar{M}$ is nonempty. Let for $x/\equiv,\, y/\equiv\, \in \bar{M}$, $x/\equiv \bar{P}\, y/\equiv$ iff $[xPy]_t = 0$. By (12) and (11) this definition is okay. We now have to show that $\bar{P}$ is reflexive and symmetric on $\bar{M}$. (12) shows that $\bar{P}$ is symmetrical. For the reflexivity observe that by (10) if $x/\equiv\, \in \bar{M}$ then $[B(x)]_t = 0$ and so by (9) and (12), $[xPx]_t = 0$.

So $(\bar{M}, \bar{P})$ is a model of $RS$. We want to show now that $(\bar{M}, \bar{P}) \models \alpha$. To show this we observe that $[E]_t = 0$ and use (23) and the same argument we used in the proof of (**) of (Lemma 22).

Thus the proof that $T_1$ is undecidable is completed.

*Remarks.* Note that in the translation $\alpha \mapsto \alpha^*$ of $RS$ in $T_1$ we have not used the ordering $<$. Also note that the structure $(S, R, o, \mathbf{D}_t)$ that was constructed is a structure of constant domains (i.e., validating $CD = HPC + \text{schema } \forall x(A \vee B(x)) \rightarrow A \vee \forall x B(x)$. Thus we get that the theory of linearly ordered abelian group in the logic $CD$ is undecidable. Also since $<$ was not in the translation, the proof presented works for the theory $T_1^*$ of abelian groups with axioms

E1–E4, S1–S5 and A1–A5. Now for the case of *HPC*, the undecidability of the theory of abelian groups with decidable equality (and hence $x \# y$ can be taken as $x \neq y$!) was proved in Section 1. However, the theory of abelian group with decidable equality with *CD* as the underlying logic is decidable. We see here, in comparison, that if we formulate the theory of abelian groups with # and decidable equality (S5), then the *CD* theory is undecidable. To summarize

THEOREM B. *In the logic CD, the theory of abelian groups formulated without # (but with decidable equality) is decidable, but when formulated with # is undecidable.*

*Algebraic Closed Fields*

The language of these theories contains $+, \cdot, =, \#, 0, 1$, (as before, $=$ is definable from #). The axioms are the following:

Axioms for the theory $T_2$ of algebraically closed fields are E1–E4, S1–S4, A1–A5, and the following:

Group M:

M1: $xy = yx$

M2: $x1 = x$

M3: $x(y + z) = xy + xz$

M4: $x \# 0 \rightarrow \exists y(xy = 1)$

M5: $x \# y \wedge z \# o \rightarrow xz \# yz$

Add: S7 to Group S

S7: $0 \# 1$

Group AC:

AC1: $(\forall x_1, \ldots, x_n) \neg\neg \exists y \left( y^{n+1} + \sum x_i y^i = o \right)$

(This is a schema; $n \in \omega$)

THEOREM C. *The theory $T_2$ of algebraically closed fields is undecidable.*

The decidability of the theory with the more natural axiom schema,

AC2: $(\forall x_1, \ldots, x_n)\exists y(y^{n+1} + \Sigma x_i y^i = 0)$ is still open.

We now turn to prove that $T_2$ is undecidable; we use similar constructions as before. Let $(M, P)$ be an infinite model of a classical reflexive and symmetric relation $P$. We construct a Kripke structure $(S, R, o, \mathbf{D}_t)$ of the theory $T_2$, called the Kripke structure associated with $(M, P)$ as follows:

(24) Let $S$ be the same as in (1)

(25) Let $R \subseteq S \times S$ be the reflexive and transitive closure of the following relation:

$$\{M\} \times S \cup \{(\{m, n\}, \{m\}) \mid \{m, n\} \in S\} \cup$$
$$\cup \{(\{m, n\}, \infty) \mid \{m, n\} \in S, m \neq n\}.$$

(26) Let $o = \{M\}$.

For each $t \in S$, we must define the classical models $\mathbf{D}_t$. These are going to be certain rings. To describe these rings we need more definitions.

(27) Let $K$ be an algebraically closed field. Let $Y$ be a set of indeterminates over $K$. We want to define a ring $K\{Y\}$ as follows. The elements of $K\{Y\}$ are all finite sums of elements from $\bigcup_{y \in Y} K[y]$, where $K[y]$ is the ring of polynomials in $y$ over $k$. So if $q \in K\{Y\}$, then $q$ has the form $\Sigma\, q_i(y_i)$, $q_i(y_i) \in K[y_i]$. Define addition on $K\{y\}$ as addition of sums, i.e., $q + q^1 = \Sigma\, q_i(y_i) + \Sigma\, q_j^1(y_j^\prime)$.

Define multiplication on $K\{Y\}$ by extending the following multiplication table to all the elements of $K\{Y\}$ through the distributive and associative laws:

$$y \cdot y^1 = \begin{cases} y^2 & \text{if} \quad y = y^1 \\ o & \text{if} \quad y \neq y^1 \end{cases}$$

for $y$, $y^1 \in Y$.

Let $y$, $y^1 \in K\{Y\}$, we can define a homomorphism $h_K(y, y^1): K\{Y\} \to K\{\{y, y^1\}\}$ by letting, for $q = \Sigma\, q_i(y_i)$, $h(y, y^1)\,(q) =$ result of substituting $o$ in $q$ for all $x \in Y - \{y, y^1\}$. Let $h_k(\infty): K\{Y\} \to K$ be defined by the substitution of $o$ for all the $y \in Y$. Let $\bar{K}\{Y\}$ be the set of all elements of the form $r = q/q^1$ (i.e., rational functions) with $q$, $q^1 \in K\{Y\}$ and $q^1$ is nonzero; i.e., $h_K(\infty)\,(q^1) \neq o$. Extend the definitions of multiplication and addition to $\bar{K}\{Y\}$ in the natural way. We can also extend $h_K(\infty)$ and $h_K(y, y^1)$ in a natural way.

We are now in a position to define the domains and models for $\mathbf{D}_t$.

(28) Proceed as follows:

(a) Let $D_o = D_{\{m, n\}}\ m \neq n = D_\infty = \bar{K}\{Y\}$, where $Y$ is a set of transcendentals of the form $Y = \{\bar{m} \mid m \in M\}$.

(b) Let $D_{\{m\}} = \overline{K(m)}\{Y - \{\bar{m}\}\}$, where $\overline{K(m)}$ is the algebraic closure of $K[\bar{m}]$.

Let us define the extensions of $+, \cdot, \#, =, 0, 1$ in $\mathbf{D}_t$.

(29) Let $+$ and $\cdot, 0, 1$ be as defined for the ring $K\{Y\}$ in (27).

(30) To define $\#$, let $r$, $r^1$ be elements of the domain; let

(a) $[r \# r^1]_t = 1$ iff $h_K(\infty)(r) \neq h_K(\infty)(r^1)$ for $t = \infty, 0, \{m, n\}, m \neq n$.

(b) $[r \# r^1]_{\{m\}} = 1$ iff $h_{K(m)}(\infty)(r) \neq h_{K(m)}(\infty)(r^1)$.

(31) To define $=$ on $\mathbf{D}_t$ let

(a) $[r = r^1]_o = 1$ iff $r$, $r^1$ are identical

(b) $[r = r^1]_\infty = 1$ iff $h_K(\infty)(r) = h_K(\infty)(r^1)$

(c) $[r = r^1]_{\{m,n\}} = 1$, for $m \neq n$, iff $h_K(\bar{m}, \bar{n})(r) = h_K(\bar{m}, \bar{n})(r^1)$

(d) $[r = r^1]_{\{m\}} = 1$ iff $h_{K(m)}(\infty)(r) = h_{K(m)}(\infty)(r^1)$.

Thus the definition of $(S, R, o, \mathbf{D}_t)$ is completed.

LEMMA 32. *The Kripke structure thus defined is a structure of the theory $T_2$ of algebraically closed fields.*

*Proof.* We check each axiom. (a) E1–E4 follow from the fact that $h$ is homomorphism. S1–S2 are simple to check. For S3, notice that if $\Sigma\, g_i(\bar{n}_i) \neq \Sigma\, g_i^1(\bar{n}_i)$ then for some $i$, $g_i \neq g_i^1$ and so $[g_i = g_i^1]_{\{\bar{n}_i\}} = 0$. Equality is defined in such a way that it amounts to substituting 0 for some of the indeterminates, e.g., $D_{\{m\}}$ is nothing but $\bar{K}(m)$. S4 also holds because $h$ is a homomorphism. A1–A5 clearly hold. So do M1–M3. To check M4, note that $x \# o$ iff $h(x) \neq o$ and then $1/x$ is also present, by construction. M5 and S7 are also easy to verify. AC1 holds because $D_{\{m\}}$ is isomorphic with $K(\bar{m})$ which is algebraically closed and $D_\infty$ is isomorphic with $K$.

Thus Lemma 32 is proved.

To proceed with the translation we need more definitions:

(33) Let $E$ be

$$(\forall x_1, x_2, x_3)\Big(\bigwedge_{i \neq j} x_i x_j = 0 \rightarrow \bigvee_j (x_j = o \vee x_j \# o)\Big).$$

(34) Let $B(x)$ be

$$\forall u, v[u \# o \vee v \# o \vee (u + v = x \wedge uv = 0 \rightarrow u = o \vee v = o] \rightarrow x \# o \vee E$$

(35) Let $x \equiv y$ be

$$\exists z(z\#o \wedge x = yz).$$

(36) Let $D(x)$ be

$$\forall y[(y\#o \leftrightarrow x\#o) \to E \vee \exists z(y = zx)] \to B(x) \vee E$$

(37) Let $xPy$ be

$$(E \to x\#o \vee y\#o \vee \forall x(x = o \vee x\#o)) \wedge (x \equiv y \to E).$$

LEMMA 38. *In the above structure:*

(a) $[E]_t = 0$ *iff* $t = o$

(b) $[B(x)]_o = 0$ *iff x has the form* $\bar{m}q(\bar{m})/r$ *with* $\bar{m} \in Y$ *and* $q(\bar{m})$ *a polynomial in* $\bar{m}$.

(c) $[D(x)]_o = o$ *iff* $x = \bar{m}a$, $\bar{m} \in Y$, $a\#o$.

(d) *For* $\bar{m}$, $\bar{n} \in Y$

*If* $a\#o$, $b\#o$ *then* $[a\bar{m}Pb\bar{n}]_o = o$ *iff* $M \models mPn$.

*Proof.* (a) For $t = o$, let $\bar{m}_i$, $i = 1$–$3$, be three distinct transcendentals; (We assumed $M$ was infinite.) Then clearly the antecedent of $E$ holds at 0 but the consequent does not. Now assume that $t \neq 0$ and $[\bigwedge_{i\neq j} x_i \cdot x_j = 0]_t = 1$. Then for $t = \{k\}$ or $t = \infty$, clearly $[\bigvee_j x_j = 0]_t = 1$, since $\mathbf{D}_t$ is a field. In the case of $t = \{n, n^1\}$ $n \neq n^1$, observe that the only elements $x$ that are not 0 are of the form $(q(\bar{n}) + q^1(\bar{n}^1))/r$. Since $[(q(\bar{n}) + q^1(\bar{n}^1))/r\#o]_t = 0$ we must have that $q = \bar{n}p$, $q^1 = \bar{n}^1p^1$, with $p$, $p^1$ polynomials in $\bar{n}$, $\bar{n}^1$ respectively. Now if $x_i = (\bar{n}p_i + \bar{n}^1p_i^1)/r_i$, then $x_1x_2 = (\bar{n}^2p_1p_2 + \bar{n}^{12}p_1^1p_2^1)/r_1r_2$ so $p_1p_2 = 0$, $p_1^1p_2^1 = 0$. Also, $p_2p_3 = p_2^1p_3^1 = p_1p_3 = p_1^1p_3 = 0$. This clearly implies that at least one of the $x_i$'s is 0.

(b) First, if $x$ has the form $\bar{m}q(\bar{m})/r$, then $[x\#o \vee E]_o = 0$. Now if $u$, $v$ are such that $(u\#o \vee v\#o]_o = 0$, then $u = \Sigma\, \bar{n}_iq(\bar{n}_i)/r$ and $v = \Sigma\bar{n}_i^1q^1(\bar{n}_i)/r^1$. We want to show that $[u + v = x \wedge u \cdot v = o \to u = o \vee v = o \vee E]_o = 1$. So assume $[u + v = x \wedge u \cdot v = o]_o = 1$. Since $[u \cdot v = 0]_o = 1$ we must have that $n_i \neq n_j^1$ for all $i$, $j$. But then since $[u + v = x]_o = 1$, either $u = x$ or $v = x$, i.e., $u = o$ or $v = o$. At any $t \neq o$, $[E]_t = 1$ and so $[u = o \vee v = o \vee E]_t = 1$. For the other direction, let $x$ be such that $[B(x)]_o = o$. By (38a) we must have that $[x\#o]_o = o$ and [consequent of $B]_o = 1$. So $x$ must have the form $\Sigma\bar{n}_iq(\bar{n}_i)/r$. If there is more than one $\bar{n}_i$ present, let $u = \bar{n}_1q(\bar{n}_1)/r$, $v = \Sigma_{i\geq 2}\bar{n}_iq(\bar{n}_i)/r$ then clearly $[u\#o \vee v\#o]_o = o$, $[u + v = x]_o = 1$, $[u \cdot v = o]_o = 1$ and $[u = o \vee v = o \vee E]_o = o$. So the antecedent of $B$ does not hold at $o$. Thus $x$ must have the form $\bar{n}_1q(\bar{n}_1)/r$.

(c) First observe that if $x$ is of the form $\bar{m}q(\bar{m})/r$ and $y\#o \leftrightarrow x\#o$, then $y$ must be of the form $\bar{m}q_1(\bar{m})/r$. Now assume that $[D(x)]_o = o$, then we show that $x \equiv \bar{m}$, for some $\bar{m}$. Now since $[D(x)]_o = o$, we get that $[B(x)]_o = o$ and therefore $x = \bar{m}^i q(\bar{m})/r$ with $q(\bar{m})\#o$ and $i \geqslant 1$. If $i = 1$, we are finished. Otherwise let $y = \bar{m}^{i-1}q/r$. Clearly $y\#o \leftrightarrow x\#o$, so we must have that either $x \equiv y$, which is impossible, or that for some $z$, $\bar{m}^{i-1}q/r = z\,\bar{m}q/r$, which again is impossible.

Now assume that $x \equiv \bar{m}$. Then certainly $[E \vee B(x)]_o = o$. To show that the antecedent of $D(\bar{m})$ holds at $o$, assume $[y\#o \leftrightarrow x\#o]_o = 1$. Then $y$ must be of the form $\bar{m}q(\bar{m})/r$ and therefore $\exists z(y = xz)$.

(d) First assume that $M \models mPn$. We distinguish two cases.

(1) $m = n$, then $[\bar{m}aP\bar{m}b]_o = 0$ since $[m\bar{a} \equiv \bar{m}b]_o = 1$.

(2) If $m \neq n$, then we have $[E]_{\{m,n\}} = 1$ and $[\bar{m}a\#o \vee \bar{m}b\#o]_{\{m,n\}} = 0$, and $[\forall x(x = o \vee x\#o)]_{\{m,n\}} = 0$. So again, $[\bar{m}aP\bar{n}b]_o = 0$.

Now assume that $[\bar{m}aP\bar{n}b]_o = 0$. If $[\bar{m}a \equiv \bar{n}b \rightarrow E]_o = 0$, then we must have $[\bar{m}a \equiv \bar{n}b]_o = 1$, which can hold only if $m = n$ and so $mPn$. If $[E \rightarrow \bar{m}a\#o \vee \bar{n}b\#o]_o = 0$, then for some $t \neq o$, $(\bar{m}a\#o)_t = 0$, $[\bar{n}b\#o]_t = 0$, $[\forall x(x = o \vee x\#o)]_t = 0$. Since $[\forall x(x = o \vee x\#o)]_t = 0$, $t \neq \infty$, and $t \neq \{n^1\}$. So $t = \{m^1, n^1\}$, $m^1 \neq n^1$. But then $[\bar{m}a\#o]_t = 0$ implies $m \in \{m^1, n^1\}$ and similarly $n \in \{m^1, n^1\}$ and so $mPn$ must hold in $M$.

Thus Lemma 38 is proved.

(39) Now observe that from (38d) it follows that if $x \equiv y$, $x^1 \equiv y^1$ then $xPy$ iff $x^1Py^1$. Thus if $\bar{M} = \{x \mid [D(x)]_o = 0\}/\equiv$ and $\bar{P}$ is defined by $x/\equiv\ \bar{P}\ y/\equiv$ iff $[xPy]_o = 0$, we get that $(M, P)$ is isomorphic to $(\bar{M}, \bar{P})$.

(40) As we did in (21), (for the theory $T_1$ of linearly ordered abelian groups), we define $\forall^*$ and $\exists^*$ by

$$(\forall^* y)\alpha = \forall y(D(y) \vee \alpha)$$

$$(\exists^* y\alpha) = \exists y((D(y) \rightarrow E) \wedge \alpha).$$

The translation $\alpha \mapsto \alpha^*$ is defined exactly as in (21), i.e., if $\alpha$ is a formula of $RS$ of the form $(Q\ldots) \wedge (\wedge\, x_iPy_i \rightarrow \vee\, u_jPv_j)$ then $\alpha^*$ is the following formula of $T_2$: $\forall u[(Q^*\ldots) \wedge (\wedge\, u_jPv_j \rightarrow \vee\, x_iPv_i \vee E) \rightarrow E \vee D(u)]$ where $P$, $D$, $E$ are those defined in (33–37); We have:

LEMMA 41. *$T_2 \vdash \alpha^*$ iff $RS \vdash \neg\alpha$.*

*Proof.* First assume that $RS \nvdash \neg\alpha$ so $\alpha$ has a structure $(M, P)$. We may assume that $M$ is infinite because $RS$ contains no equality. Construct the structure $(S, R, o, \mathbf{D}_t)$ associated with $(M, P)$. Observe, by following the lines of argument of the proof of (22) that $[\alpha^*]_o = o$. The proof of (22) used (14), (18) and (20). Our proof now will use (38)–(39), which were defined for the same purpose.

For the other direction, let $(S, R, o, \mathbf{D}_t)$ be a structure in which $\alpha^*$ is false. Obtain a structure of $\alpha$ by following the line of the proof following (23). The reflexivity and symmetry of $\bar{P}$ follow from (37).

Thus Theorem C is proved.

We turn now to the theory $T_3$ of real closed fields. The language of $T_3$ contains the symbols $+, \cdot, =, \#, 0, 1$ and $>$. The axioms are the following:

Axioms for the theory $T_3$ of real closed fields are E1–E4, S1–S4, A1–A5, M1–M5, S7 and Groups P and RC.

P1: $x \# o \rightarrow x > o \vee (-x) > 0$

P2: $x > o \wedge y > 0 \rightarrow x + y > 0$

P3: $x > 0 \wedge y > 0 \rightarrow x \cdot y > 0$

P4: $\neg(0 > 0)$

RC1: $x \# o \rightarrow \neg\neg\exists y(x = y^2 \vee -x = y^2)$

RC2: $(\forall x_o, \ldots, x_{2n})\neg\neg\exists y\left(y^{2n+1} + \sum x_i y^i = o\right).$

We do not know whether $T_3$ is decidable or not. Neither do we know whether the theory with RC3 and RC4 is undecidable, where

RC3: $x \# o \rightarrow \exists y(x = y^2 \vee -x = y^2)$

RC4: $(\forall x_o, \ldots, x_{2n})\exists y\left(y^{2n+1} + \sum x_i y^i = 0\right).$

*Linear Order*

We begin with the theory $T_4$ of dense linear ordering. This theory was first proved undecidable by Smorynski by interpreting in it the intuitionistic theory of one monadic letter. In Section 2 we show that the theory of decidable linear ordering is undecidable (here $\#$ can be defined from $<$).

The language of $T_4$ contains $<$, $\#$, $=$, and the following axioms: E1–E4, S1–S5, O1, O3, O5, O6, S6 and O7, O8, O9 below:

O7: $\exists y(x<y)$

O8: $\exists y(y<x)$

O9: $\exists y(x<z \rightarrow x<y \wedge y<z)$.

THEOREM D. *$T_4$ is undecidable.*

We sketch the *proof*; we use the same method we used bofore.

Let $(M, P)$ be a countably infinite model of a reflexive and symmetric relation. We can assume that $M$ is the set of rational numbers. We construct a Kripke structure $(S, R, o, \mathbf{D}_t)$ called the Kripke structure associated with $(M, P)$ as follows:

(42) $S=\{M\} \cup\{\infty\} \cup\{\{x, y\} \mid x P y\}$, let $o=\{M\}$.

(43) $R=\{M\} \times S \cup S \times\{\infty\} \cup\{(t, t) \mid t \in S\}$.

Let $K$ be the rational numbers. Let $\mathbf{D}_t=K \times M$. We have to define $<$, $\#$, $=$ on $\mathbf{D}_t$, $t \in S$. We proceed as follows:

(44) $[(r, m)=(r^1, m^1)]_t=1$ iff $r=r^1$, $m=m^1$.

(45) $[(r, m)<(r^1, m^1)]_o=1$ iff $m<m^1$.

(46) $[(r, m)<(r^1, m^1)]_\infty=1$ iff $m<m^1$ or if $m=m^1$ then $r<r^1$.

(47) $[(r, m)<(r^1, m^1)]_{\{n, n^1\}}=1$ iff $m<m^1$ or if $m=m^1 \in \{n, n^1\}$ then $r<r^1$.

Let $x \# y$ mean $x<y \vee y<x$.

LEMMA 48. *All axioms of $T_4$ hold in the structure.*

LEMMA 49. *$[x \# y]_o=o$ is an equivalence relation in the structure.*

*Proof.* Transitivity follows from S4.

LEMMA 50. *Let $E$ be the sentence $\exists x \forall y(x=y \vee x \# y)$, then $[E]_t=o$ iff $t=o$.*

LEMMA 51. *If* $xPy$ *is defined as* $E \to (\forall z(z = x \vee z \# x) \leftrightarrow \forall z(z = y \vee z \# y))$ *then* $[(r, m)P(r^1, m^1)]_o = o$ *iff* $mPm^1$.

LEMMA 52. *The relation* $x \equiv y$ *defined below is an equivalence relation on* $D_o$ *and furthermore,* $x \equiv x^1 \wedge y \equiv y^1 \wedge xPy \to x^1Py^1$, *where* $x \equiv y$ *is* $\forall z(z \# x \leftrightarrow z \# y)$. *In fact, the set* $\bar{M}$ *of equivalence classes with the relation* $\bar{P}$ *(induced by* $[xPy]_o = o$*) is isomorphic to* $(M, P)$.

Now let $\alpha$ be a sentence of $RS$ in the form $(Q\ldots) \wedge (\wedge\, x_iPy_i \to \vee\, u_jPv_j)$. Let $\alpha^*$ be $(Q\ldots) \wedge (\wedge\, u_jPv_j \to \vee\, x_iPy_i \vee E) \to E$.

We claim that

$$T_4 \vdash \alpha^* \quad \text{iff} \quad RS \vdash \neg\alpha.$$

*Remark.* This translation and construction shows that the apartness relation itself (E1–E4, S1–S5) is undecidable. Simply take the same constructions for a language without <.

## 5. Undecidability of some finite extensions of the theory of abelian groups

DEFINITION 1. Let $T$ be the theory of abelian groups with decidable equality formulated in Heyting's predicate calculus *HPC*; that is, the language of $T$ has 0, +, = and the axioms of $T$ are

(a) $x + y = y + x$

(b) $(x + y) + z = x + (y + z)$

(c) $x + 0 = x$

(d) $\forall x \exists y(x + y = 0)$

(e) $x = y \vee \neg x = y$

(f) $x = x$

(g) $x = y \to y = x$

(h) $(x = y) \wedge (y = z) \to x = z$

(k) $x = y \to (A(x) \to A(y))$, $A$ any wff.

In Section 2 it was shown that the Theory $T$ is undecidable. It is of interest to examine the decision problem for various finite extensions of $T$. We prove the following:

THEOREM 2. *Let $G_m$ be the extension of $T$ with the axiom $\forall x(mx = 0)$. Then $G_m$ is decidable iff $m$ is the product of pairwise different prime numbers.*

For comparison, in the next chapter we show:

THEOREM 3. *Let $G_m^*$ be the extension of $G_m$ with the axiom*

$$\forall x\Big(\bigwedge_{1\leqslant n\leqslant m}(\exists y(x=ny)\vee\neg\exists y(x=ny))\Big)$$

*Then $G_m^*$ is decidable.*

In this section we prove that if there exists a prime $p$ such that $p^2 | m$, then $G_m$ is undecidable. Our method is to faithfully interpret in $G_m$, the classically undecidable theory $RS$ of a reflexive and symmetric relation. We can assume that $m = p^2$, since $G_{p^2}$ is a finite extension of $G_m$, and if $G_{p^2}$ is undecidable, so is $G_m$.

DEFINITION 4. Let $(\mathbf{A}, <)$ be a countable structure of $RS$. We shall construct a Kripke structure $(S, \leqslant, \emptyset, \mathbf{G}_t) \in \mathbf{M}_m$, called the Kripke structure associated with $(\mathbf{A}, <)$. Assume that $\mathbf{A} \subseteq$ set of natural numbers and that $\langle , \rangle$ is a coding function for pairs of natural numbers.

(a) Let $S = \{0\} \cup \{(\langle k, n\rangle) | k, n \in \mathbf{A}, k < n\}$.

(b) To define $\mathbf{G}_t$, $t \in S$, let $p^2 | m$. Let $C_k$ denote the cyclic group of order $k$. Let $\mathbf{G}_\emptyset = \Sigma_{a\in A}\mathbf{G}_a^\emptyset$, with $\mathbf{G}_a^\emptyset = \mathbf{C}_p$, $a \in \mathbf{A}$.

$$\mathbf{G}_{(\langle k,n\rangle)} = \sum\nolimits_{a\in\mathbf{A}} \mathbf{G}_a^{(\langle k,n\rangle)},$$

$$\text{with}\quad \mathbf{G}_a^{(\langle k,n\rangle)} = \begin{cases} C_p & a \neq k, n \\ C_{p^2} & a = k \quad\text{or}\quad a = n. \end{cases}$$

(c) We have now to indicate in what manner is $\mathbf{G}_\emptyset$ identified as a subgroup of $\mathbf{G}_{(\langle k,n\rangle)}$. The isomorphism is

$$f \in \mathbf{G}_\emptyset \mapsto f^* \in \mathbf{G}_{(\langle k,n\rangle)}$$

$$\text{defined by}\ :\ f^*(a) = \begin{cases} f(a) & a \neq k, n \\ pf(a) & a = k \quad\text{or}\quad a = n; \end{cases}$$

for $a \in \mathbf{A}$; i.e., we identify $C_p$ of $\mathbf{G}_\emptyset$ with the subgroup (of $C_{p^2}$), generated by $p$.

Thus the Kripke structure associated with $(\mathbf{A}, <)$ has been defined.

DEFINITION 5. We define a list of formulas of $G_m$ to be used in the syntactical translation of $RS$ into $G_m$, $m = p^2$.

(a) $E =_{\text{def}} \exists x(px \neq 0)$.

(b) $x \equiv y =_{\text{def}} \bigvee_{1 \leq n < p} (ny = x \vee nx = y)$

(c) $E^1 =_{\text{def}} E \wedge \forall u, v(pu \neq 0 \wedge pv \neq 0 \rightarrow pu \equiv pv)$.

(d) $D(x) =_{\text{def}} \neg(E^1 \wedge \exists u(x = pu)) \vee x = 0$.

(e) $xRy =_{\text{def}} \neg(E \wedge \exists u(x = pu) \wedge \exists u(y = pu))$.

DEFINITION 6. Let $(S, \leqslant, \emptyset, \mathbf{G}_t)$ be a Kripke structure. Let $A(x_1, \ldots, x_n)$ be a formula and let $a_1, \ldots, a_n \in \mathbf{G}_t$. We denote the truth value of $A$ at the point $t$ under the indicated assignment by $[A(a_1, \ldots, a_n)]_t$.

LEMMA 4. *In the structure associated with* $(\mathbf{A}, <)$ *we have*:

(a) $[E]_t = 1$ *iff* $t \neq \emptyset$.

(b) $[E^1]_t = 1$ *iff* $t = (\langle n, n \rangle)$, *for some* $n \in \mathbf{A}$.

(c) $[D(x)]_\emptyset = 0$ *iff for some* $a_x \in \mathbf{A}$ *we have that for all* $b \in \mathbf{A}$, $(x(b) \neq 0$ *iff* $b = a_x)$.

(d) *Let* $x, y \in \mathbf{G}_\emptyset$ *be such that* $[D(x) \vee D(y)]_0 = 0$. *Then* $[xRy]_0 = 0$ *iff* $a_x < a_y$.

*Proof.* (a) and (b) are obvious from the construction. (c) if $x$ is such that for some $a_x$, $x(b) \neq 0$ iff $b = a_x$, for all $b \in \mathbf{A}$, then clearly in $(\langle a_x, a_x \rangle)$, $E^1$ holds, and so does $\exists y(x = py)$. It is also clear that $x \neq 0$. Now suppose $[D(x)]_\emptyset = 0$. Then $x \neq 0$ and for some $n \in \mathbf{A}$, $[\exists y(x = py)]_{(\langle n, n \rangle)} = 1$. But then it is clear that for $b \neq n$, $py(b) = 0$. Thus for all $b \in \mathbf{A}$, $x(b) \neq 0$ iff $b = n$. (d) Let $x$, $y$ be such that $[D(x) \vee D(y)]_\emptyset = 0$. Then by construction $[\exists u(x = pu) \wedge \exists u(y = pu)]_t = 1$ iff $t$ has the form $(\langle a_x, a_y \rangle)$ and $(\langle a_x, a_y \rangle) \in S$ iff $a_x < a_y$.

LEMMA 5. *For* $x_i$, $y_i$, $i = 1, 2$, *such that* $[D(x_i) \vee D(y_i)]_\emptyset = 0$ *and* $[x_i \equiv y_i]_\emptyset = 1$ *we have that* $[x_1 R x_2]_\emptyset = [y_1 R y_2]_\emptyset$.

*Proof.* Follows from Lemma 4(d).

LEMMA 6. **A** *can be identified with the set* $\{x \in \mathbf{G}_\emptyset \mid [D(x)]_\emptyset = 0\}|\equiv$ *and* $<$ *can be identified with the relation* $[xRy]_\emptyset = 0$.

*Proof.* Follows from Lemmas 4 and 5.

We are now in a position to describe the translation from the theory $RS$ of a reflexive and symmetric relation $<$ to the theory $G_m$. First, a definition:

Let $\forall^*$, $\exists^*$ be the quantifiers of $G_m$ defined by:

$$\forall^* yA =_{\text{def}} \forall y(D(y) \vee A)$$
$$\exists^* yA =_{\text{def}} \exists y(A \wedge (D(y) \rightarrow E)).$$

Now let $\alpha$ be a sentence of $RS$. Write $\alpha$ in the form

$$(*) \qquad \alpha = (Q \ldots) \wedge (\wedge x_i < y_i \rightarrow \vee\, u_j < v_j)$$

where $Q$ is a string of quantifiers. Let $\alpha^*$ be the following sentence of $G_m$:

$$\alpha^* = \forall u\{[(Q^* \ldots) \wedge (\wedge\, u_j R v_j \rightarrow \vee\, x_i R y_i \vee E)] \wedge$$
$$\wedge\, [\forall x(xRx \rightarrow D(x) \vee E)] \rightarrow [E \vee D(u)]\}$$

LEMMA 7. *$G_m \vdash \alpha^*$ iff $RS \vdash \neg\alpha$.*

*Proof.* (1) Assume $RS \nvdash \neg\alpha$.

Let $(\mathbf{A}, <)$ be a structure of $\alpha$. Construct the Kripke structure $(S, \leqslant, \emptyset, \mathbf{G}_t)$ associated with $(\mathbf{A}, <)$. In this structure $[\alpha_1]_\emptyset = 1$, where $\alpha_1 = (Q^* \ldots) \wedge (\wedge\, u_j R v_j \rightarrow \vee\, x_i R y_i \vee E)$.This follows from Lemmas 5 and 6, and from the details of the proofs of Lemmas 8 and 13 of Section 1. We can therefore show that also $[\alpha^*]_\emptyset = 0$.

(2) Assume that $G_m \nvdash \alpha^*$.

Let $(S, \leqslant, \emptyset, \mathbf{G}_t) \in \mathbf{M}_m$ be a structure where $\alpha^*$ is false. We can assume that for some $u \in \mathbf{G}_\emptyset$, $[\alpha_1]_\emptyset = 1$ and $[E \vee D(u)]_\emptyset = 0$.

Let $A = \{u \in \mathbf{G}_\emptyset \mid [D(u)]_\emptyset = 0\}$. Now since $[\exists y(py \neq 0)]_\emptyset = 0$, we get that the relation $[x \equiv y]_\emptyset = 1$ is an equivalence relation on $A$. So let $\mathbf{A} = A|\equiv$.

Lemma 5 also holds for $\equiv$ and therefore we can define $<$ on $A$ by $x|\equiv\, <\, y|\equiv$ iff $[xRy]_\emptyset = 0$. Now an argument similar to the one given for (1) above, and in the proofs of Lemmas 8 and 13 of Section 1 establishes that $(\mathbf{A}, <) \models \alpha$.

COROLLARY. *If $m$ is not a product of pairwise different prime numbers, then $G_m$ is undecidable.*

## 6. Undecidability of Second Order Propositional Calculus

For notation and terminology, see Chapter 9.

To show that $C2I$ is undecidable, we interpret in it the undecidable classical theory of a reflexive and symmetric relation. The interpretation is given semantically and is faithful.

Let $(M, <)$ be a countably infinite structure of a reflexive and symmetric relation. We shall use this structure to define a $C$-structure of $C2I$ as follows:
Let $S = \{M\} \cup M \cup \{\{a, b\} \subseteq M \mid a < b\}$.
Let $R = \{M\} \times S \cup \{(a, \{a, b\}) \mid a < b\} \cup \{(t, t) \mid t \in S\}$.
Let $O = M$.

Let D be the set of all subsets X of $S$ such that $t \in X \wedge tRs \rightarrow s \in X$. Let our language have enough propositional variables in it so that we can take $g$ to be a function such that $g(f) =$ empty set, and for all $X \in D$ there exists some $u \in V$, $g(u) = X$.

We now define certain formulas of the language of $C2I$. Recall that $\neg x$ is $x \rightarrow f$.

$$(1) \qquad E = (\forall x_1, x_2, x_3)(x_3 \rightarrow [((x_2 \rightarrow x_1) \rightarrow x_2) \rightarrow x_2] \rightarrow x_3)) \rightarrow x_3.$$

$$(2a) \qquad N(x) = \forall u, v((x \rightarrow u \vee v) \rightarrow (x \rightarrow u) \vee (x \rightarrow v)) \wedge (\neg x \rightarrow E).$$

$$(2b) \qquad D(x) = N(x) \wedge (x \rightarrow E) \wedge \forall y[N(y) \wedge (y \rightarrow E) \wedge (x \rightarrow y) \rightarrow (y \rightarrow x)].$$

$$(3) \qquad xPy = \neg(x \wedge y).$$

Let $\alpha$ be any statement of the reflexive and symmetric relation $<$, write $\alpha = (Q\ldots) \wedge (\wedge x_i < y_i \rightarrow \vee u_j < v_j)$. Where $(Q\ldots)$ is a string of quantifiers. We translate $\alpha$ into $\alpha^*$, in the language of $C2I$ as follows

$$\alpha^* = \exists x D(x) \wedge \{(Q^* \ldots) \wedge [\wedge u_j P v_j \rightarrow \vee x_i P y_i \vee E] \rightarrow \rightarrow E \vee \exists x(D(x) \wedge \neg x)\},$$

where $Q^*$ is the same string of quantifiers relativized to

$$D(x) (\text{i.e., } \forall^* x A = \forall x(D(x) \rightarrow A) \text{ and } \exists^* x A = = \exists x(D(x) \wedge A)).$$

Observe the following:
(a) In the model associated with $(M, <)$ we have:

(1) $\|E\|_g = S - \{M\}$.

(2a) $\|N(x)\|_g = S$ iff for some $a_x \in S$, $g(x) = \{t \mid a_x Rt\}$.

(2b) $\|D(x)\|_g = S$ iff for some $a_x \in M$, $g(x) = \{t \mid a_x Rt\}$.

(3a) $\|E \vee \exists x(D(x) \wedge \neg x)\|_g = S - \{M\}$.

(3b) For $x$, $y$ such that $D(x)$ and $D(y)$ hold at $M$ we have that: $xPy$ is false at $M$ iff $a_x < a_y$ holds at $(M, <)$.

*Proof of Observations.* (1) Notice that if $t \notin \|((x_3 \to u) \to x_3) \to x_3\|_g$ than there must be an $s \neq t$ such that $tRs$ and $s \in \|x_3\|_g$ and $s \notin \|u\|_g$. Similarly if $u = ((x_2 \to x_1) \to x_2) \to x_2$, there must be an $s' \neq s$ such that $sRs'$ with $s' \in \|x_2\|_g$ and $s' \notin \|x_1\|_g$.

Thus since it takes 3 linearly ordered points to falsify $E$ we get $\|E\|_g = S - \{M\}$.

(2a) Suppose that for some $S_0 \subseteq S$, a set of pairwise $R$-incomparable elements we have, $\|x\|_g = \{t \mid \exists s \in S_0(sRt)\}$. Then if $\bar{\bar{S}}_0 > 1$, decompose $S_0$ into nonempty $S_1$, $S_2$. There must be $y_i$, such that $g(y_i) = \{t \mid \exists s \in S_i(sRt)\}$. Certainly, $\|x \to y_1 \vee y_2\|_g = S$. $\|(x \to y_1) \vee (x \to y_2)\|_g \neq S$.

(2b) Notice that if $\|D(x)\|_g = S$, then for some $a_x \in S$, $g(x) = \{s \mid a_x Rs\}$. Now since $\|x \to E\|_g = S$, $a_x \neq M$. Can $a_x$ be an end point? i.e. $a_x = \{c, d\}$, $c$, $d \in M$? If this is the case let $y$ be such that $g(y) = \{t \mid dRt\}$. Then clearly, $\|N(y) \wedge (y \to E) \wedge (x \to y)\|_g = S$, but $\|y \to x\|_g \neq S$. Thus we cannot have $\|D(x)\|_g = S$. Therefore $a_x \in M$. The other direction of (2b) is simple.

(3a) Notice that for each $t \neq M$, $t \in \|E\|_g$ and so $t \in \|F\|_g$, where $F = E \vee \exists x(D(x) \wedge \neg x)$. Now $M \notin \|E\|_g$, and if for some $x$, $M \in \|D(x)\|_g$, then by (2b) $\|x\|_g = \{t \mid a_x Rt\}$, $a_x \in M$, and therefore $M \notin \|\neg x\|_g$. Thus $M \notin \|F\|_g$.

(3b) Follows from the construction and (1), (2a, b) (3a).

(4) We now claim that $\alpha$ is valid in $(M, <)$ iff $\|\alpha^*\|_g \neq S$. To show this observe that by (3a), $\|\alpha^*\|_g \supseteq S - \{M\}$. So $\|\alpha^*\|_g \neq S$ holds only if $\|$antecedent of $\alpha^*\|_g = S$, as $\|F\|_g = S - \{M\}$. So let us examine the antecedent of $\alpha^*$. Clearly $\|\exists x D(x)\|_g = S$. So we must check $\|(Q^* \ldots) \wedge (\ldots)\|_g$. Let us call:

$$w = \wedge (\wedge\, x_i < y_i \to \vee\, u_j < v_j)$$

$$w^* = \wedge (\wedge\, u_j P v_j \to \vee\, x_i P y_i \vee E).$$

We want to state Lemma 5, that will be used also in future argument. Let $w_1 = Q_1 w$, $w_1^* = Q_1^* w^*$ be two subformulas of $\alpha$ and of the part $(Q^* \ldots) \wedge (\ldots)$ of the antecedent of $\alpha^*$. Suppose we substitute in $w_1^*$ variables $x_1, \ldots, x_n$ (for all its free variables) such that $\|D(x_i)\|g = S$. Then we know that there exist elements $a_{x_i} \in M$ as in observation (2b). We claim that $\|w_1^*(x_1, \ldots)\|_g = S$ iff $(M, <) \models w(a_{x_1}, \ldots)$. We must also remember that for any $b \in M$, there exists an $x$ such that $\|D(x)\|g = S$ and $b = a_x$. So let us write $x$ for $a_x$, for the sake of simplicity.

LEMMA 5. *For the above convention of notation we have*:

(a) $(M, <) \models w_1$ *implies* $\|w_1^*\|_g = S$

(b) $(M, <) \not\models w_1$ *implies* $\|w_1^*\|_g = S - \{M\}$.

*Proof*. By induction. (1) We check for an arbitrary conjunct of $w$ of the form $\wedge\, x_i < y_i \to \vee\, u_j < v_j$. This conjunct holds in $(M, <)$ iff [by (3b)] the following implication (I) holds, where:

(I): If for all $j$, $\|u_j P v_j\|_g = S$ then for some $i$ $\|x_i P y_i\|_g = S$. Now since $\|E\|_g = S - \{M\}$ (*I*) holds iff $\|\wedge\, u_j P v_j \to \vee\, x_i P y_i \vee E\|_g = S$. Observe that since $E$ is inner conjunct in the antecedent of the implication, the $\|\cdots\|_g$ value is always $= S - \{M\}$.

(2) Since the lemma holds for each conjunct of $w$ it holds for $w$.

(3) Assume the lemma holds for $w_1$, and look at $(\forall x) w_1(x)$. $(M, <) \models \forall x w_1(x)$ iff for all $b \in M$, $(M, <) \models w_1(b)$ and by the induction hypothesis and the construction of the intuitionistic structure, iff for all $x$ such that $\|D(x)\|_g = S$, $\|w_1(x)\|_g = S$. Now since $\|w_1(x)\|_g \supseteq S - \{M\}$ (this by the induction hypothesis) we can continue: iff $\|\forall x (D(x) \to w_1^*(x)\|_g = S$ iff $\|\forall^* x w_1^*\|_g = S$. One can verify that $\|\forall^* x w_1(x)\|_g \supseteq S - \{M\}$ always.

(5) The case of $\exists$ is treated similarly.

Thus Lemma 5 is proved and therefore (4) follows.

(6) Now observe that $C2I \vdash \alpha^*$ iff $RS \vdash \neg\alpha$ where $RS$ is the theory of reflexive and symmetric relation with infinitely many elements.

To prove this observe that one direction follows from 4. If $RS \nvdash \neg\alpha$, then let $(M, <)$ be a structure where $\alpha$ is true and so by construction, the structure associated with $(M, <)$ is a structure of $C2I$ where $\alpha^*$ is false. Now for the other direction, let $(S, R, o, \mathrm{D}, g)$ be a structure where $\alpha^*$ is false (i.e., $\|\alpha^*\|_g \neq S$). Then there must be a $t \in S$ such that $t \in \|$antecedent of $\alpha^*\|_g$ and $t \notin \|F\|_g$. Let $M = \{g(x) \mid t \in \|D(x)\|_g\}$. Clearly $M$ is non empty. Define $<$ on $M$ by $x < y$ iff $t \notin \|xPy\|_g$. $<$ is reflexive and symmetric. To show that $(M, <) \models \alpha$, observe that $t \notin \|E\|_g$ and that $t \in \|(Q^* \ldots)w^*\|_g$. An induction argument for $w_1^*(x_1, \ldots, x_n)$, $x_i \in M$, $w_1^*$ as in (4) above, shows that $(M, <) \models w_1$ if $t \in \|w_1^*\|_g$.

Thus $(M, <) \models \alpha$. Though the structure may be finite it is always elementarily equivalent to an infinite structure.

## NOTES

Section 6 on the undecidability of 2nd order intuitionistic propositional calculus is based on the paper by the author in the *Archiv für Math. Logic* 1974. Sobolev 1977 pointed out that the original definition of $N(x)$ (Definition 2a in my paper of 1974) allows for the case $x = f$. Thus in the definition of $N(x)$ in this book (Section 6) the conjunct $(\neg x \to E)$ is added and the case $x = f$ is excluded. Sobolev 1977 suggests another way of correcting the gap. He changes the definition of $D(x)$ to handle the case $x = f$.

## CHAPTER 15

# DECIDABILITY RESULTS

There are several methods for proving the decidability of a logical system. The first useful method for propositional calculi is via the finite model property. A theorem of Harrop states:

THEOREM 0. *If a logical system x has the f.m.p. and is finitely axiomatizable then it is decidable.*

*Proof.* Since we can recursively enumerate both the theorems and non-theorems of $x$.

As corollaries we get, since $h$ has the f.m.p., that $h$ is decidable and so are all the extensions of $h$ of previous chapters that were shown to have the f.m.p. and are finitely axiomatizable. Also since every finitely axiomatizable extension of $h^+$ has the f.m.p., every such extension is decidable.

There are decidable intermediate logics that lack the f.m.p., and their decidability can be shown by the second method of proving decidability, called the reduction method, based on a theorem of Rabin. The idea is as follows:

Let $T$ be a given theory of Heyting's predicate calculus. We know that Heyting's predicate calculus is complete for the Kripke interpretation. We choose a class **M** of Kripke structures for which $T$ is complete, i.e., all axioms of $T$ are valid in any structure of the class and whenever $A$ is not a theorem of $T$, $A$ is false in some structure of **M**. Perhaps a class of structures **M** can be chosen such that it can be described in a convenient way (convenient in the sense of the purpose mentioned below) in a classical language (possibly second order with additional relation symbols). Call this language $L^*$. Let $\Theta^*$ be a theory of $L^*$ such that one is able to associate with any $A$ of the language of $T$ an $A^*$ of the language of $L^*$ that actually says that '$A$ is true in all structures of **M**,' and such that:

$\Theta^* \vdash A^*$ (classically) iff $T \vdash A$ (intuitionistically). Then, under these circumstances, if $\Theta^*$ is decidable, so is $T$.

A very useful possible $\Theta^*$ is the monadic second order theory of $\aleph_0$-successor functions (denoted by $S\omega S$) and proved decidable by Rabin. The theorems below are obtained by choosing the appropriate classes **M** and presenting the translation: $A$ goes to $A^*$.

First let us quote Rabin's theorem. Let $\omega^*$ be the set of all finite sequences of natural numbers with $\Lambda$ the empty sequence. Define:

(a) For each $n$, let $f_n, g_n, d_n, e_n$ be the functions with:

$$d_n(t) = t_*(4n)$$
$$e_n(t) = t_*(4n+1)$$
$$f_n(t) = t_*(4n+2)$$
$$g_n(t) = t_*(4n+3)$$

Where $t$ is arbitrary and $_*$ denotes concatenation of sequences, and $n$ is any natural number.

(b) Let $t <_1 s$ denote the relation $s = d_n(t)$ for some $n$. Let $t <_2 s$ denote the relation $s = e_n(t)$ for some $n$, and similarly define $<_3$, $<_4$, using $f_n$ and $g_n$ respectively. Let $\leqslant_i$ denote the reflexive and transitive closure of $<_i$.

THEOREM (*Rabin*). *The monadic 2nd order theory SwS of the following model is decidable*: $(\omega^*, \leqslant, \leqslant_i, d_n, e_n, f_n, g_n, <, \Lambda)$, *where* $n = 1, 2, 3, \ldots, 1 \leqslant i \leqslant 4$, *and* $<$ *denotes the lexicographic ordering of* $\omega^*$ *and* $\leqslant$ *denotes* $\bigcup_{1 \leqslant i \leqslant 4} \leqslant_i$.

Notice that the notion of 'finite set' is definable in *SwS*. $Y$ is finite iff $Y$ is well ordered by both $<$ and its converse. Equality is also defined as $x = y$ iff $\forall C(x \in C \leftrightarrow y \in C)$.

A third method of proving decidability is a syntactical one: express the system as a Gentzen system or other convenient formulation and obtain decidability by checking this particular formulation. Some of the original decidability results were given in this way. We do not use this method in this book.

## 1. Decided monadic predicates

THEOREM 1. *The intuitionistic theory of* $\aleph_0$ *decidable monadic predicates with decidable equality is decidable. (The axioms are* $\forall x(x = y \vee x \neq y)$ *and* $\forall x(P_n(x) \vee \neg P_n(x))$ *for each monadic* $P_n$.)

To prove Theorem 1, we need a class $M_1$ of structure for which, e.g., $T_1$, the first order theory of Theorem 1 is complete. This class $M_1$ is the following (the reader can verify the completeness himself using Henkin type completeness proof):

The structure of $M_1$ are of the form $(S, R, \Lambda, \mathbf{A}_t)$, where $S \subseteq \omega^*$,

$\Lambda \in S$, in the empty sequence and the following holds:

(1) $S \subseteq \{y \mid \Lambda \leqslant_1 y\} \cup \{\Lambda\}$.

(2) $y \in S \wedge x \leqslant_1 y \rightarrow x \in S$.

(3) $R = \leqslant_1 \upharpoonright S \cup \{(y, y) \mid y \in S\}$.

(4) $\mathbf{A}_x$ is an ordinary classical model with domain $A_x$, for the language with equality $=$ and $\aleph_0$ unary predicates $P_0, P_1, \ldots$. We assume that $a = b$ iff $a$ is the same element as $b$, for any $a, b \in A_t$. We can assume this since equality is decidable (this is an axiom of $T_1$). The structure fulfils the requirements of a Kripke structure, i.e.,

(5) $tRs \rightarrow A_t \subseteq A_s$.

The fact that this is a structure of $T_1$ corresponds to the conditions:

(6) $[P(a)]_t = 0$ and $tRs$ imply that $[P(a)]_s = 0$ for atomic $P$.

We may also assume (and this assumption, unlike (6) and the assumption about $=$, can be imposed on any class of Kripke structures without changing its theory) that

(7) $tRs_1 \wedge tRs_2 \wedge \forall s(sRs_1 \wedge sRs_2 \rightarrow sRt) \rightarrow$
$\rightarrow (A_{s_1} - A_t) \cap (A_{s_2} - A_t) = \emptyset$ (empty set).

We adopt assumption (7) because of the fact that if a Kripke structure fulfils (7) then condition (6) on a predicate $P_n$ is equivalent to the assumption that

(8) There exists a set $\Pi_n \subseteq \bigcup_{x \in S} A_x$ such that, for all $x$ and $a$, $[P(a)]_x = 1$ iff $a \in \Pi_n$.

Let us put $B = \bigcup_{x \in S} A_x$.

Let us now turn to the proof of Theorem 1 and begin with showing that the class $M_1$ of structures can be defined in the language of $SwS$. Clearly the set $S$, the relation $R$ and $\Lambda$ can be defined in $SwS$, simply by taking a set variable (let us denote it by $S$ as well) such that (1) and (2) hold for $S$, and defining $R$ in terms of $S$ and $\leqslant_1$, via condition (3). We now have to define the domains $A_x$, for each $x \in S$. To do that let us define $B$ first. For $B$ we can take any set variable (denoted by $B$ as well) such that the following formula $F(B)$ of $SwS$ holds for $B$. ($F$ depends on $S$ as well.)

(9) $F(B)$ iff $[\forall y(y \in B \rightarrow \exists z(z \in S \wedge z \leqslant_2 y)] \wedge [\exists u(u \in B \wedge \Lambda \leqslant_2 u]$.

Now to define $A_x$, for $x \in S$, we use a formula $G(x, y, A)$, with set variable $A$, and element variables $x$ and $y$ which will mean $y \in A_x$ ($G$, of course, depends on $B$).

(10) $y \in A_x$ iff $\exists z[(z \leqslant_1 x \vee z = x) \wedge z \leqslant_2 y \wedge y \in B]$.

Conditions (9) and (10) say that for $x \in S$, the elements of $\{y \mid x \leqslant_2 y\}$ that are in $B$ are the new elements that were put into $A_x$ and are not in $A_u$, where $u$ is the predecessor of $x$ in $S$.

Equality is defined as

(11) $x = y$ iff $\forall C(x \in C \leftrightarrow y \in C)$ where $C$ is a set variable.

Let us now translate the expression: "$\beta(a_1, \dots, a_n)$ holds at $x \in S$," where $a_i \in A_x$ are elements of the domain. This we do by induction: First, with each $P_n$ of the language we associate a set variable $\Pi_n$. Denote the translation by *:

(12) $([P_n(a)]_x = 1)^* = a \in A_x \wedge a \in \Pi_n \wedge \Pi_n \subseteq B$.

(13) $(a = b)^* = \forall C(a \in C \leftrightarrow b \in C)$.

(14) $([\alpha \wedge \beta]_x = 1)^* = ([\alpha]_x = 1)^* \wedge ([\beta]_x = 1)^*$.

(15) Similarly for $\vee$ (like in (14)).

(16) $([\alpha \to \beta]_x = 1)^* = \forall y[(x = y \vee x \leqslant_1 y)$
$\to (([\alpha]_y = 1)^* \to ([\beta]_y = 1)^*)]$.

(17) $([\exists a\alpha(a)]_x = 1)^* = (\exists v \in A_x)([\alpha(v)]_x = 1)^*$ where all substitution are properly done.

(18) $([\forall a\alpha(a)]_x = 1)^* = \forall y \forall v[xRy \wedge v \in A_y \to ([a(v)]_y = 1)^*]$.

One can easily see that the following lemma holds:

LEMMA 19. *For a closed $\beta$ with predicates $P_1, \dots, P_n$, $\beta$ is a theorem of $T_1$ iff the following is a theorem of SwS:*
($\forall S$ *such that* (1) *and* (2) *hold*)($\forall B$ *such that* $F(B)$)($\forall \Pi_1, \dots, \Pi_n \subseteq B$)($[\beta]_\Lambda = 1)^*$.

*Proof.* Show by induction that $([\alpha(a)]_y = 1)^*$ actually says that the value of $\alpha(a)$ at $y$ is truth.

Now since *SwS* is a decidable theory, $T_1$ must be decidable.

## 2. Unary functions with decided equality

THEOREM 1. (a) *The following intuitionistic theory of one unary function with decidable equality is decidable*:

(1) *Axioms for decidable equality.*

(2) $\forall x \exists y_1, \dots, y_n(\bigwedge_{i=1}^{n} x = f(y_i))$.

(3) $$(\forall x)(\forall y_0, \ldots, y_n)(\wedge_{i=0}^{n}(x = f(y_i) \rightarrow \vee_{i \neq j} y_i = y_j).$$

(b) *If we drop from* (a) *the condition that every point x has exactly n point* $y_1, \ldots, y_n$ *such that* $x = f(y_1), \ldots, x = f(y_n)$, *then the resulting theory is undecidable.*

*Proof.* We prove Theorem 1 for the special case of a first order theory with the axiom that any $y$ has a unique predecessor. We assume this to simplify matters. Since we gave the full translation in Section 1, we shall be more brief here and describe only how the semantics can be expressed in *SwS.*

Let $(S, R, \Lambda, \mathbf{A}_t)$ be a structure of the theory $T_2$ of Theorem 1. We can assume again, as in Section 1, that assumptions (1), (2), (3), (5), (6) and (7) hold. Since equality is decidable we can assume it to be 'being the same element' in the structure. Now what do the structures $\mathbf{A}_x$, $x \in S$, look like?

At the structure $\mathbf{A}_0$, we have one or many related chains of the form

$$\ldots, f^{-2}(x), f^{-1}(x), x, f(x), f^2(x), \ldots$$

some of which are finite cycles (e.g., if $f(x) = x$) some are infinite chains. In any higher structure we may have more elements and therefore more chains, but one thing is certain, since any element *must have* exactly one predecessor, no new element is related to an element that appears in a lower structure. We need this in order to be able to describe the structure in *SwS,* and without it, the theory (i.e., the theory without the axiom about the existence of exactly one predecessor) is undecidable. If we require that every element has exactly $n$ predecessors we get chains of the form

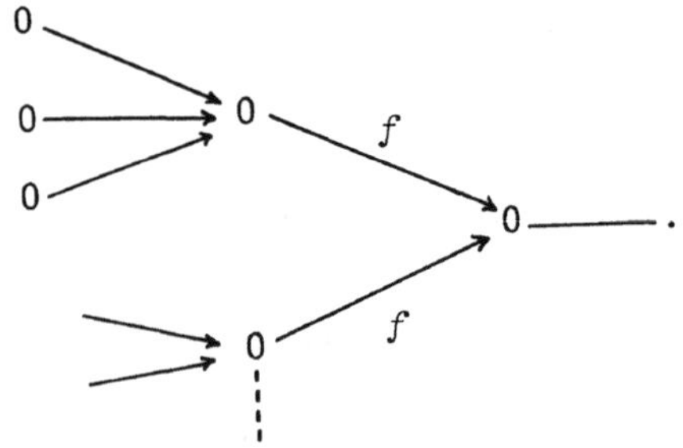


We need this kind of axiom in order that, again, no new element has any relation with an old element; if we do not require this, the theory is undecidable, as we shall show later, using methods of the previous chapter.

Now to define the interpretation take $S$ and $R$ as before.

Let $G_x$ for $x \in S$ be the following set:

$$G_x = \{e_1^n(e_2^k(x)) \mid n, k \text{ natural numbers}\} \cup \{e_0^n(e_2^k(x)) \mid n, k \text{ natural numbers}\} \cup \{e_0^k(y) \mid \text{for } k \leqslant n \text{ for } n \text{ such that } e_0^n(y) \in D_x, \text{ where } D_x \text{ is a given subset of } \{e_0^n(y) \mid e_3(x) \leqslant_2 y\} \text{ such that if } e_0^n(y) \in D_x \wedge e_0^m(y) \in D_x \text{ then } m = n\}.$$

(This can be expressed in the second order language by saying 'if $z, z^1$ are of the form $e_0^n(y)$ for $y$ such that $e_3(x) \leqslant_2 y$ then $z = z^1$'.)

We define $f$ on $G_x$ as follows: On elements of the form $e_0^n(e_2^k(x))$ or $e_1^n(e_2^k(x))$, $f$ goes as in the diagram:

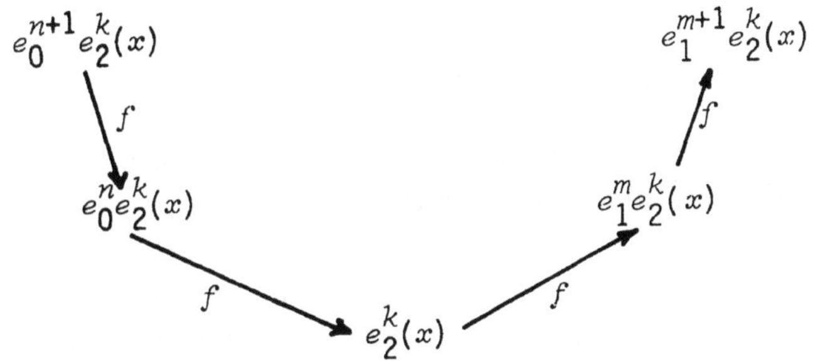


On elements of the form $e_0^n(y)$ for $e_3(x) \leqslant_2 y$, we define $f$ to be a cycle:

$$f(e_0^k(y)) = e_0^{k+1}(y) \quad \text{and for} \quad e_0^n(y) \in D_x$$
$$\text{let} \quad f(e_0^n(y)) = e_0(y).$$

The set $B$ is a subset of $\cup G_x$ with the properties:

$$e_0^n(y) \in B \cap D_x \rightarrow e_0^m(y) \in B \qquad \text{for } m \leqslant n,$$

$$e_2^n(x) \in B \rightarrow e_0^m(e_2^k(x)) \in B \wedge e_1^m(e_2^k(x)) \in B \qquad \text{for all } m, k.$$

We thus get that $B$ is closed under $f$.

Now that we have $B$, $S$ and $R$, we can continue to define $\mathbf{A}_x$, $x \in S$, and the translation as before. To show that the theory of Theorem 1 is complete for the class of structures just defined, observe the following: (1) Let $(S, R, 0, \mathbf{A}_t)$ be a model of a sentence in the form $\gamma = (Q \ldots) \wedge (\wedge \alpha_i \rightarrow \vee \beta_j)$, where $\alpha_i, \beta_j$ are atomic, then $[\gamma]_t = 1$ implies

$\mathbf{A}_t \models \gamma$ (classically). From this we conclude that all Kripke structures of the theory of Theorem 1 have the property that $\mathbf{A}_t$ is a classical model of a unary function. Now since equality is decidable ($x = y \vee x \neq y$) we conclude that $A_t \subseteq A_s$ (for $tRs$) is an isomorphism into. The class of structures that we defined in *SwS* is exactly the class of all structures of this kind.

Thus Theorem 2 is proved.

If we relax the requirements about the predecessors of $x$ the resulting theory is undecidable, because we can interpret a unary predicate in it, e.g., the domains are those elements $x$ of the form $f(x) = x$, $P(x)$ holds iff for at least for two different $y$'s we have $f(y) = x$. The HPC theory of unary predicate is undecidable.

## 3. Monadic fragment again

We have shown that the intuitionistic theory of one monadic letter is undecidable. We can also show that:

THEOREM 1. *There exists an extension of intuitionistic logic (Dummett LC with constant domains, in fact) where the theory of one monadic letter is decidable but the theory of two monadic letters is undecidable but is axiomatizable.*

THEOREM 2. *There exists an extension of intuitionistic predicate logic (in fact, defined by the class of Kripke structures with $S = \omega$) such that the theory of one monadic letter and decidable equality is decidable of two letters and decidable equality is undecidable and of (say)* 10 *monadic letters and decidable equality, is not arithmetical. (For the last, we use ideas of Scott.)*

*Proof of Theorem* 1. Let us now prove that there exists an extension of the intuitionistic predicate logic for which the theory of one monadic letter is decidable while the theory of two is not.

The extension is known as constant domains LC and is with the axioms $\forall x(\alpha \vee \beta(x)) \to (\alpha \vee \forall x \beta(x))$, $x$ is not free in $\alpha$; and $(\alpha \to \beta) \vee (\beta \to \alpha)$. This is complete for the class of all Kripke structures with linear sets of possible worlds and with the same domain for all worlds. We shall show later that the theory on one monadic letter in this system is decidable.

Let us now show that the theory with two monadic letters of this fragment is undecidable. Rogers proved that the classical theory of two equivalence relations is undecidable. We shall faithfully interpret this undecidable theory in the monadic fragment of LC.

Let $P_1, P_2$ be the two monadic predicates of the fragment.

Define $x \equiv_i y$ as $P_i(x) \leftrightarrow P_i(y)$. As we have in LC an infinite linear set of possible worlds we can obtain this way an infinite number of equivalence classes and therefore any classical equivalence relation can be represented by a $P_i$. We can also arrange that $\exists x P_i(x)$ be false at the origin of the Kripke structure but true at all higher points. Let $\gamma = (Q\ldots) \wedge (\bigwedge_j x_j \equiv y_j \rightarrow \bigvee_i u_i \equiv v_i)$, where $\equiv$ can be either $\equiv_1$, or $\equiv_2$ and $Q$ is a string of quantifiers. This is the general form of a classical sentence in the theory of two equivalence relations. Let $\gamma^*$ be the following sentence in the monadic LC

$$\Big[(Q\ldots) \wedge \Big(\bigwedge_j (P(x_j) \leftrightarrow P(y_j)) \rightarrow \bigvee_i (P(u_i) \leftrightarrow P(v_i))$$
$$\vee \exists x P_1(x)\Big)\Big] \rightarrow \forall x P_1(x),$$

where $P$ is $P_1$ or $P_2$ according to what $\equiv$ is.

LEMMA 3: $\vdash_{CE} \neg\gamma \leftrightarrow \vdash_{LC} \gamma^*$, *where* CE *is the classical theory of two equivalence relations.*

We leave the proof to the reader, see Chapter 14.

Let us now prove that the theory of one monadic letter of constant domain LC is decidable.

The set $T_1 = \{x \mid \Lambda \leq_1 x\}$ is densely ordered by $<$ and therefore every countable ordered set can be embedded in $(T_1, <)$.

Let $A \subseteq T_1$ such that it has a first element. Let $T_2^x$ be $\{y \mid x \leq_2 y\}$. Let $B \subseteq \bigcup_{x \in A} T_2^x$.

Then $A$ and $B$ define a model of a unary predicate with $B$ its domain, $A$ its set of possible worlds and $[P(y)]_x = 1$ iff $\exists z(z < x \wedge y \in T_2^z \wedge z \in A)$.

Since all this can be expressed in Rabin's theory *SwS*, we get the decidability result.

We turn to Theorem 2, we sketch the proof. Let $CLC_\omega$ be the intermediate logic defined by the class of all Kripke structures with constant domains and with $\omega$ as the set of possible worlds. The monadic fragment of this system with equality and one individual constant is not axiomatizable as arithmetic can be interpreted in it. Let $\gamma$ be a sentence of arithmetic with successor $'$, 0, addition $\oplus$, and multiplication $\odot$, then there exists an $\gamma^*$ such that $\models_\omega \neg\gamma$ iff $\models_{\mathrm{CLC}_\omega} \gamma^*$. We now work our way towards constructing $\gamma^*$. We assume that the language of $\mathrm{CLC}_\omega$ contains equality $=$, the individual constant 0 and the unary predicates $D$, $N$, $P_1$, $P_2$, $P_3$, $Q_1$, $Q_2$ and $Q_3$. Consider the following formulae:

$F_0$ is $\forall x(Nx \to N0)$,

$F_1$ is $(\forall xy)(x = y \vee x \neq y)$,

$F_2$ is $\forall x \neg\neg Nx \wedge \neg\forall x(Nx \vee \neg Dx)$,

$F_3$ is $\forall xy((Nx \leftrightarrow Ny) \to (x = y \vee N0))$,

$F_4$ is $\forall x(Dx \vee \neg Dx) \wedge \exists x Dx$.

We need the following definitions:

$$x < y =_{\mathrm{def}} x \neq y \wedge (Ny \to Nx),$$

$$S(x, y) =_{\mathrm{def}} x < y \wedge \forall z(x < z \to (Nz \to Ny)),$$

$$R_\oplus(x, y, z) =_{\mathrm{def}} Dx \wedge Dy \wedge Dz \wedge \exists u(\neg Du \wedge (P_1u \leftrightarrow P_1x) \wedge (P_2u \leftrightarrow P_2y) \wedge (P_3u \leftrightarrow P_3z)),$$

$R_\odot(x, y, z) =_{\mathrm{def}}$ like $R_\oplus$ with $Q_i$ replacing $P_i$.

Now let

$F_5$ be $\forall xy[N0 \vee \exists z R_\oplus(x, y, z)]$,

$F_6$ be $\forall x, y, z, z^1(R_\oplus(x, y, z) \wedge R_\oplus(x, y, z^1) \to (N0 \vee z = z^1))$,

$F_7$ be like $F_5$ with $R_\odot$ replacing $R_\oplus$,

$F_8$ be like $F_6$ with $R_\odot$ replacing $R_\oplus$.

Now let

$F_9$ be $\forall u, x, y, z(N0 \vee (R_\oplus(x, 0, x) \wedge R_\odot(x, 0, 0)) \wedge [S(y, z) \wedge R_\odot(x, y, u) \wedge R_\oplus(x, z, v) \to (N0 \vee S(u, v))]$

$$\wedge [S(y, z) \wedge R_\odot(x, y, v) \wedge R_\odot(x, z, u) \\ \to (N0 \vee R_\oplus(v, x, u))]).$$

We can now define the translation. Let $(Q \ldots) \wedge (\wedge \alpha_i \to \vee \beta_i)$ be a sentence $\gamma$ of arithmetic with $\oplus$, $\odot$, $'$, and 0, and with $Q$ a string of quantifiers and $\alpha_i$ and $\beta_i$ atomic propositions. Let $\gamma^*$ be

$$\wedge F_i \wedge (Q^D \ldots) \wedge (\wedge \alpha_i^* \to [\vee \beta_i^* \vee N0]) \to \exists x Nx,$$

where $Q^D$ is the same string of quantifiers relativized to $D$ and $\alpha_i^*$ is the translation of the atomic by

$R_\oplus(x, y, z)$ if $\alpha_i$ is $x \oplus y = z$,

$R_\odot(x, y, z)$ if $\alpha_i$ is $x \odot y = z$,

$S(x, y)$ if $\alpha_i$ is $y = x^1$.

LEMMA 4. $\models_\omega \neg \gamma$ *iff* $\models_{\mathrm{CLC}_\omega} \gamma^*$.

We now sketch the proof of the lemma. First let us assume that $\gamma$ is true in the natural numbers and let us construct a model of $\mathrm{CLC}_\omega$ in which $\gamma^*$ is false. Let the set of possible worlds be $\{-1, 0, 1, 2, \ldots\}$ in the order of magnitude and let the common domain $A$ be

$$A = \omega \cup \{t_{x_1, x_2, x_3} \mid x_3 = x_1 + x_2\} \cup \{s_{x_1, x_2, x_3} \mid x_3 = x_1 \cdot x_2\},$$

where $t_{x_1, x_2, x_3}$ and $s_{x_1, x_2, x_3}$ are special indexed symbols.

Define the Kripke structure as follows:

(1) Equality is the real equality. Let 0 be $0 \in \omega$.

(2) $[D(x)]_n = 1$ iff $x \in \omega$.

(3) For $x \in A$ let $[Nx]_i = 1$ iff $x \notin \omega \wedge i \geq 0$ or $x \in \omega \wedge x \leq i$.

(4) To define $P_i$ let, for each $t_{x_1, x_2, x_3}$, $[P_i(x_i)]_n = [P_i(t_{x_1, x_2, x_3})]_n = 1$ for each $x_i \leq n$, $i = 1, 2, 3$.

(5) Similar definition for $Q_i$ using $s_{x_1, x_2, x_3}$.

We now prove some lemmas.

LEMMA a. *For $x, y \in D$ we have*

$$[x < y]_{-1} = 1 \quad \textit{iff} \quad x < y.$$

*Proof.* By definition.

LEMMA b. *For $x, y \in D$,* $[S(x, y)]_{-1} = 1$ *iff* $y = x + 1$.

*Proof.* By definition.

LEMMA c. *For* $x_1, x_2, x_3 \in D$, $[R_\oplus(x_1, x_2, x_3)]_{-1} = 1$ *iff* $x_3 \equiv x_1 + x_2$.

*Proof.* Clearly if $x_3 = x_1 + x_2$ then $t_{x_1, x_2, x_3}$ fulfils the definition of $R_\oplus$ at $-1$, by the definition of the $P_i$'s. Assume now that $[R_\oplus(x_1, x_2, x_3)]_- = 1$, then, for some $a \in A - \omega$, we have $[P_i(x_i) \leftrightarrow P_i(a)]_{-1} = 1$. $a$ must be of the form $t_{k,l,m}$ as these are the only elements in $A - \omega$ for which any $P_i$ is true. Now since $[P_1(x_1) \leftrightarrow P_1(a)] = 1$, we must have $x_1 = k$. Similarly we get that $x_2 = l$ and $x_3 = m$. So $x_3 = x_1 + x_2$. A similar lemma holds for $R_\odot$.

LEMMA d. $[\wedge F_i]_{-1} = 1$.

*Proof.* By inspection. Notice the role that $N0$ plays in the formula $[N0]_m = 1$ for $m > -1$.

LEMMA e. $(Q^D \ldots) \wedge (\wedge \alpha_i^* \rightarrow \vee \beta_i^* \vee N0)$ *holds at* $-1$.

*Proof.* The reader is referred to Chapter 14 for this type of argument. In the outline observe the following: Since $N0$ is true from 0 upwards it is sufficient to check the quantifiers on the world $-1$. But there $-S$, $R_\oplus$, $R_\odot$ have their true meaning and so the formula is true, since $\gamma$ is true for the natural numbers. Now since $[\exists x Nx]_{-1} = 0$ we get a countermodel for $\gamma^*$.

We now turn to show the other direction, that is, if $\gamma^*$ is false in some $\text{CLC}_\omega$-Kripke structure, then $\gamma$ is true for the natural numbers. We can assume that the set of possible worlds is $\omega$ and that the antecedent of $\gamma^*$ is true at 0 and the consequent is false. Let us now check what information we get from the fact that the antecedent of $\gamma^*$ holds at 0.

(1) Since $F_1$ holds we conclude that equality is decidable, i.e., is like the actual equality.

(2) From $F_4$ we get that $D$ is decidable and $\{x \mid [Dx]_0 = 1\}$ is nonempty.

(3) From $F_2$ we conclude that for every element $x$ of $D$ there exists a $n \in w$ such that $[Nx]_n = 1$ and that for each $n \in w$ there is a $y \in D$ such that $[Ny]_n = 0$.

(4) Let $D_n \subseteq D$ be the set of all $x$ such that $n$ is the *first* number such that $[Nx]_n = 1$. $F_3$ says that $D_n$ can contain *at most one* element. (In fact equality is a congruence relation but since equality is decidable we can take the quotient model.)

Thus the elements of $D$ can be enumerated $0 = d_0, d_1, d_2, d_3, \ldots$ in a way that $D_{n_i} = \{d_i\}$ and $i < j$ iff $[Nd_j \rightarrow Nd_i]_0 = 1$.

(5) One can verify that

$$[d_i < d_j]_0 = 1 \quad \text{iff} \quad i < j,$$

$$[S(d_i, d_j)]_0 = 1 \quad \text{iff} \quad j = i + 1.$$

Now $R_\oplus$ and $R_\odot$ induce two three place relations on $D$ at 0. $F_5$ to $F_8$ say that, e.g., the relation $[R_\oplus(x, y, z)]_0 = 1$ is a function. Similarly for $R_\odot$. $F_9$ says that indeed it is $\oplus$ and $\odot$ on $\{d_i\}$, e.g.,

$$[R_\oplus(d_i, d_j, d_k)]_0 = 1 \quad \text{iff} \quad k = i + j.$$

Now from $[(Q^D \ldots) \wedge (\wedge \alpha_i^* \to \vee \beta_j^*)]_0 = 1$, we get that $\gamma$ (note that $[N0]_0 = 0$) holds on the natural numbers.

## 4. Some abelian groups

We want to show the decidability of the theory $G_m^*$ of groups of Definition 12.5.2. We need some further notions.

(a) Let $N_1^* = \{t \in w^* \mid \Lambda \leqslant_1 t\}$, $N_t^2 = \{s \in w^* \mid t \neq s \wedge t \leqslant_2 s\}$.

(b) Let $q_1, \ldots, q_n$ denote all the possible divisors of $m$.

(c) Let $C, X, Y, D$ range over subsets of $w^*$.

Recall that $G_m^*$ is complete for a class of structures $\mathbf{M}_m^*$ of the form $(S, R, 0, G_t, I_t)$ as mentioned in (12.5.3). We want to express this interpretation in $SwS$. We go by stages:

(0) To express the notion that **S** is a set of possible worlds, we replace $w$ by the isomorphic $N_1^*$ (as we need the rest of $\omega^*$ for other purposes). Let '**S** is a set of possible worlds' mean in $SwS$ the following:

$$(\emptyset \subsetneq \mathbf{S} \subseteq N_1^*) \wedge \forall s, t(s \in \mathbf{S} \wedge t \leqslant_1 s \to t \in \mathbf{S}).$$

(1) $R$ is replaced, of course, by $\leqslant_1$.

(2) We now want to express the set of indices $I_t$. We define $J^t$, an auxiliary set by: $J_0 = I_0$. If $t, s \in S$ and $t$ is an immediate successor of $s$, then $J_t = I_t - I_s$. So we have that $I_t = \bigcup_{s \leqslant_1 t} J_s$. The only requirement of $J$ is, of course, $J_\emptyset \neq \emptyset$. So given a set variable **S** expressing **S** as (0) above, (i.e., for which '**S** is a set of possible worlds' holds) we define in $SwS$ the notion '**J** is a set of indices for **S**' by saying: '$\mathbf{J} \cap N_\emptyset^2 \neq 0$ and $\forall t \in \mathbf{J} \exists s \in \mathbf{S}(s \leqslant_2 t \wedge s \neq t)$'.

(3) The set $I_t$, for $t \in \mathbf{S}$ can be defined by a formula $I(t)$ of $SwS$ as follows: $\beta \in J(t)$ (or $J(t, \beta)$) is: $\exists s \in \mathbf{S}(s \leqslant_1 t \wedge s \leqslant_2 \beta \wedge \beta \in \mathbf{J})$.

(4) To define $G^t_\beta$, for $\beta \in I(t)$ we have to provide the information of (a) what is the order of $G^t_\beta$, and (b) what are the elements of $G^t_\beta$. To do this we introduce the set variable **P** and the functions $f_i$, $1 \leqslant i \leqslant n$ (recall that $q_1, \ldots, q_n$ are all the divisors of $m$). For $\beta \in I(t)$, we let $f_i(\beta) \in \mathbf{P}$ exactly when the order of $G^t_\beta$ is $q_i$. We let $g_1(\beta), \ldots, g_{q_i}(\beta)$ be the elements of $G^t_\beta$. We consider $g_{q_i}(\beta)$ as the zero of $G^t_\beta$. So we can express the fact that **P** defines $G^t_\beta$, by writing a formula $\Psi =$ '**P** defines the groups' of *SwS* saying in *SwS* that '**P** contains only elements of the form $f_i(\beta)$, for $\beta \in \mathbf{J}(t)$ and for each such $\beta \in \mathbf{J}(t)$, exactly one such $f_i(\beta) \in \mathbf{P}$, $1 \leqslant i \leqslant n$'.

(5) Given a set variable **Y**, we want to define in *SwS* the notion that '**Y** is an element of $G_{t_0}$, for $t_0 \in \mathbf{S}$'. To do this we say that 'For any $\beta \in \bigcup_{t \in \mathbf{S}} I(t)$ such that $f_i(\beta) \in \mathbf{P}$, **Y** contains exactly one of $g_1(\beta), \ldots, g_{q_i}(\beta)$, and also the set **C** of all elements $\beta$ such that $\beta \in \bigcup_{t \in \mathbf{S}} I(t)$ and $f_i(\beta) \in \mathbf{P} \wedge g_{q_i}(\beta) \in \mathbf{Y}$ is *finite*', and $\mathbf{C} \subseteq I(t_0)$.

(6) For $\beta \in I(t)$, $f_i(\beta) \in \mathbf{P}$, the elements of $G^t_\beta$ are $g_1(\beta), \ldots, g_{q_i}(\beta)$, with $g_{q_i}(\beta)$ considered the o of the group and $g_1(\beta)$ the generator and $g_j(\beta)$ the $j$-th element. Now given two elements **X** and **Y** of $G_t$ we can define the unique element of $G_t$ denoted by $\mathbf{X}+\mathbf{Y}$ (actually one defines a 3-place formula $\oplus(\mathbf{X}, \mathbf{Y}, \mathbf{Z})$) which is the sum of **X** and **Y**. We simply have to give a definition by cases (of $\oplus(\mathbf{X}, \mathbf{Y}, \mathbf{Z})$) for each $\beta \in I(t)$ and each case of $f_i(\beta) \in \mathbf{P}$ and what $g_?(\beta)$ happens to be in **X** and **Y** and say for this case what $g_?(\beta)$ should be in **Z**. We can therefore say that there are formulas of *SwS* saying '**Y** is an element of $G_t$)' and '(**Y** is the sum of $\mathbf{X}_1$ and $\mathbf{X}_2$)'.

(7) Let $\Psi$ be any formula of $G^*_m$. Let $(\mathbf{S}, \mathbf{J}, \mathbf{P})$ define a structure of $G^*_m$ in *SwS* (see (1)–(6) above). We want to define a formula $\Psi^\#(t)$ of *SwS* saying that in this model $[\Psi]_t = 1$. We do this by induction.

(a) For atomic $\Psi(y_1, \ldots, y_n)$ simply let $\Psi^\#$ say that '$\mathbf{Y}_1, \ldots, \mathbf{Y}_n$ are elements of $G_t$' and 'write same *equation* appearing in $\Psi$ using $\oplus$ and $=$ of *SwS*'.

(b) $(\Psi_1 \wedge \Psi_2)^\# = \Psi^\#_1 \wedge \Psi^\#_2$.

$(\Psi_1 \vee \Psi_2)^\# = \Psi^\#_1 \vee \Psi^\#_2$.

(c) $(\Psi_1 \to \Psi_2)^\# = \forall s \in S(t \leqslant_1 s \wedge \Psi^\#_1(s) \to \Psi^\#_2(s))$.

(d) $(\exists y \Psi(y))^\# = \exists \mathbf{Y}$('**Y** *is an element of* $G_t$' $\wedge \Psi^\#(\mathbf{Y})$).

(e) $(\forall y \Psi(y))^\# = (\forall s \in \mathbf{S})\forall \mathbf{Y}$('**Y** *is an element of* $G_s \wedge t \leqslant_1 s$' $\to \Psi^\#(s, \mathbf{Y})$).

LEMMA. *For any* $\Psi$ *of* $G_m^*$, $\Psi$ *holds in all structures of* $\mathbf{M}_m^*$ *iff* $SwS \models (\forall \mathbf{S}, \mathbf{J}, \mathbf{P})$ ('$\mathbf{S}$ *is a set of possible words*' $\wedge$ '$\mathbf{J}$ *a set of indices*' $\wedge$ '$\mathbf{P}$ *defines the groups*' $\rightarrow \Psi^{\#}(\Lambda)$).

*Proof.* Follows from (1)–(7) above.

COROLLARY. $G_m^*$ *is decidable.*

## 5. Linear order

**Exercise.** Show that the theory of linear order of Theorem 12.3.8 is decidable.

*Hint.* The interpretation is directly expressible in *SwS*.

## NOTES

Based on the author's papers. Used two key results of M.O. Rabin, on the decidability of *SwS*, the monadic second order theory of $\aleph_0$-tree with $\aleph_0$ successors, and the second is the Rabin–Scott method of interpretation.

The undecidability of the theory of two monadic letter in *H* was proved by S. Kripke. Maslov-Minc-Orevkov gave a syntactical proof of the undecidability on one letter in *H*. Lifshits handled the decision problem for pure equality. C. Smorynski obtained undecidability results for e.g., dense-linear order.

Decidability for the various propositional calculi follow from the fact that they have f.m.p. F.m.p. was obtained by McKinsey–Tarski for *h* and using filtration by K. Segerberg obtained decidability for many other systems.

# REFERENCES


Aczel, P. H. G., 1968: 'Saturated intuitionistic theories', in H. A. Schmidt, K. Schutte and H. J. Thiele, eds., *Contributions to Mathematical Logic*, Amsterdam.

Anderson, J. G., 1966: *Some Results in Super-Constructive Propositional Calculi*, Thesis, Newcastle upon Tyne.

Anderson, A. R., and Belnap, N., 1975: *Entailment*, Princeton.

Beth, E. W., 1956, 'Semantic construction of intuitionistic logic', Mededelingen der Koninklijke Nederlandse Akademie van Wetenschappen, Afd. Letterkunde, Nieuwe Reeks **19**, 357–388.

Bowen, K. B., 1971: 'An extension of the intuitionistic propositional calculus', *Indag. Math.*, pp. 287–293.

Bull, R. A., 1962: 'The implicational fragment of Dummett's LC', *J. S. L.* **27**, 189–194.

Bull, R. A., 1966: 'MIPC as the formalisation of an intuitionist concept of modality', *J. S. L.* **31**, 609–616.

Curry, H. B., 1963, *Foundations of Mathematical Logic*, McGraw-Hill.

van Dalen, D., 1974, 'A model for HAS, a topological interpretation of the theory of species of natural numbers', *Fund. Math.* **82**, 167–174.

van Dalen, D., 1974: 'Choice sequences in Beth Models', 'Notes on Logic and Computer Science', *Logos* **20** (October).

van Dalen, D. and Gordon, C. E., 1971: 'Independence problems in subsystems of intuitionistic arithmetic', *Proc. Neth. Acad. Sciences*, Series A, **74**, 448–456.

Diego, A., 1966: *Sur les algèbres de Hilbert*, Paris, Gauthiers-Villars.

Dummett, M. A., 1959: 'A propositional calculus with a denumerable matrix', *J. S. L.* **24**, 96–107.

Dyson, V. H. and Kreisel, G., 1961, Technical Report No. 3, Stanford University.

Fine, K., 1970: 'An intermediate logic without the finite model property', unpublished.

Fine, K., 1972: 'A logic containing S4 without F.M.P.', in *Conference on Mathematical Logic*, Vol. 70, Ed. W. Hodges, Springer, pp. 98–102.

Fine, K., 1974: 'An ascending chain of S4 logics', *Theoria* **XL**, 110–116.

Fitting, M. C., 1969: *Intuitionistic Logic Model Theory and Forcing*, Amsterdam.

Gabbay, D. M., 1969, 'Montague type semantics for non-classical logic', *Technical Report*, U. S. Air Force, pp. 1–25.

Gabbay, D. M., 1969, 'Semantic proof of Craig's Theorem for intuitionistic logic and extensions, Part I and Part II', *Proceedings of the 1969 Logic Colloquium in Manchester*, North-Holland Publishing Company, pp. 391–410.

Gabbay, D. M., 1970: 'Decidability of the Kreisel-Putnam System', *J. S. L.*, **35**, 431–437.

Gabbay, D. M., 1971: 'A general theory of filtration', *J. P. L.*, **1**, pp. 29–34.

Gabbay, D. M., 1972: 'Applications of trees to intermediate logic I', *J. S. L.* **37**, pp. 135–138.

Gabbay, D. M., 1972: 'Decidability of some intuitionistic predicate calculi', *J. S. L.* **37**, 579–587.

Gabbay, D. M., 1972: 'Model theory for intuitionistic logic', *Z. f. Math. Logik u. Grundlagen d. Math.* **18**, 49–54.

Gabbay, D. M., 1972, 'Sufficient conditions for the undecidability of intuitionistic theories with applications', *J. S. L.* **37**, pp. 375–384.

Gabbay, D. M., 1973, 'Survey of decidability results in non-classical logic', *Proceedings of the 4th International Congress on Logic, Methodology, and Philosophy of Science*, North-Holland Publishing Co., pp. 29–43.
Gabbay, D. M., 1973: 'Undecidability of intuitionistic theories of algebraic and real closed fields', *J. S. L.*, **38**, pp. 86–92.
Gabbay, D. M., 1974: 'A Generalization of the concept of intensional semantics with applications', *Philosophia*, **4**, 251–270.
Gabbay, D. M., 1974: 'On 2nd order intuitionistic propositional calculus with full comprehension', *Archiv f. Math. Logik u. Grund.* **16**, 177–186.
Gabbay, D. M., 1975: 'Decidability results in non-classical logics I', *Annals of Mathematical Logic*, **8**, 236–307.
Gabbay, D. M., 1975: 'The decision problem for finite extensions of the intuitionistic theory of Abelian Groups', *Studia Logica* **34**, 59–67.
Gabbay, D. M., 1976, 'Completeness properties of Heyting's predicate calculus with respect to RE models', *J. S. L.*, **41**, 81–95.
Gabbay, D. M., 1976: *Investigations in Modal and Tense Logics*, D. Reidel.
Gabbay, D. M., 1977: 'Undecidability of intuitionistic theories formulated with the apartness relation', *Fundam. Mathem.* **97**, 57–69.
Gabbay, D. M., 1977: 'A note on Intuitionistic connectives', *Studia Logica*, **36**, 127–141.
Gabbay, D. M., 1976: 'A note on Kreisel's notion of validity in Post systems', *Studia Logica*, **35**, 285–295.
Gabbay, D. M., 1977: 'Craig's theorem for intuitionistic logic', Part III, *J. S. L.* **42**, 269–272.
Gabbay, D. M., 1977: 'A new version of Beth semantics', *J. S. L.* **42**, 306–309.
Gabbay, D. M., 1978: 'What is a classical connective', *Z. f. Math. Logik*, **24**, 37–44.
Gabbay, D. M. and deJongh, D., 1974, 'Sequences of Decidable Finitely Axiomatizable intermediate logics with the disjunction property', *J. S. L.* **39**, 67–79.
Glivenco, V. I., 1929: 'Sur quelques points de la logique de Brouwer', *Bull. Acad. Royal de Belgique, Classe de Sciences* **15**, 183–188.
Gödel, K., 1932: 'Zum intuitionistischen Aussangenkalkul', *Akademie der Wissenschaften in Wien, Mathematisch-naturwissenschaftliche Klasse* **69**, 65–66.
Görnemann, S., 1971: 'A logic stronger than intuitionism', *J .S. L.* **36**, 249–261.
Grzegorczyk, A., 1972, 'An approach to logical calculus', *Studia Logica*, **30**, 33–43.
Gurevitz, Y., 1977: 'Intuitionistic logic with strong negation', *Studia Logica* **XXXVI**, 49–59.
Harrop, R., 1956: 'On disjunctions and existential statements in Intuitionistic systems of logic', *Math. Annalen* **132**, 347–361.
Harrop, R., 1958: 'On the existence of finite models and decision procedures for propositional calculi', *Proc. Camb. Phil. Soc.* **54**, 1–13.
Harrop, R., 1965: 'Some structure results for propositional calculi', *J. S. L.*, **30**, 271–292.
Heyting, A., 1956: *Intuitionism*, Amsterdam.
Horn, A., 1962: 'The separation theorem of intuitionist propositional calculus', *J. S. L.*, **27**, 391–399.
Hosoi, T., 1967: 'On intermediate logics I', *Journal Faculty of Science, Univ. of Japan*, **XIV**, 293–312.
Hosoi, T., 1969: 'On intermediate logics II', *Journal Faculty of Science, Univ. of Tokyo* **16**, 1–12.

Hosoi, T. and Ono, H., 1973: 'Intermediate propositional logics (a survey)', *Journal of Fouda College* **5**, 67–82.

Jankov, V. A., 1968: 'Constructing a sequence of strongly independent super-intuitionistic propositional calculi', *Sov. Math. Dok.*, **9**, 806–807.

Jaskowski, V. A., 1936: 'Recherches sur la logique', *Intuitioniste Actes du Congres International de Philosophie Scientifique*, **6**, 806–807.

deJongh, D. H. J., 1970: 'The maximality of intuitionistic predicate calculus with respect to Heyting's arithmetic' (abstract) *J. S. L.*, **35**, 606, paper to appear in *Compositia Mathematica.*

deJongh, D. H. J. and Smorynski, C., 1976: 'Kripke models and the intuitionistic theory of species', *Annal Math. Logik* **9**, 157–186.

Kleene, S., 1962: 'Disjunction and existence under implication in elementary intuitionistic formalisms', *J. S. L.*, **27**, 11–18.

Kreisel, G. and Putnam, H. 1957: 'Eine unableitungsbeweismethode fur den Intuitionistischen aussagenkalkul', *Archiv f. Math. Logik*, **3**, 74–78.

Kripke, S., 1965: 'Semantical analysis of intuitionist logic I' in J. N. Crossley and M. A. E. Dummette (eds.), *Formal Systems and Recursive Functions*, North-Holland.

Kripke, S., (unpublished), 1968: 'Semantical analysis for intuitionistic logic II'.

Kuznecov, A. V., and Gerciu, V. J., 1970: 'Superintuitionistic logics and finite approximability', *Sov. Math. Dokl.*, **11**, 1614–1619.

Leivant, D., 1971: 'A note of translation of C into Γ, Amsterdam, technical report.

Lemmon, E. and Scott, D., 1966: *Intensional logics*, published 1977 as *An Introduction to Modal Logic* by E. J. Lemmon with D. Scott, edited by K. Segerberg. Blackwell, Oxford.

Lifshits, V. A., 1969: 'Problem of decidability for some constructive theories of equality', in A. O. Slisenko (ed.), *Studies in Constructive Mathematics and Mathematical Logic I*, New York.

Maslov, S., Minc, G. E. and Orevkov, V. P., 1965: 'Unsolvability in the constructive predicate calculus of certain classes of formulas containing only monadic predicate variables', *Sov. Dok.*, **6**, 918–920.

McCullough, D. P., 1971: 'Logic Connectives for Intuitionistic propositional logic', *J. S. L.* **36**, 15–21.

McKay, C. G., 1968: 'The decidability of certain intermediate propositional logics', *J. S. L.*, **33**, 258–264.

Medvedev, T., 1966: 'Interpretation of logical formula by means of finite problems', *Sov. Doklady*, **7**, 857–860.

Moschovakis, J. R., 1973: 'A topological interpretation of second order intuitionistic arithmetic', *Compos. Math.* **26**, 261–275.

Mulvey, C., 1974: 'Intuitionistic algebra and representations of rings and $C^*$ algebras by continuous sections', *Mem. Amer. Math. Soc.* **148**, 3–57.

Nagashima, T., 1966: 'An extension of the Craig-Schutte interpolational theorem', *Am. Japan Assoc, of Phil. Soc.*, **3**, 12–18.

Nishimura, I., 1960: 'On formulas of one variable in intuitionistic propositional calculus', *J. S. L.*, **25**, 327–331.

Pogovzelski, W., 1968: 'On the scope of the classical deduction theorem', *J. S. L.*, **33**, 77–81.

Prawitz, D., 1971: 'Some results for intuitionistic logic with second order quantification rules', in *Intuitionism and Proof Theory*. Eds. A. Kino, J. Myhill, and R. E. Vesley. North-Holland, pp. 259–270.

Prior, A. N., 1964: 'Two additions to positive implication', *J. S. L.*, **29**, 31–32.

Rabin, M. O., 1965: 'A simple method of undecidability proof and some applications', Y. Bar-Hillel (ed.), *Logic Methodology and Philosophy of Science*, Amsterdam, 58–68.

Rabin, M. O., 1969: 'Decidability of second order theories and automata on trees', *T. A. M. S.* **141**, 1–35.

Rawiowa, H., 1974: *Algebraic Approach to Non-classical logic*, Amsterdam.

Rauszer, C., 1974: 'Semi-boolean algebras and their application to intuitionist logic with dual operation', *Fund. Math.*, **83**, 219–249.

Schutte, D., 1962: 'Der Interpolationssatz der intuitionistischen Pradikatenlogic', *Math. Ann.*, **148**, 192–200.

Schutte, D., 1968: *Vollstandige systeme modaler und intuitionistischer Logik*, Berlin.

Scott, D., 1968: 'Extending the topological interpretation to intuitionistic analysis', *Compositio Mathematica* **20**, 194–210.

Scott, D., 1970: 'Extending the topological interpretation to intuitionistic analysis II', in A. Kino, J. Myhill and R. E., Vesley, (eds.), *Intuitionism and Proof Theory*, Amsterdam, 235–255.

Scott, D., 1974: 'Completeness and axiomatizability in many valued logic', Proceedings of Tarski Symposium. Ed. L. Henkin *et al. AMS*, 411–435.

Segerberg, K., 1968: 'Propositional logics related to Heyting's and Johansson's', *Theoria*, **34**, 26–61.

Segerberg, K., 1974: 'A proof of a conjecture of McKay', *Fund. Math.* **81**, 267–270.

Smorynski, C., 1972: 'The undecidability of the 2nd order intuitionistic propositional calculus', (unpublished).

Smorynski, C., 1973: *Investigations of Intuitionistic Formal Systems by Means of Kripke Models*, Thesis, Univ. of Illinois.

Smorynski, C., 1973: 'Elementary intuitionistic theories', *J. S. L.*, **38**, 122–135.

Smorynski, C., 1975: 'Consistency and related metamathematical properties', Report 75, 02, Amsterdam.

Smullyan, R., 1961: *Theory of Formal Systems*, Princeton.

deSwart, H., 1976: 'Another intuitionistic completeness proof', *J. S. L.*, **41**, 644–662.

Thomason, R. H., 1968: 'On the strong semantical completeness of the intuitionistic predicate calculus', *J. S. L.* **33**, 1–7.

Thomason, R. H., 1969: 'A semantical study of constructive falsity', *Zeitschrift für Math. Logic Grund.* **15**, 247–257.

Troelstra, A. S., 1965: 'On intermediate propositional logics', *Indagaciones*, 141–152.

Troelstra, A. S., 1976: 'Completeness and validity for intuitionistic predicate logic', Report 76.05, Amsterdam.

Urquhart, A., 1975: 'Implicational formulas intuitionistic logic', *J. S. L.*, **40**, 661–665.

Veldman, W., 1976: 'An intuitionistic completeness theorem for intuitionistic predicate logic', *J. S. L.*, **41**, 159–166.

Wronski, A., 1972: 'Intermediate logics and the disjunction property', unpublished.

## ADDITIONAL REFERENCES

Van Dalen, D., 1979: 'Another semantics for intuitionistic logic and some metamathematical uses', Univ. of Utrecht, reprint 133.

Dummett, M., 1977: *Elements of Intuitionism*, OUP.

Smorynski, C., 1973: 'Applications of Kripke Models' in *Metamathematical Investigations of Intuitionistic Arithmetic and Analysis*, ed. A. S. Troelstra, Springer-Verlag.

Löb, M. H., 1976, 'Embedding first order predicate logic in fragments of intuitionistic logic', *J. S. L.*, **41**, 705–718.

de Swart, H., 1977: 'An intuitionistically plausible interpretation of intuitionistic logic', *J. S. L.*, **42**, 564–578.

Troelstra, A. S., 1978: 'Some remarks on the complexity of Henkin-Kripke models', *Proc. Koninklijke Nederlandse Akademie van Wetenschappen*, Amsterdam, Series A, **81**, 296–302.

Maximova, L. L., 1977: 'Craig interpolation theorem and amalgamable varieties', *Soviet Math. Dok.*, **18**, 1550–1553.

Komorie, Y., 1979: 'Logics not Holding Craig Interpolation', to appear.

Zucker, J., 1978: 'Interpolation for fragments of the propositional calculus', preprint ZW 116/78, Amsterdam.

Goad, C. A., 1978: 'Monadic infinitary propositional logic: a special operator', *Reports on Mathematical Logic*, **10**, to appear.

Kreisel, G., 1977: 'Monadic operators defined by means of propositional quantification in intuitionistic logic, addendum to the paper Constructivist approaches to logic', *Proceed. Conference on Modern Logic*, Rome, *Encyclopedia Italiana.*

Sobolev, S. K., 1977: 'On the intuitionistic propositional calculus with quantifiers', *Mathematical Notes*, **22**, 528–532.

Renardel de Lavalette, G. R., 1979: *The Interpolation Theorem* (Report), Amsterdam.

Grayson, R., 1980, *Topological Models*, Bar-Ilan, Israel.

Fourman, M. and Hyland, J., 1979: 'Sheaf models for analysis', in *Springer Lecture Notes No. 753*, pp. 280–302.

M. Fourman and Scott, D. 1979: 'Sheaves and Logic', in *Springer Lecture Notes No. 753*, pp. 302–402.

Scott, D., 1979: 'Identity and existence in intuitionistic logic', in *Springer Lecture Notes No. 753*, pp. 660–697.

# INDEX

# SYNTHESE LIBRARY

Studies in Epistemology, Logic, Methodology,
and Philosophy of Science



1. J. M. Bochénski, *A Precis of Mathematical Logic.* 1959.
2. P. L. Guiraud, *Problèmes et méthodes de la statistique linguistique.* 1960.
3. Hans Freudenthal (ed.), *The Concept and the Role of the Model in Mathematics and Natural and Social Sciences.* 1961.
4. Evert W. Beth, *Formal Methods. An Introduction to Symbolic Logic and the Study of Effective Operations in Arithmetic and Logic.* 1962.
5. B. H. Kazemier and D. Vuysje (eds.), *Logic and Language. Studies Dedicated to Professor Rudolf Carnap on the Occasion of His Seventieth Birthday.* 1962.
6. Marx W. Wartofsky (ed.), *Proceedings of the Boston Colloquium for the Philosophy of Science 1961-1962.* Boston Studies in the Philosophy of Science, Volume I. 1963.
7. A. A. Zinov'ev, *Philosophical Problems of Many-Valued Logic.* 1963.
8. Georges Gurvitch, *The Spectrum of Social Time.* 1964.
9. Paul Lorenzen, *Formal Logic.* 1965.
10. Robert S. Cohen and Marx W. Wartofsky (eds.), *In Honor of Philipp Frank.* Boston Studies in the Philosophy of Science, Volume II. 1965.
11. Evert W. Beth, *Mathematical Thought. An Introduction to the Philosophy of Mathematics.* 1965.
12. Evert W. Beth and Jean Piaget, *Mathematical Epistemology and Psychology.* 1966.
13. Guido Küng, *Ontology and the Logistic Analysis of Language. An Enquiry into the Contemporary Views on Universals.* 1967.
14. Robert S. Cohen and Marx W. Wartofsky (eds.), *Proceedings of the Boston Colloquium for the Philosophy of Science 1964-1966. In Memory of Norwood Russell Hanson.* Boston Studies in the Philosophy of Science, Volume III. 1967.
15. C. D. Broad, *Induction, Probability, and Causation. Selected Papers.* 1968.
16. Günther Patzig, *Aristotle's Theory of the Syllogism. A Logical-Philosophical Study of Book A of the Prior Analytics.* 1968.
17. Nicholas Rescher, *Topics in Philosophical Logic.* 1968.
18. Robert S. Cohen and Marx W. Wartofsky (eds.), *Proceedings of the Boston Colloquium for the Philosophy of Science 1966-1968.* Boston Studies in the Philosophy of Science, Volume IV. 1969.

19. Robert S. Cohen and Marx W. Wartofsky (eds.), *Proceedings of the Boston Colloquium for the Philosophy of Science 1966-1968*. Boston Studies in the Philosophy of Science, Volume V. 1969.
20. J. W. Davis, D. J. Hockney, and W. K. Wilson (eds.), *Philosophical Logic*. 1969.
21. D. Davidson and J. Hintikka (eds.), *Words and Objections. Essays on the Work of W. V. Quine*. 1969.
22. Patrick Suppes, *Studies in the Methodology and Foundations of Science. Selected Papers from 1911 to 1969*. 1969.
23. Jaakko Hintikka, *Models for Modalities. Selected Essays*. 1969.
24. Nicholas Rescher *et al.* (eds.), *Essays in Honor of Carl G. Hempel. A Tribute on the Occasion of His Sixty-Fifth Birthday*. 1969.
25. P. V. Tavanec (ed.), *Problems of the Logic of Scientific Knowledge*. 1969.
26. Marshall Swain (ed.), *Induction, Acceptance, and Rational Belief*. 1970.
27. Robert S. Cohen and Raymond J. Seeger (eds.), *Ernst Mach: Physicist and Philosopher*. Boston Studies in the Philosophy of Science, Volume VI. 1970.
28. Jaakko Hintikka and Patrick Suppes, *Information and Inference*. 1970.
29. Karel Lambert, *Philosophical Problems in Logic. Some Recent Developments*. 1970.
30. Rolf A. Eberle, *Nominalistic Systems*. 1970.
31. Paul Weingartner and Gerhard Zecha (eds.), *Induction, Physics, and Ethics*. 1970.
32. Evert W. Beth, *Aspects of Modern Logic*. 1970.
33. Risto Hilpinen (ed.), *Deontic Logic: Introductory and Systematic Readings*. 1971.
34. Jean-Louis Krivine, *Introduction to Axiomatic Set Theory*. 1971.
35. Joseph D. Sneed, *The Logical Sstructure of Mathematical Physics*. 1971.
36. Carl R. Kordig, *The Justification of Scientific Change*. 1971.
37. Milic Capek, *Bergson and Modern Physics*. Boston Studies in the Philosophy of Science, Volume VII. 1971.
38. Norwood Russell Hanson, *What I Do Not Believe, and Other Essays* (ed. by Stephen Toulmin and Harry Woolf). 1971.
39. Roger C. Buck and Robert S. Cohen (eds.), *PSA 1970. In Memory of Rudolf Carnap*. Boston Studies in the Philosophy of Science, Volume VIII. 1971.
40. Donald Davidson and Gilbert Harman (eds.), *Semantics of Natural Language*. 1972.
41. Yehoshua Bar-Hillel (ed.), *Pragmatics of Natural Languages*. 1971.
42. Sören Stenlund, *Combinators, λ-Terms and Proof Theory*. 1972.
43. Martin Strauss, *Modern Physics and Its Philosophy. Selected Papers in the Logic, History, and Philosophy of Science*. 1972.
44. Mario Bunge, *Method, Model and Matter*. 1973.
45. Mario Bunge, *Philosophy of Physics*. 1973.
46. A. A. Zinov'ev, *Foundations of the Logical Theory of Scientific Knowledge (Complex Logic)*. (Revised and enlarged English edition with an appendix by G. A. Smirnov, E. A. Sidorenka, A. M. Fedina, and L. A. Bobrova.) Boston Studies in the Philosophy of Science, Volume IX. 1973.
47. Ladislav Tondl, *Scientific Procedures*. Boston Studies in the Philosophy of Science, Volume X. 1973.
48. Norwood Russell Hanson, *Constellations and Conjectures* (ed. by Willard C. Humphreys, Jr.). 1973.

49. K. J. J. Hintikka, J. M. E. Moravcsik, and P. Suppes (eds.), *Approaches to Natural Language.* 1973.
50. Mario Bunge (ed.), *Exact Philosophy – Problems, Tools, and Goals.* 1973.
51. Radu J. Bogdan and Ilkka Niiniluoto (eds.), *Logic, Language, and Probability.* 1973.
52. Glenn Pearce and Patrick Maynard (eds.), *Conceptual Change.* 1973.
53. Ilkka Niiniluoto and Raimo Tuomela, *Theoretical Concepts and Hypothetico-Inductive Inference.* 1973.
54. Roland Fraissé, *Course of Mathematical Logic* – Volume 1: *Relation and Logical Formula.* 1973.
55. Adolf Grünbaum, *Philosophical Problems of Space and Time.* (Second, enlarged edition.) Boston Studies in the Philosophy of Science, Volume XII. 1973.
56. Patrick Suppes (ed.), *Space, Time, and Geometry.* 1973.
57. Hans Kelsen, *Essays in Legal and Moral Philosophy* (selected and introduced by Ota Weinberger). 1973.
58. R. J. Seeger and Robert S. Cohen (eds.), *Philosophical Foundations of Science.* Boston Studies in the Philosophy of Science, Volume XI. 1974.
59. Robert S. Cohen and Marx W. Wartofsky (eds.), *Logical and Epistemological Studies in Contemporary Physics.* Boston Studies in the Philosophy of Science, Volume XIII. 1973.
60. Robert S. Cohen and Marx W. Wartofsky (eds.), *Methodological and Historical Essays in the Natural and Social Sciences. Proceedings of the Boston Colloquium for the Philosophy of Science 1969-1972.* Boston Studies in the Philosophy of Science, Volume XIV. 1974.
61. Robert S. Cohen, J. J. Stachel, and Marx W. Wartofsky (eds.), *For Dirk Struik. Scientific, Historical and Political Essays in Honor of Dirk J. Struik.* Boston Studies in the Philosophy of Science, Volume XV. 1974.
62. Kazimierz Ajdukiewicz, *Pragmatic Logic* (transl. from the Polish by Olgierd Wojtasiewicz). 1974.
63. Sören Stenlund (ed.), *Logical Theory and Semantic Analysis. Essays Dedicated to Stig Kanger on His Fiftieth Birthday.* 1974.
64. Kenneth F. Schaffner and Robert S. Cohen (eds.), *Proceedings of the 1972 Biennial Meeting, Philosophy of Science Association.* Boston Studies in the Philosophy of Science, Volume XX. 1974.
65. Henry E. Kyburg, Jr., *The Logical Foundations of Statistical Inference.* 1974.
66. Marjorie Grene, *The Understanding of Nature. Essays in the Philosophy of Biology.* Boston Studies in the Philosophy of Science, Volume XXIII. 1974.
67. Jan M. Broekman, *Structuralism: Moscow, Prague, Paris.* 1974.
68. Norman Geschwind, *Selected Papers on Language and the Brain.* Boston Studies in the Philosophy of Science, Volume XVI. 1974.
69. Roland Fraissé, *Course of Mathematical Logic* – Volume 2: *Model Theory.* 1974.
70. Andrzej Grzegorczyk, *An Outline of Mathematical Logic. Fundamental Results and Notions Explained with All Details.* 1974.
71. Franz von Kutschera, *Philosophy of Language.* 1975.
72. Juha Manninen and Raimo Tuomela (eds.), *Essays on Explanation and Understanding. Studies in the Foundations of Humanities and Social Sciences.* 1976.

73. Jaakko Hintikka (ed.), *Rudolf Carnap, Logical Empiricist. Materials and Perspectives.* 1975.
74. Milic Capek (ed.), *The Concepts of Space and Time. Their Structure and Their Development.* Boston Studies in the Philosophy of Science, Volume XXII. 1976.
75. Jaakko Hintikka and Unto Remes, *The Method of Analysis. Its Geometrical Origin and Its General Significance.* Boston Studies in the Philosophy of Science, Volume XXV. 1974.
76. John Emery Murdoch and Edith Dudley Sylla, *The Cultural Context of Medieval Learning.* Boston Studies in the Philosophy of Science, Volume XXVI. 1975.
77. Stefan Amsterdamski, *Between Experience and Metaphysics. Philosophical Problems of the Evolution of Science.* Boston Studies in the Philosophy of Science, Volume XXXV. 1975.
78. Patrick Suppes (ed.), *Logic and Probability in Quantum Mechanics.* 1976.
79. Hermann von Helmholtz: *Epistemological Writings. The Paul Hertz/Moritz Schlick Centenary Edition of 1921 with Notes and Commentary by the Editors.* (Newly translated by Malcolm F. Lowe. Edited, with an Introduction and Bibliography, by Robert S. Cohen and Yehuda Elkana.) Boston Studies in the Philosophy of Science, Volume XXXVII. 1977.
80. Joseph Agassi, *Science in Flux.* Boston Studies in the Philosophy of Science, Volume XXVIII. 1975.
81. Sandra G. Harding (ed.), *Can Theories Be Refuted? Essays on the Duhem-Quine Thesis.* 1976.
82. Stefan Nowak, *Methodology of Sociological Research. General Problems.* 1977.
83. Jean Piaget, Jean-Blaise Grize, Alina Szeminska, and Vinh Bang, *Epistemology and Psychology of Functions.* 1977.
84. Marjorie Grene and Everett Mendelsohn (eds.), *Topics in the Philosophy of Biology.* Boston Studies in the Philosophy of Science, Volume XXVII. 1976.
85. E. Fischbein, *The Intuitive Sources of Probabilistic Thinking in Children.* 1975.
86. Ernest W. Adams, *The Logic of Conditionals. An Application of Probability to Deductive Logic.* 1975.
87. Marian Przelecki and Ryszard Wójcicki (eds.), *Twenty-Five Years of Logical Methodology in Poland.* 1977.
88. J. Topolski, *The Methodology of History.* 1976.
89. A. Kasher (ed.), *Language in Focus: Foundations, Methods and Systems. Essays Dedicated to Yehoshua Bar-Hillel.* Boston Studies in the Philosophy of Science, Volume XLIII. 1976.
90. Jaakko Hintikka, *The Intentions of Intentionality and Other New Models for Modalities.* 1975.
91. Wolfgang Stegmüller, *Collected Papers on Epistemology, Philosophy of Science and History of Philosophy.* 2 Volumes. 1977.
92. Dov M. Gabbay, *Investigations in Modal and Tense Logics with Applications to Problems in Philosophy and Linguistics.* 1976.
93. Radu J. Bogdan, *Local Induction.* 1976.
94. Stefan Nowak, *Understanding and Prediction. Essays in the Methodology of Social and Behavioral Theories.* 1976.
95. Peter Mittelstaedt, *Philosophical Problems of Modern Physics.* Boston Studies in the Philosophy of Science, Volume XVIII. 1976.

96. Gerald Holton and William Blanpied (eds.), *Science and Its Public: The Changing Relationship*. Boston Studies in the Philosophy of Science, Volume XXXIII. 1976.
97. Myles Brand and Douglas Walton (eds.), *Action Theory*. 1976.
98. Paul Gochet, *Outline of a Nominalist Theory of Proposition. An Essay in the Theory of Meaning*. 1980.
99. R. S. Cohen, P. K. Feyerabend, and M. W. Wartofsky (eds.), *Essays in Memory of Imre Lakatos*. Boston Studies in the Philosophy of Science, Volume XXXIX. 1976.
100. R. S. Cohen and J. J. Stachel (eds.), *Selected Papers of Léon Rosenfeld*. Boston Studies in the Philosophy of Science, Volume XXI. 1978.
101. R. S. Cohen, C. A. Hooker, A. C. Michalos, and J. W. van Evra (eds.), *PSA 1974: Proceedings of the 1974 Biennial Meeting of the Philosophy of Science Association*. Boston Studies in the Philosophy of Science, Volume XXXII. 1976.
102. Yehuda Fried and Joseph Agassi, *Paranoia: A Study in Diagnosis*. Boston Studies in the Philosophy of Science, Volume L. 1976.
103. Marian Przelecki, Klemens Szaniawski, and Ryszard Wójcicki (eds.), *Formal Methods in the Methodology of Empirical Sciences*. 1976.
104. John M. Vickers, *Belief and Probability*. 1976.
105. Kurt H. Wolff, *Surrender and Catch: Experience and Inquiry Today*. Boston Studies in the Philosophy of Science, Volume LI. 1976.
106. Karel Kosík, *Dialectics of the Concrete*. Boston Studies in the Philosophy of Science, Volume LII. 1976.
107. Nelson Goodman, *The Structure of Appearance*. (Third edition.) Boston Studies in the Philosophy of Science, Volume LIII. 1977.
108. Jerzy Giedymin (ed.), *Kazimierz Ajdukiewicz: The Scientific World-Perspective and Other Essays, 1931-1963*. 1978.
109. Robert L. Causey, *Unity of Science*. 1977.
110. Richard E. Grandy, *Advanced Logic for Applications*. 1977.
111. Robert P. McArthur, *Tense Logic*. 1976.
112. Lars Lindahl, *Position and Change. A Study in Law and Logic*. 1977.
113. Raimo Tuomela, *Dispositions*. 1978.
114 Herbert A. Simon, *Models of Discovery and Other Topics in the Methods of Science*. Boston Studies in the Philosophy of Science, Volume LIV. 1977.
115. Roger D. Rosenkrantz, *Inference, Method and Decision*. 1977.
116. Raimo Tuomela, *Human Action and Its Explanation. A Study on the Philosophical Foundations of Psychology*. 1977.
117. Morris Lazerowitz, *The Language of Philosophy. Freud and Wittgenstein*. Boston Studies in the Philosophy of Science, Volume LV. 1977.
118. Stanislaw Leśniewski, *Collected Works* (ed. by S. J. Surma, J. T. J. Srzednicki, and D. I. Barnett, with an annotated bibliography by V. Frederick Rickey). 1980. (Forthcoming.)
119. Jerzy Pelc, *Semiotics in Poland, 1894-1969*. 1978.
120. Ingmar Pörn, *Action Theory and Social Science. Some Formal Models*. 1977.
121. Joseph Margolis, *Persons and Minds. The Prospects of Nonreductive Materialism*. Boston Studies in the Philosophy of Science, Volume LVII. 1977.
122. Jaakko Hintikka, Ilkka Niiniluoto, and Esa Saarinen (eds.), *Essays on Mathematical and Philosophical Logic*. 1978.
123. Theo A. F. Kuipers, *Studies in Inductive Probability and Rational Expectation*. 1978.

124. Esa Saarinen, Risto Hilpinen, Ilkka Niiniluoto, and Merrill Provence Hintikka (eds.), *Essays in Honour of Jaakko Hintikka on the Occasion of His Fiftieth Birthday.* 1978.
125 Gerard Radnitzky and Gunnar Andersson (eds.), *Progress and Rationality in Science.* Boston Studies in the Philosophy of Science, Volume LVIII. 1978.
126. Peter Mittelstaedt, *Quantum Logic.* 1978.
127. Kenneth A. Bowen, *Model Theory for Modal Logic. Kripke Models for Modal Predicate Calculi.* 1978.
128. Howard Alexander Bursen, *Dismantling the Memory Machine. A Philosophical Investigation of Machine Theories of Memory.* 1978.
129. Marx W. Wartofsky, *Models: Representation and the Scientific Understanding.* Boston Studies in the Philosophy of Science, Volume XLVIII. 1979.
130. Don Ihde, *Technics and Praxis. A Philosophy of Technology.* Boston Studies in the Philosophy of Science, Volume XXIV. 1978.
131. Jerzy J. Wiatr (ed.), *Polish Essays in the Methodology of the Social Sciences.* Boston Studies in the Philosophy of Science, Volume XXIX. 1979.
132. Wesley C. Salmon (ed.), *Hans Reichenbach: Logical Empiricist.* 1979.
133. Peter Bieri, Rolf-P. Horstmann, and Lorenz Krüger (eds.), *Transcendental Arguments in Science. Essays in Epistemology.* 1979.
134. Mihailo Marković and Gajo Petrović (eds.), *Praxis. Yugoslav Essays in the Philosophy and Methodology of the Social Sciences.* Boston Studies in the Philosophy of Science, Volume XXXVI. 1979.
135. Ryszard Wójcicki, *Topics in the Formal Methodology of Empirical Sciences.* 1979.
136. Gerard Radnitzky and Gunnar Andersson (eds.), *The Structure and Development of Science.* Boston Studies in the Philosophy of Science, Volume LIX. 1979.
137. Judson Chambers Webb, *Mechanism, Mentalism, and Metamathematics. An Essay on Finitism.* 1980.
138. D. F. Gustafson and B. L. Tapscott (eds.), *Body, Mind, and Method. Essays in Honor of Virgil C. Aldrich.* 1979.
139. Leszek Nowak, *The Structure of Idealization. Towards a Systematic Interpretation of the Marxian Idea of Science.* 1979.
140. Chaim Perelman, *The New Rhetoric and the Humanities. Essays on Rhetoric and Its Applications.* 1979.
141. Wlodzimierz Rabinowicz, *Universalizability. A Study in Morals and Metaphysics.* 1979.
142. Chaim Perelman, *Justice, Law, and Argument. Essays on Moral and Legal Reasoning.* 1980.
143. S. Kanger and S. Öhman (eds.), *Philosophy and Grammar. Papers on the Occasion of the Quincentennial of Uppsala University, Sweden.* 1980.
144. Tadeusz Pawlowski, *Concept Formation in the Humanities and the Social Sciences.* 1980.
145. Jaakko Hintikka, David Gruender, and Evandro Agazzi (eds.), *Theory Change, Ancient Axiomatics, and Galileo's Methodology. Proceedings of the 1978 Pisa Conference on the History and Philosophy of Science.* 1980. (Forthcoming.)
146. Jaakko Hintikka, David Gruender, and Evandro Agazzi (eds.), *Probabilistic Thinking, Thermodynamics, and the Interaction of the History and Philosophy of Science. Proceedings of the 1978 Pisa Conference on the History and Philosophy of Science.* 1980. (Forthcoming.)

147. Uwe Mönnich, *Aspects of Philosophical Logic.* 1980. (Forthcoming.)
149. Evandro Agazzi, *Modern Logic - A Survey. Historical, Philosophical, and Mathematical Aspects of Modern Logic and its Applications.* 1980.
150. A. F. Parker-Rhodes, *The Theory of Indistinguishables. A Search for Explanatory Principles below the Level of Physics.* 1981. (Forthcoming.)

MIX
Papier aus verantwortungsvollen Quellen
Paper from responsible sources
FSC® C105338

If you have any concerns about our products,
you can contact us on
**ProductSafety@springernature.com**

In case Publisher is established outside the EU,
the EU authorized representative is:
**Springer Nature Customer Service Center GmbH**
**Europaplatz 3, 69115 Heidelberg, Germany**

Printed by Libri Plureos GmbH
in Hamburg, Germany